The Way of 'Q'

Alfred de Grazia

ALFRED DE GRAZIA

The Way of 'Q'

Essentials of the
Quantavolution Paradigm

An Abridgement of the
Quantavolution Series by

Anne-Marie de Grazia

With an Introduction by
Anne-Marie de Grazia

Metron Publications

ISBN: 978-1-60377-100-9
Library of Congress Control Number:2019949855

Metron Publications P.O.Box 1205
PRINCETON, NJ, 0

About the Author

Alfred de Grazia was born December 29th, 1919 in Chicago. His father was a clarinettist and orchestra conductor who had immigrated from Sicily. His mother was born in Chicago, also of Sicilian ascent. Alfred was admitted to University of Chicago age not yet 15, on a half-scholarship, where he studied political science. He was drafted into the Army in 1942 and participated in six campaigns in WWII as a liaison officer in Africa and Europe, one of the first OSS officers specialized in psychological warfare. By the end of the war, he was the head of Psychological Warfare for the 7th Army. He then resumed his studies, taught at University of Indiana, and obtained his PhD in Political Science at University of Chicago in 1948. His thesis was published to acclaim in 1951 by Alfred Knopf as *Public and Republic: Political Representation in America.* In 1952 he published the 2 volume textbook *Elements of Political Science: Political Behavior* & *Political Organization.* He taught at University of Minnesota, Brown University, Stanford University. From 1958 at New York University. He published a dozen more books in political science and created the magazine *The Behavioral Scientist,* as well as *The Universal Reference System,* the first ever computerized reference system in the Social Sciences. He met and befriended Immanuel Velikovsky in 1962 and published *The Velikovsky Affair* first in his magazine, then as a book. Having retired early from New York University, he took to writing and publishing the ten book *Quantavolution Series* between 1977 and 1984. He divided his life between the United States, Europe, and his property on the Greek island of Naxos, in the

Aegean Sea. He continued to be active for the sake of a better world, writing *Kalos: What Is To Be Done With Our World?* and investigating the Bhopal chemical catastrophe, on which he wrote the first book: *A Cloud Over Bhopal.* He continued writing poetry and short plays. In 1997, he created the *grazian-archive.com* website, destined to contain his complete life's output, and to serve as a model for his Living Archive Project, to preserve and make available the archives of deceased producers of culture. He was a professor at University of Bergamo in Italy from 2002 to 2005. He died age 94 in La Ferté-Bernard, France, on July 13th, 2014. He had been made a Chevalier of the French Légion d'Honneur and was made posthumously a Distinguished Member of the Regiment of Special Operations at Fort Bragg, N.C. One month after his death, sensational archaeological finds of Neanderthal artefacts were made in the garden of his property in Naxos, Greece. He had seven children by his first wife, Jill Oppenheim, was shortly married to Nina Mavridis, then lived for 37 years with his third wife, Anne-Marie (Ami), a French writer.

Websites:

https://grazian-archive.com
https://kalos.co
http://metron-publications.org
http://www.q-ency.com

THE QUANTAVOLUTION SERIES

Chaos and Creation - *An Introduction to Quantavolution in Natural and Human History* (1981)

God's Fire - *Moses and the Management of Exodus* (1982)

The Lately Tortured Earth - *Exoterrestrial Forces and Quantavolutions in the Earth's Sciences* (1982)

Homo Schizo One - *Human and Cultural Hologenesis* (1983)

Homo Schizo Two - *Human Nature and Behavior* (1983)

The Divine Succession - *A Science of Gods Old and New* (1983)

Solaria Binaria - *Origins and History of the Solar System* (1983)

The Disastrous Love-Affair of Moon and Mars - *Celestial Sex, Earthly Destruction and Dramatic Sublimation in Homer's Odyssey* (1984)

The Burning of Troy *and other works in Quantavolution and Scientific Catastrophism* (1984)

The Cosmic Heretics - *A Personal History of Attempts to Establish & Resist Theories of Quantavolution and Catastrophes in the Natural & Human Sciences,* 1963-1983 (1984)

The eleventh book:
The Iron Age of Mars - *Speculations on a Quantavolution* (2009)

The Way of 'Q' was conceived as an easy to read, quick tour around the ten volumes of the Quantavolution Series, with some prolonged stops at the main ideas of the theory. For references and further study, *Q-Ency, the Encyclopedia of Quantavolution,* available at http://www.q-ency.com/ should prove to be useful, as well as those of the texts with are published, with footnotes and bibliographies, on the Internet. As well as, of course, the ten published books, available in paper as well as in electronic form.

TABLE OF CONTENTS

INTRODUCTION

BORN ON 'Q'-STREET

The last guests are gone. Tony has taken off at dawn for Milan airport, to fly back to Los Alamos and to his NASA lab. After lunch, and a heated discussion over whether space plasma is of any relevance to rock art all over the world, Vladimir has driven Rens in our car to the local airport, whence he is to fly back on Ryanair to London. Alfred and I then drive Vladimir to the train station - in a few hours he will take off for Sofia, Bulgaria, and his office at the Academy of Science. Kisses, fond farewells. Thus ends the first Colloquium on Quantavolution, held at the University of Bergamo, Italy. Some twenty-five quantavolutionists have sat together around glass tables for several warm spring days in a beautiful Gothic room, squeezed between two medieval cloisters in the newly restored Convent of San Agostino, which stands like a prow at the end of the Upper town of Bergamo, the *Città Alta* - the Lombard plain in front, the Alps in back.

These have been heady and exciting days. Germans questioning the validity of measuring early geological and historical time by radiochronometry. Victor Clube from Oxford examining the historic or prehistoric records of catastrophic cometary approaches of Earth. The demise of the dinosaurs, of course, one extinction among many. Greek professors introducing their new field of "geomythology:" the traces left on Earth's surface by violent events reported as ancient myths. How repeated planetary convulsions may have made us human. Quantavolution is the science of such changes.

Quantavolution has been my life with this extraordinary man, Alfred de Grazia. The word was born in the first year of our life together, in Washington, DC, where we dwelled for a while, while Alfred was consulting for Government agencies. He was writing "The Lately Tortured Earth," the first of the ten books of what was to become the "Quantavolution Series." "Quanta," in physics, standing for a discrete (i.e., distinct, non-continuous) natural unit, or packet, of energy, charge, angular momentum, or other physical property. "Volution," a turn, down or up, a jump, an event as opposed to a regular, continuous process.

Quantavolution posits that the major changes in nature occur in rapid, intense bursts of energy rather than accruing at steady, incremental slow rates. A Quantavolution is a disruption of the processes of steady change, accelerating or transforming them. Many people had trouble with the word, arising often from a deeply ingrained reluctance to admit the existence of an unsettled universe. It was tough to get it accepted. It was met with blank stares, it was derided, it was criticized, in the end - one quarter of a century later - it was proving itself unmatchable. It was born, as I said, in Washington, D.C. in a small brick house in Georgetown and, in what can only be termed an instance of poetic coincidence, the house was located on 'Q' Street... At the time, Liz Taylor and Senator Warner lived in back of us, on P...

It is hard to imagine today how foreign to the way of thinking that prevailed then in the sciences were the notions of Quantavolution. The only establishment scientist endowed with a sizeable audience who was then pushing at the edge of the all-powerful paradigm of uniformitarian change was the Paleontologist Stephen Jay Gould, at Harvard. He called his concept of rapid change intervening between long periods of constancy "punctuated equilibrium". His views raised controversy and he got his share of flak from the scientific establishment, but he had adroitly managed to bring his opponents to fight him on the battlefield of the media, where he enjoyed considerable advantage, through his extraordinarily engaging, mild-mannered personality and a regular column in the magazine of the New York Museum of Natural History. He was able to score significant successes and to break a first breach in the wall, from within. A few broke out after him. Quantavolution broke in. Although he never espoused Alfred's paradigm, Gould agreed with isolated elements of it and there was a friendly exchange between them.

One might bring back the revolution in established scientific thinking some fifteen years earlier, to the emergence of chaos theory in mathematics in the early sixties. Incrementalism, slow change, uniformitarianism were in essence linear thinking, whether applied to cosmology, geology, evolution, or the social sciences. With the major turn into non-linear studies in mathematics, which itself had been made necessary by the even earlier breakthroughs of quantum mechanics in fundamental physics, another big stride was made, although this revolution had taken place without bloodshed, as there were no overriding ideological interests tied in with linear mathematics, and the dominant ideology, which held the natural sciences in such a tight grip, paid little attention to mathematics and fundamental physics, and saw no danger coming from those quarters. Still, minds shift, and when they do, they are not wont to turn back... Changes of paradigm are necessarily preceded by such shifts - changes in the frame of mind, what one might call, in the fashion in thinking...

Quantavolutions happen all the time, at all levels, through the universe. With tremendous, disturbing bursts of energy wandering loose through the cosmos, even to the interior of our cells. Galaxies crash into each other, stars explode - it is happening right now... At the scale of the universe, a large meteorite crashing into a planet is a crude, insignificant event. Far less significant than a mutation in a mouse's litter might turn out to be for the species of mice, or for the order of mammals.

It is interesting to note that, despite the fact that slow change, at least in the form of the cyclical and seasonal, was available to man for observation all through history, and was indeed given intense attention, man explained the world in terms of fast change, sudden catastrophes, violent events, and mostly lived in dread of them, and organized in anticipation of them. The idea of evolution, of uniformitarian change, occurring unthreateningly, and over a long time, emerged in England at the very beginning of the Nineteenth Century, after the French Revolution, and in reaction to it. It spread out from there and may well be termed an ideology - replacing the faltering religious faiths of the Western world - which helps to explain the fierceness with which its tenets were defended whenever challenged. Political revolution, as it were, was thus outlawed, on grounds of non-conformity with the laws of nature. The rash of political revolutions which continued to agitate Europe during the rest of the century only contributed to tighten the hold of the new ideology. The natural sciences were increasingly pushed into the descriptive, the pedantic, and the myopically specialized. Overarching explanations, reaching into different fields of the increasingly atomized sciences, became taboo. Theoretical thinking wasted away. Evidences of large scale, sudden catastrophic changes were explained away, or suppressed and, after a while, became simply non-existent to the eye trained not to see them, and their study was postponed for over a century and a half.

Alfred de Grazia eschewed the term "catastrophism" in favor of "quantavolution." But that was because there was something wrong with the word. For, however "catastrophic," the violent, sudden changes visited upon ourselves, our Earth, and our Solar system have actually been of limited destructiveness, as they have permitted us to exist and to survive to this day.

The balance, until now, between the creative and the destructive aspects of quantavolutions on Earth, is mightily positive from our point of view, and from the point of view of life itself, as we have maintained ourselves splendidly up to this point - from man to microbe - and are indeed thriving. If life exists on these galaxies now in the process of crashing into each other - and we are fast approaching a consensus in the scientific world according to which life exists virtually everywhere - the processes of destruction are probably far more radical and, for those forms of life that would be equivalent to the mammalian or the reptile, probably thoroughly fatal. (At least one third of the stars are now thought to be able to support systems where life could develop. We can logically imagine that every day, somewhere, a world of intelligent beings might be annihilated.)

The more measured quantavolutions which Earth has undergone have created us humans and we would not exist as we are, were it not for their occurrence. The catastrophic extinction of the dinosaurs was a boon to the mammals and in the end, to you and me (if you and me consider it a boon to have been born, a simple event which is equivalent, for each of us, to winning odds which are far greater than winning the biggest jackpot ever, at the biggest lottery ever, with the smallest possible stake.) Quantavolutions occur rapidly, wreak their transformations within a short time and their effects are profound and usually irreversible. They occur in a universe of tremendous plasticity and complexity, endowed with practically infinite capacities of response, which it has "learned" precisely at the school of these mega-changes.

Alfred de Grazia's reflections on rapid change had started a long time before he moved to 'Q' Street and settled on the term "Quantavolution." In fact, it happened a decade and a half before, when he became a close friend and supporter of Immanuel Velikovsky, who lived down the street from his house in Princeton. He was then a professor of political science of impeccable credentials, one of the top political scientists of his generation, with a round dozen of books published, very much an establishment scientist himself, founder of a magazine, *The American Behavioral Scientist,* which is still going strong in the stable of Sage Publications, along with the many journals it has sired. He held teaching appointments at University of Chicago, University of Minnesota, Brown

University, Harvard, Columbia, Stanford University, New York University.

The first meeting of the two scholars was regularized into frequent tête-à-tête and phone calls over the years whenever Alfred was in Princeton. He had taken little notice of the controversy surrounding Velikovsky when it had happened, in the early fifties, and by the time he met Velikovsky, this controversy had considerably subsided. Except for a few diehard supporters, it was pretty much history.

What aroused his indignation before all, and moved him to action, was the scandalous way in which Velikovsky had been treated by several leaders of the scientific establishment, who had threatened his first publisher, Macmillan, with boycott if it did not give up the publication of his book, *Worlds in Collision.* This outrageous threat had had its intended effect: the book was dropped from the list of Macmillan, and owed it only to its success as a best-seller to be picked up smoothly by another publisher, Doubleday.

Alfred de Grazia stunned his readership by devoting an entire issue of *The American Behavioral Scientist* to "The Velikovsky Affair," forcefully raising the central question, the one of utmost importance indeed to political and social science - what he called "the Reception System of Science," for which the treatment dealt to Velikovsky by the scientific establishment represented indeed an ideal case.

The controversy was restarted, stronger than before, both regarding the behavior of establishment science towards new ideas, and, inevitably, as regards the substance of Velikovsky's ideas and theories. Alfred later told the whole story, as well as that of the revived movement around Velikovsky, in his book *The Cosmic Heretics.* When Alfred convened a college at Valais, Switzerland, Immanuel came along in due course, bringing with him his extended family. Alfred established a special studio on civilizations and catastrophes for him there.

Many things happened between them before I arrived on the scene. They had their quarrels, always quickly subdued by their rare bond of two men who could and did respect each other. Velikovsky was still living in Princeton when I met Alfred and we visited him often in his house on Hartley Avenue, close to Lake Carnegie. We had lunch there in summer, on the new wooden deck which he had added to the side of his house, drinking Israeli wine. He maintained a strong connection to Israel, of which his father, emigrating from Lvov after the Communist

Revolution, had been one of the earliest settlers. The house on Hartley Avenue struck me as "Central-European," it resembled the home of old people I visited as a child in Alsace, with heavy furniture and rooms kept dark to protect the upholstery. There was an atmosphere of heavy coziness and I was old enough to feel a nostalgia for it, although I myself liked to live in airy, sunny places.

Sixteen months after I first met him, eighteen years into their friendship, Velikovsky died, at his home, unexpectedly. That day, when Velikovsky's widow, Elisheva, tried to call Alfred to announce him the great man's passing, we had been working together at Princeton University's Firestone Library, doing research on Alfred's book on Moses and the Exodus: *God's Fire*. Particularly, in the Egyptology section, on the drawing of an extraordinary device, boat shaped, with two winged creatures facing each other on top of a box, in which Alfred saw the Egyptian prototype of the Ark of the Covenant - an electrical capacitor, an accumulator of static electricity, an early and more sophisticated form of what would be rediscovered in the XVIII century as the Leyden jar. It produced sparks, and transmitted charge which could shock, and even kill, as was clear from several passages in the Bible relevant to the Ark of the Covenant. It would have been particularly effective - beyond anything a Leyden jar could produce in our times - in the highly electrified environment of the time of the Exodus, which was created by electromagnetic conditions caused by close passages of a wandering, comet-like planet Venus - the linchpin of Velikovsky's theories.

The early seventies witnessed the publication of three books by Alfred in political science. From then on, it was all 'go' for Quantavolution. Alfred had retired early from New York University in order to write the books which would become the Quantavolution Series, of which *God's Fire* was the second. The ten books took approximately another seven years to write and to produce - the first seven years of my life together with Alfred. Needless to say, this task required his full time and the full engagement of his tremendous work capacity - from five o'clock in the morning, day after day, year after year, until late at night - and a good deal of mine, as well. Our main source of income was Alfred's early retirement pension and we led a frugal, yet highly satisfying life - thanks, in great part, to our unusual living circumstances: Alfred had acquired, years before, a large piece of land facing the Aegean sea and the sunset on a promontory on the Greek Island of Naxos, then still undiscovered by mass tourism, a setting of

Eden-like simplicity, where he had built a cabin of rough stones - without running water, without electricity, without telephone. No taxes. Wonderful fresh food. Restaurants where a hearty meal with wine cost less than a dollar.

We didn't have a car - there was no road to speak of, either, on our promontory, called Stelida. Alfred used to drive to town on a motorcycle until the day, early on - the very day he finished the first draft of an article on Moses which he was already thinking of expanding into a book (it would be *God's Fire*) - when he broke his leg trying to get his motorcycle unstuck from a muddy spot. The road was so poor that we did not ride it on the motorcycle together - I had taken a short cut on foot and was waiting for him farther down the road. After I found him, I ran to the closest farm and, with scant knowledge of Greek and passionate gesticulation, got the farmer to pick him up on his tractor and drive him into the main town of the island, where there were no medical facilities besides a competent and overworked young generalist. After this, I forbade him the motorcycle and I walked to town twice a week - three miles around the mountain and along deserted beaches, with my knapsack on my back, to buy food. Having no electricity, we had no refrigerator, either. It was probably the happiest time in our lives.

For six months of the year, we went back to live in America, picking up with the same work schedule, in a different environment. The cabin in Naxos was not winterized, and of course Alfred needed to work in good academic libraries. Among his New York University post-retirement privileges, he had been allowed to keep an apartment on the 25th floor of the Silver Towers, in the heart of Greenwich Village, with what was then probably the most beautiful view of New York: the perfect unobstructed Midtown skyline - one could see a sliver of both the Hudson and the East River - and a bit of New Jersey and a bit of Brooklyn. It would have been the perfect place to live if, due to restricted financial circumstances made necessary for him to be free to write the books, Alfred had not been forced to sublet it part-time, the first year, to an only too-eager close relative who, living quite comfortably in New Jersey himself, decided to use it as a luxurious Manhattan pied-à-terre. The part-time arrangement became quickly, and unilaterally, full-time and, preferring to keep the family peace, we ended up eking out the right to a few short days of residence every fortnight from this reluctant relative. Even on the day of our wedding, April 23rd, 1983, it was not free for us to use.

We were married at the Church of Saint Luke's in the Fields, actually then a basement because it had burned to the ground, but where the Rev. Chia Ballantine, Alfred's niece-in-law, officiated, and then jumped in our wedding clothes on the PATH train to return to New Jersey, on to Trenton, where we had bought a small old brick row-house with a backyard to plant tomatoes in a mixed neighborhood. I am reminded of Truman Capote - who derived a feeling of basic security from knowing that, whatever happened to him financially, he would be able to live creatively in a marginal urban neighborhood. The house on Centre Street, Trenton, was narrow, had two floors, a basement where Al's archive was kept, an attic where some quantavolutionist or other sometimes slept. It was not close to any usable academic library, but we made do with frequent drives to nearby Princeton, where Alfred's mother still lived, and our bi-monthly trips to Manhattan, where we would work in the nearby Bobst Library and at New York Public Library - what joy that was, too...

These were therefore the places where the ten books of the Quantavolution Series were mostly written, 'Q' Street in Georgetown, DC, the Greek stone cabin, the greatest-view-in-Manhattan, the Trenton slum - over seven years... If books, and ideas, are a product of places as well as of minds, Quantavolution is a product of these places and of that mind, Al de Grazia's.

India's leading environmentalist, Dr. Rashmi Mayur, reviewed *Chaos and Creation* for the million readers of *The Hindu,* dated May 31, 1981, and asked Alfred whether the planet was in peril from a new quantavolution. "More dangerous is the threat of man's self-destruction. Nuclear explosions can reduce India's huge people to Manu and his crew once more, but there is no guarantee that they would be wise. Only a responsible world-government can insure us against this disaster." Ever interested in world conflicts, Alfred was in India pushing the cause of the kind of world union he had devised in his basic book on the subject: *Kalos: What is to be Done with Our World?*

Earl Milton I met in Washington in 1978 - he came for a month from Lethbridge, Alberta, to talk about Alfred's theory of 'Solaria Binaria,' among other things. Lethbridge University had made Velikovsky a doctor *honoris causa.* The model of Solaria Binaria had been first expounded in a chapter of Volume One of the Quantavolution Series, *Chaos and Creation,* which had not yet been published. Earl, a professor of astronomy and a member of the Royal Astronomical Society of Canada, was enthusiastic with the proposed new paradigm of the solar system and before long the pair decided to do a separate, co-authored volume with Ralph Juergen, already committed. This brought for Alfred the benefit of Earl's expert knowledge and his original mind, which was intent on pushing as far as it could go - even to the breaking point - the electromagnetic dynamics included in the theory - in fact, doing away with the gravity model altogether. So went gravity, to which, in *Chaos and Creation,* Alfred had still been holding onto - so to speak - with the help of electro-magnetic forces. The result was a highly original mind-blower which Prof. Vladimir Damgov, a top-ranking physicist and member of the Bulgarian Academy of Science wondered, in a lecture twenty years later, why it had not been widely discussed in scientific journals. Earl and Alfred met again in Princeton, later that year. But the book needed a long stretch of intimate discussions and collaboration for which the stone cabin on the Aegean Sea was to offer the perfect haven. Earl Milton took several months off from his teaching and brought along his wife Joan and his little son Davin, who was then two and half years old.

We met up with them in London - it was the first time that Earl Milton and his family were travelling outside North America. They had gone straight to Westminster Abbey, "the oldest place we have ever been to." Both Alfred and Earl took part in a meeting of the Society for Interdisciplinary Studies, founded by British supporters of Velikovsky's work. We then flew together to Athens and took the eight hour ferry-boat ride to Naxos. We settled Earl and Joan and Davin in one of the small, vine and flower-covered bungalows built among the rocks by our friend Juergen Hecht, a German geologist, which he rented out to foreigners who had found their way to the remote spot. The bungalows had no electricity, either, the water came from a cistern which collected the winter rain, the nearest telephone was a crank-up model at the taverna down by the beach of Agios Prokopios. The beach was then vast, white and empty, used mostly by straggles of beautiful, penniless students from Germany and Scandinavia, a few members of the tailing-off hippie-movement of America, occasional Australians, all nudists by conviction. Some of them lived right on the beach, in bamboo huts. Juergen Hecht's bungalow had a large, vine-shaded terrasse, with a view to the sea and the mountains, a kitchen where Joan cooked meals, a small bathroom, it cost five US dollars a day. Earl and Joan and Davin stayed for two months - May and June 1980.

From the bungalows, if you walked West, the nearest house was Alfred's cabin, almost a mile away and after that, only the hill and sea. Every morning, in the cool hours, Earl walked over to our place. Al would be waiting for him with their coffee. And work would start with the first word of greeting - for this was a blissful time - life was work, all interaction between them was work, almost inadvertently, as well as it was life at its happiest and most exciting. It seems to me that I have never been in proximity of a human being so evenly elated over such a long stretch of time as Earl Milton was, working with Alfred at Stelida. They would go at it all morning, then the three of us would have lunch on the patio, then we would have a nap, Al with me, Earl on a couch in the "study" where they worked, which was a longish, oddly shaped house that Al had built next to the cabin and which we still call "the annex." Then they would work again, until evening, ouzo and whisky time - when Joan and Davin would come over for dinner, or we would all together walk a mile or two to one of the tavernas, coming home in the dark of night, along unlit paths, groping in the dark with weak flashlights (the quality of the batteries was poor), unless the moon was up, bright as day. Or Joan would invite us for dinner at the bungalow.

The weather was perfect all that time, the air was dry, the sub-desert Mediterranean vegetation produced no pollens or irritants, a fact which Earl, who much suffered from asthma, appreciated immensely. He had just turned forty, he was tall and athletic, he enjoyed the sun and the simple life - his mind was working at full speed. With balls and spheres, and strings and rulers, he and Alfred would play around in the annex, demonstrating the developing planetary configurations of Solaria Binaria. I would hear them speak loudly, enthusiastically, and laugh a lot of the time. The available library resources were scant, but included the full set of the *Encyclopedia Britannica,* for which Alfred professed contempt, kept in its own little oak chest. The nights were mild and dark - no street-lights, no clouds. Often Earl told us he had spent hours after dinner, watching the sky.

Sometime in May, we took a trip to the island of Thira-Santorini, three hours away by slow boat from Naxos, an awesome site, an island the largest part of which was blown off in a volcanic explosion dated to around 1450 BC, an event of such proportions that it has been suspected variously to be at the origins of the Atlantis legend, of the Plagues of Exodus and of the destruction, through concomitant earthquakes, tsunamis, gas-clouds and ash fall-outs, of the Minoan civilization of

Crete. In its present shape, it is semi-circular, the sea having invaded the caldera, with two small islets in the middle of it which emerged only three centuries ago and constitute some of the youngest land to be trodden on Earth. It is one of the great sites on the planet, a place of shattering beauty. We took a late afternoon ferry boat leaving from Naxos, planning to enter the caldera of Thira at sunset, when it appears in full glory. We did not make it...

A bad wind blew up. The ferry made a routine call at the neighboring island of Ios, and decided it would stay over for the night in the protected little harbor there. There was not much going on in Ios at the time, except for some notorious hippie-beaches on the opposite side of the island. On Ios, a tenacious legend insists, Homer died and lay buried. Our party's enthusiasm was only heightened by this forced halt in the homeric site: Alfred had already almost finished *The Disastrous Love Affair of Moon and Mars,* which was to be part of the ten books on Quantavolution, and which is a catastrophic, astronomic interpretation of the Homeric Song of Demodocus, such as it appears in the Odyssey, one of the most ancient song pieces taken up in the epic. To Alfred, it recorded the planetary instabilities of the age immediately preceding the Trojan War, and the interactions of a catastrophically unstable, des-orbited Mars with Venus, Earth and the Moon. We left the boat and went for a long walk in the hills, then returned on the ferry to spend an uncomfortable night: as the trip had been supposed to last only three hours, we had not taken cabins... The boat left sometime in the middle of the night and reached Thira at dawn, on a rough sea, passing, as it entered the caldera, the stately 'Achille Lauro,' an Italian cruise-boat, which would be a few years later the target of a notorious Palestinian terrorist attack (remembered in an opera: *The Death of Klinghoffer*).

Earl Milton was deeply moved by the experience of this trip to Thira - it was the only time he went there in his life - by the awesome sight of the caldera, the excursion to the excavations of the Late Bronze Age city of Akrotiri, which had been buried by volcanic ashes, and particularly by a boat trip to the "new lands," barely emerged, of the volcanic islands, where the soil was hot and sulfur bubbled up under his hands as he dug into it. When he died, nineteen years later, in 1999, we learned from his devoted second wife, Ann, that he had asked for his ashes to be spread on this island, in the middle of the caldera.

This, then, was where and how *Solaria Binaria,* the co-authored book by Alfred and Earl, was spun out from a chapter in *Chaos and Creation.* A year later, Rosemary Burnard started typesetting the manuscript in her house in East London. It was slow work. Every week, batches of type-set pages arrived. Earl proved a slow and demanding editor. Alfred and I brought the book to a printer in Manhattan. It was the seventh of the ten books to be put into print. Almost simultaneously with *The Disastrous Love Affair of Moon and Mars* and *The Burning of Troy.* Trips to printers had become routine. The arrival of cartons of books too, which we stored in the basement of the little Trenton house.

After that, there were left to finish, and publish, *The Divine Succession,* which rounded off the theoretical model of Quantavolution by moving it to its ultimate and inevitable conclusion, into the theological sphere, through what Alfred called theotropy, the dynamics working in opposition to entropy - creation, as against destruction. The last, autobiographical volume, *The Cosmic Heretics,* told the story of his unique friendship with Velikovsky and his involvement with the catastrophist movement. *The Cosmic Heretics* came out in the Spring of 1984. The last of the Ten Volumes, the achievement of the Herculean task, the Labors Accomplished.

All ten books were published by our Metron Publications - a company created and owned by Alfred since the 1950's, and which had published products as distinguished as *The American Behavioral Scientist,* and the "Universal Reference System," the first computerized index in the social sciences in America - or in the world, for that matter - created in 1964, another achievement of Alfred which had been ahead of its time.

The books were difficult to handle, to market, to send off. During the years when they had been coming out one by one, we had advertised them, sold them, filled orders, out of the Trenton house. During the months we spent and worked in Naxos, Sigmund Kardas used to take over. Sigmund Kardas, who lived in a fine house in a Trenton neighborhood which bordered on the terrifying, taught science in one the state's most difficult schools and owed to his heroic life-style a financial independence allowing him to pursue his multiple and offbeat intellectual interests. He had lived in Spain, he still owned a house on Teneriffe, but did not care much for travelling. When the Quantavolution Series was complete and published, we took off for Naxos and Sigmund took over, for the following years, the business of selling and expediting the books.

Later that year, after the chemical catastrophe which caused thousands of deaths in Bhopal, India, we went there, together with Rashmi Mayur. Alfred wrote his next book, *A Cloud Over Bhopal,* about the catastrophe, on behalf of the victims and the survivors. With the Ten Books behind him, he could move on, hardly giving them the personal promotion they deserved and required. But this direction was not new to him, rather a return to the sources for he had been, after all, a political scientist with a major reputation. After Bhopal, he went on to write his autobiography, three large books, *The Babe,* about his childhood as a bandmaster's son on Chicago's Near North Side, in the age of Jazz, *The Student,* about his studies at University of Chicago, where he had entered on a half-tuition scholarship at the age of fifteen (and worked-off the other half of his tuition working as a busboy), *The Taste of War,* about his experience in five campaigns of WWI in Africa and Europe.

It was followed by a large history of the United States, from 1492, or rather, from geological times to the millennium, and by a book of poems, *Twentieth Century Fire Sale,* his second.

In 1997, Alfred created his Grazian-Archive website with the help of our friends from Princeton, Syamala and Krishna Jonnalagadda, on which his enormous output in all fields was progressively published and which within a few years developed to a million and a half visitors a year. The Daily Average of Data Transferred from Alfred de Grazia's archive to visitors to the archive, in the month of April 2005, was 310,544 kilobytes. The total number of successful requests of the same month came to 185,353.. The Ten Books were published as a Quantavolution CD during that same year.

With all this, it has been clear to me for over a decade that reading ten thick, tightly printed books (654,198 words!) which, moreover, challenged so many notions commonly held by an educated American, was an almost impossible demand to make of readers.

There was a need, I felt, for a book that would bring together the main ideas of the Quantavolution Series, leaving out much, perforce, but keeping the essential. There are few major problems of contemporary science that are not addressed directly or by inference in the Grazian Corpus. The extent of the coverage cannot be explored even in the many pages to follow. I have drawn nearly nothing, even by mention, of the special, highly differentiated articles of *The Burning of Troy,* little from *The Disastrous Love Affair of Moon and Mars,* certainly a myth/reality epic *in extremis explicatus.* Or of the *Cosmic Heretics* (with its intriguing technique of the double-ego author). *Lately Tortured Earth* is easily the most holistic and adventurous of the geography and geology of the latter twentieth century. *The Divine Succession* is a stunning compendium and logical analysis of the totality of religion, with a catechism worthy of recital at college chapel meetings periodically. The *Cosmic Heretics* tells of all these books, going far beyond what I can divulge here, and to it ought be appended the famous *Velikovsky Affair,* a work which Alfred, once he was convinced of Velikovsky's cause and scientific achievement, produced with the cordial and dedicated support of Livio Stecchini and Ralph Juergens, a classic of the history of science, forever being cited, although narrowly treated, as if only the historical pages were read and not the records of the history of science inscribed by Dr. Stecchini, nor his own contribution to the social psychology and behavior of the scientific establishment. It could truly be said of the book, as befitted Alfred's temperament and past performances, that it forced the issue to the doors of the enemy.

I tried constantly to preserve the deep, organic unity of the whole, which was produced in one enormous and practically uninterrupted drive, a tour de force, a true Herculean cycle, all integrated, all at the avant-garde of scientific investigation and still comfortably far-out, reacting immediately to new advances which conventional science, due to its own obligations, could only afford to consider and absorb slowly, striking out boldly yet always stopping in time to reflect upon the advance and making the necessary adjustment to the whole in order to receive it. Thus ending up in a new paradigm of many parts from cosmology to geology to psychology to history to theology. I followed the process daily, and in endless conversations. Never once did I catch Al trying to sweep under the rug a contradiction or an embarrassing fact - a sin sometimes exposed in Velikovsky's own work - never did he fudge a citation, or take from it only what was useful to his purpose and silence the part of the message which did not support him. Nor did he have an agenda beyond the scientific - like proving, for instance, that Moses' monotheism owed nothing to Akhenaton's religious revolution - the religious and nationalistic aspects of such questions (for which he had a famously keen nose, going back to his early studies in the sociology of knowledge) were mightily indifferent to him...

He followed doggedly his line of thinking and any contradiction and new input had to be addressed, and any new explanation satisfactorily fitted into the whole, because the coherence of the whole was more important than the immediate usefulness or relative importance of its parts. There would be no peace until contradictions were fitted according to the rules of pragmatic logic, and to this the 'Q' series, whether right or dead wrong in many of its parts, owes its terrific conceptual unity. He wanted to prove nothing, promote nothing, except for his own case of trying to get recognition of a scientific truth which was being ignored and neglected - precisely for reasons of ideology and history and motives the non-scientificity of which had come to be forgotten. First of all, as only a social scientist could do - he revealed systematically the disabilities of what was then conventional science, and announced peremptorily: "The King is blind!"

Denouncing the opacity of the science establishment over the years would have been a useful activity. But, having glimpsed the substance of the truths which science was depriving itself of, he set out after them and would not pause until he had pursued them and cornered them at their logical end in a new complex of paradigms. He created a model, reaching from the constitution of the solar system as an evolving catastrophic binary, all the way to the dawn of human awareness in a minuscule delay in instinct, which made us suddenly incapable of acting with the innocence and security of the beasts, and instead made us create all else in an attempt to recapture the wholeness and simplicity of our minds, to reclaim and endlessly transcend the wholeness and health of our lost animal selves, and set us on the path of discovery of which the 'Q' paradigm should be, like all science, one of the crude, early stepping-stones.

Anne-Marie de Grazia

Naxos, 2005

The little red house on Q-Street, Georgetown, Washington, D.C.

Part One

HOMO SCHIZO

Most scholars believe that man has progressed since his original appearance on earth. Probably so, but what a strange kind of progress, hardly understood, and often showing a negative balance of what we, the present result of this progress, must consider the "bad" over the "good."

Some scholars believe that man is a rational animal. In limited ways he is, but, again, what a strange kind of rationality, more ape-like than are other traits of humans that are specifically called "non- rational." For, to preview an argument that comes later, man is continually seeking ways to reestablish the uninterrupted flow of instinctive responses enjoyed by his forebears, and this is the homologue of "rationality." When Descartes wrote of animals as machines, he was obviously unaware that the precise "rationality" of man, which he, of all philosophers, elevated to awesome status, was just this homologue of the machine and animal.

So constrained and confused is whatever is called human rationality, that I prefer to call mankind by the name homo schizo, that is, *homo sapiens schizotypus,* rather than *homo sapiens.* Humans were created and are born *schizotypical,* with a set of traits to be distinguished in this book. They were from the first, and are now, more schizophrenic than otherwise. What is called "rational" is a derivation out of *schizotypicality.*

Might all types of known hominids and proto-humans have been of the species *homo sapiens (schizotypus)* in physiology and culture? Might these and all modern races have appeared during the past 14,000 years? Might man have originated *hologenetically* in the holocene period, by *quantavolution?* Such is the line of questioning and argument to be followed here; outrageous as it may be to conventional theory, it may be also productive.

It can be asserted and defended that in the past fourteen thousand years, a disorder befell the solar system that terrorized and transmutated the sensible biosphere, changed the atmosphere, cleaved and ravaged the crust of the earth, altered drastically the sky and surface waters and destroyed or severely damaged every civilization up to the seventh century before the present era.

During these fourteen thousand years, it can be argued, human groups spent one-third of their time in an environment of natural and social chaos and suffered intense physical and mental stress. Continents were fractured; mountain ranges rose; crustal material was exploded into space; cataclysms of water, ashes, oils, gases, and fire rained from the skies; ice ages came quickly and avalanched, not melted, into oblivion. Oceans were created; seas were drained; floods raged in every direction up to the very mouths of highly placed caves; climate altered in a day and the atmosphere was deprived, enriched, and poisoned on numerous occasions.

We posit that no single mile of the surface of the world can be bored for its actual stratigraphic column without discovering it to be at some points a catastrophic column. No matter what part of the destruction can be assigned to the ages before man, some part of it has to be attributed to the ages of man.

Settlements and civilizations everywhere, from the Arctic Sea to the Tropics, from Spitzbergen to Tiahuanacu, are now, upon exhumation, shown to have been the victims of such events. It appears, therefore, that every hypothesis trying to explain the means of humanization must be developed within the historical bounds of natural catastrophe.

By what means did hominid become man? By electrochemical means, and suddenly. Was the change large or small? The change was substantially minute, but profound in its consequences. When did it happen? Recently -- about one thousand reproductive generations ago, which comes to about 260 memorial generations. What role did great natural forces play? They precipitated and perpetuated the change. Did culture spring up with, or did it lag behind, the human transformation? Culture sprang up with the gestalt of human creation. How many symptoms of mental illness are innate in man? All of them. How many cultures are "sick"? All of them, but the sickness is "normal." Can homo schizo aspire to become *homo sapiens?* One can aspire to a fiction, but cannot achieve it. Occasionally, a person, or even a group, can reach a delicate equilibrium, which can be called "reasonable," thus becoming *homo sapiens schizotypus.* Anything more than that is most uncertain.

These answers are tentative, as must be many scientific propositions. They may appear far-fetched, rightly so, because they must be brought in from faraway fields

CHAPTER ONE

HOMINIDS IN HOLOGENESIS

Australopithecus had certain human qualities. He was adequately supplied with cranial matter. Specimens exceeding the minimal brain size known for normal humans have been discovered. His brain-body build proportions were modern. His size was that of many millions of modern people. His dentition was close to modern man's, far removed from the apes. He was bipedal and held his head high (higher than we do, said Louis Leakey). He was social. He used tools. He built enclosures. He was right-handed. It appears that his brain was hemispherically asymmetric, which introduces additional human potentials.

"McKinley," Wolpoff says, "demonstrated that Australopithecus (gracile and robust) followed a 'human' model of short birth spacing," and Mann showed that "the rate of australopithecine development and maturation were delayed, as in modern man, rather than accelerated, as in modern chimpanzees." (Based upon the timing of molar eruption.) There are no signs yet of his having had speech, but no evidence to the contrary; Louis Leakey thought he had a human palate. There have been few indications yet of his having been religious and artistic. There are signs of his having used fire.

Opposing the theory that australopithecus was human stands largely the thesis that he is anatomically too different from modern man. To the forgoing response may be added the following: we do not know what are the limits of variation within the single species, or how the principal distinction employed -- that interbreeding be impossible -- would apply here. It is of significance that Johanson had persistent doubts about classifying his fossil hominid, "Lucy."

There is much diversity among the australopithecines themselves to fuel controversy; several attempts have been made to call new specimens by new species names. Generally, anthropologists have wanted to join them together as a single species, if only to avoid barren disputation and to make it easier to sum up the primordial situation in textbooks.

Australopithecine was connected with *homo erectus* in time and with the Acheulian-Chellean culture at Olduvai, which culture extends into the Terrafine of North Africa and is found also at Swanscombe and Steinheim, with practically modern man.

We would like to know whether *homo erectus,* too, might be human. If so, how did he come to be created? And when was he created? And if five million years old, or three, or one, or one-half, or one-tenth -- to cite various estimates -- why then should he have evolved so slowly until the Upper Paleolithic, which is variously reckoned at from 50,000 to 10,000 years ago?

Homo erectus cannot be dismissed from the motley ranks of modern man. He had large supraorbital ridges but so have some modern individuals and so, too, Neanderthal man, who had a cranium larger than modern man and a culture. *Homo erectus* had a low skull, yet possessed the cranial capacity of the smaller skulls found among ourselves. And amongst ourselves, cranial size has little or no relationship to average intelligence and competence, or perhaps even to extreme intelligence.

Le Gros Clark, seeking to prove gradual evolution, wrote that by "the end of the Middle Pleistocene, the hominid skull had attained a degree of development very similar to modern man; indeed, except for the rather strongly developed supraorbital ridges, some of the cranial remains of this date are hardly to be distinguished from modern man." Probably *homo erectus* and *homo sapiens* were contemporaries. Inasmuch as fire making was also assigned to Peking man, whether 400,000 or four million years ago, it would seem that humans have been allowed an inordinately long time to sit around fires in a mental funk.

To allow quantavolution in a short time, one must agree that some part of evolution might be systemic, that is, permit a set of crucial human changes to occur together in the same moment and perhaps by the same instant mutation. The issue has been hotly argued. A plurality of biologists are point-by-point evolutionists; very few are saltationists, quantavolutionists or systemists; many are puzzled over the great variety of points to be covered over time, no matter how long, and yet unready to accept "successful monsters" as the answer.

Geologists bought evolutionary time to preserve themselves from alternative catastrophic hypotheses. Whereupon the biologists and anthropologists, together with the geologists, were persuaded of radiochronometry by geo-physicists. The potassium-argon test claims validity over a time span of a billion years and more, beginning at 100,000 years or less before the present. Its favorite rock for testing is erupted volcanic material, ashes and lava. It establishes a constant rate of decay of the isotope potassium-40 into the isotope argon-40 (40K to 40A). Then it measures the amount of 40K and 40A in a rock sample and, by the proportion of the two, determines the "age" of the rock, hence of fossils embedded in the rock. A high proportion of Argon-40 signifies an old age.

Unfortunately for its validity, and despite the brilliant technical theory and achievements represented in its applications, the 40K to 40A test suffers from a defect common to radioactive elements in nature. The elements migrate. In consequence, the proportions change, giving illusory ages. Rocks can both acquire and lose both elements or either alone.

The Upper Paleolithic period falls between the claimed periods of competence of potassium-argon dating and radiocarbon dating. The most careful work on this period is therefore dependent on sedimentary dating in large part, and this cannot get around the possibilities of periods of flood and torrents, laying down blanket after blanket of clay and gravel to create illusions, in today's peaceful landscape, of the passage of much time. This is no new problem. In repeated discussions at the Dordogne cave and shelter sites with French scientists who have excavated and are responsible for them, I have been unable to accept their meticulous reconstructions as valid.

In the end, they rely nowadays upon carbon dating, which although it often upsets their expectations, at least keeps them in the Paleolithic period rather than moving them into more recent times. That radiocarbon dating, which is based upon measuring a ratio involving the diminishing amount of carbon14 isotopes discoverable in organic remains, can be erratic, owing to atmospheric, species, and soil transformations, has already been the subject of investigation. Recently, changes in the Earth's geomagnetic field have been added to the several conditions that alter radiocarbon dating. Unfortunately, the usefulness of radiocarbon dating decreases exponentially as we move into the periods of the Neolithic and beyond, when the need for a dating instrument becomes increasingly acute.

Chapter Two

A SURPRISING COLLAPSE OF TIME

But nothing stands in the way of objectively and empirically explaining the whole set of fossil hominids that rift excavations extending from Syria to Southeast Africa have produced as a short term occurrence under catastrophic conditions. The same is true of Peking Man and of all other hominid and protohuman finds. The Olduvai Gorge hominids and homo can be readily brought into the Holocene period.

Consider how rapidly man changes, physiologically and culturally, under present-day observation and from our earliest direct knowledge, which is Upper Paleolithic and Neolithic. To thereupon add five million years (or 100,000 "memorial" fifty-year generations) of mental and cultural evolution to a substantially completed anatomical structure would reduce to absurdity the uniformitarian theory of the evolution of modern mankind. Or else man would have evolved, and been destroyed, time and time again, never being extinguished. (But of course this would be another form of universal catastrophic theory).

So much time is not needed, if man is evolving on a consistent anatomical base. More time is now defeating to evolutionary theory; the evolutionists do not yet appreciate that they have crawled out farther and farther on a limb which may suddenly and soon break off at the trunk. For instance, could humans and hominids have lived for millions of years without having reached the Americas, where elephants, camels, horses and other mammals already abounded? Would they have waited until 100,000 years ago to descend upon Australia?

Nor can evolutionists cease to stretch time and beat retreat to shortened time. If the time is drastically shortened for paleoanthropology, the radiodating techniques collapse. Then all which depends upon the techniques - prehistory, paleontology, geophysics, geology, climatology, etc. - will come under assault.

Against the background of this stupendous reversal in prospect, other conclusions about fossil man pale. The australopithecines existed alongside *homo erectus* and other types of man, as well as many kinds of ape. The vanished hominids were destroyed by, or adapted to a dominant strain of the human race, in conjunction with natural catastrophes. We shall consistently maintain that *homo sapiens schizotypus* (catastrophized *homo sapiens*) reduced his live, physiologically compatible brethren, whether *australopithecus,* or *homo erectus,* or *homo sapiens,* to subjection, or he exterminated them.

CHAPTER THREE

MECHANICS OF HUMANIZATION

Many processes, external as well as internal, must necessarily have concurred to bring about humanization. Let us examine the most salient:

1. ANCIENT CATASTROPHES

Legends everywhere carry stories of great numbers of people reduced to a few survivors. They are obsessive tales, repeated over thousands of years. Psychiatrists will readily deem them to be founded upon strong memories, a possibility that most historical scholars are loath to admit. The English prehistorian Childe says that the population of England in paleolithic times numbered only in the hundreds, during neolithic times in some thousands. Were then all the people of England mustered to drag the neolithic megaliths of Stonehenge into position, and then reassembled in Ireland and Brittany for the same tasks?

Mircea Eliade, whose research on myths worldwide is justly renowned, is most impressed by the obsession of all peoples with the earliest times of creation, by the permeation of the totality of their cultures with the same obsession that great and terrible events occurred; yet he has not ventured to say that anything at all happened then. Are the legends mere fantasies of primeval poets, primordial Dante's, whose plots no later poet could improve upon?

Dobzhansky needs only one sentence in a large book on human evolution to dismiss obsessions of creation as a "natural" reluctance of people to conceive of infinity.

For all the people who have ever lived, and who claim by a kind of culturally transmitted history that their ancestors arose in large numbers and were wiped out in cycles of catastrophes and revival, no place is to be reserved in science. The Holocene period itself embraces many more fundamental natural events than were once accredited to it, as the latest editions of the *Encyclopedia Britannica* point out. A start has been made in assigning quantavolutions to it, but the major allocations cannot occur until chronological methods are criticized and reformed.

Even though their character as "ages" is not yet defined, the occurrences of catastrophes of continental, if not global, scope may have occurred in half a dozen sections of the Holocene period -- approximately 2700, 3400, 4400, 5700, 11500, and 14000 years ago. For the moment, and so that argument may proceed along its main line, the extent of catastrophism of the past fourteen thousand years can be barely sketched. It would help conceptually to regard all expressions of natural forces of a destructive character witnessed by modern humanity as but the flattened tails of negatively exponential curves of catastrophism.

2. THE HUMANIZING FACTOR

In his treatise on *The Ghost in the Machine*, Arthur Koestler has placed the origins of human "mis-behavior" in a malfunctioning relation of the limbic system to the cerebral region of the brain. The basic reptilian and mammalian control and response systems are located below and behind the cerebrum, which is grossly "over-developed" in man. The rational and constructive inclinations of the uniquely human cerebrum, he thinks, may be frustrated all too often by the more instinctive, unconscious, and irrational animal systems. Human behavior, as a result, is prone to contradictions, rage and aggressiveness, destructiveness, and madness.

Even while admitting that a specialization is occurring in the human central nervous system that can bring about schizoid behavior from a lack of perfect coordination, we must say that the problem is incorrectly stated and may explain why Koestler did not arrive at the focal center of human nature. The problem is not one of "mis-behavior" but simply of behavior, both "bad" and "good, "normal" and "abnormal." *Pari passu,* there is no "malfunctioning," but only "functioning." We do not turn off a spigot marked "rational" and turn on the spigot labelled "irrational."

Once we brush aside this specious and decrepit Aristotelianism and scholasticism, Koestler's work becomes valuable. For now it becomes possible to seek a mechanism of delayed instinct between the automatic and cognitive specialization of the brain, which, in conjunction with other sources of delayed, diffused, and over-loaded responses, may explain the self-awareness, existential fear, and profuse displacements of the human being.

The bilateral structure of the brain, providing two hemispheres, had been fashioned long before the advent of humans, probably on some quantavolutionary occasion between two ages. A division of functions between the hemispheres may have come only with the origination of mankind. The skull case tends to warp to conform to the concentration of functions in the brain; and external asymmetry conforms to the internal asymmetry. Such asymmetry, implying human specialization, may characterize most or all hominids. Ornstein asserts that hemispheric specialization (asymmetry, that is) appears to be unique to humans. Handedness in favor of the right hand, and language, are dominated by the left hemisphere. Asymmetry in the language region is, for instance, discoverable on the skull of "Arago XXII" coming from Tautavel, France. This specimen is classified as *homo erectus* and assigned an age of 450,000 years by uranium-thorium and electron-spin resonance tests. (Source: Musée de l'Homme, Paris.)

Besides governing right hand and body movements and language, the left hemisphere is specialized in analysis and mathematical functions. It is also assertive and, in observed behavior and experiments, tends to dominate decision-making. The right hemisphere of the cerebrum initiates and supports activities of the left side of the body, and pursues non-verbal and holistic forms of thought and appraisals of experience. It is described as artistic and analogical in its ways of processing the external world for internal consumption and action. Thomas Parry has surmised that a relation exists between ancient catastrophism and a take-over of internal and external behavioral leadership by the right hemisphere of the brain on the occasion of traumatic experiences.

Each hemisphere alone can convey to the whole person the possibility of physical and mental survival. Each is in constant touch with the other through the medium of the *corpus callosum* which carries millions of connecting links between them. The severance of this membrane has permitted direct observation of the individuality of the two hemispheres. It leaves a still "normal" person "with two separate minds, that is, with two separate spheres of consciousness."

If the key to humanization is a general delay of instinctive response with a consequent choice-factor introduced into a wide range of behavioral decisions, then a possible source of the delay lies in the corpus callosum and/or any drug that can inhibit the full and complete communication or near-identity of action of the two hemispheres. If, for example, fatigue and exhilaration both produce schizoid symptoms, some quantitative measure of interaction between the cerebral hemispheres may define the normal schizotypical state of the hemispheric relationship; the norm itself would be genetically and/or socially induced on a continuing basis, providing typical human behavior.

The closest that we can come to distinguishing a key factor in humanization is an instinct-delay system operating in the brain but serviced by the body's electrical and hormonal system. This could be called the humanizing factor. In baseball language, the animal in us has been forced to touch base several times before completing the circuit and scoring, and so the human ball game is on. A close investigation of instinct-delay emerges, with the theory that it is an effect of the specialization of the brain, with consequent transmission delays in the coordination of the total brain and organism, and that there is an overcrowding of consciousness because of a spillover of repeatedly insistent messages taking alternative routes for delivery or ending up in functional cul-de-sac.

I shall try to formulate the process of instinct delay here in a manner that will assist in seeking the mechanism of humanization. Instinct-delay *(D)* is a function or ratio of the product of the mass of the brain tissue requiring service *(M),* and the specialization thereof *(S),* to the product of access facilities (the number of receptors or docks and the number of routes pursued by messages) *(A)*, the input of electro-chemical signals *(E)*, and the velocity of the work of transmission *(V)*. That is, $D = f(MS/AEV)$. When all other variables are held constant: **1.** If the mass of the brain increases, instinct-delay *(D)* will usually increase. **2.** If specialization of the brain increases, *D* will usually increase. **3.** If the number of message docks and the number of routes to them increase, *D* will usually decrease. **4.** If the input of electro-chemical signals increases, *D* will usually decrease. **5.** If the signals move faster, with less impedence, *D* will usually decrease.

However, the variables are not entirely independent, although we do not know the extend of their interdependence. Thus, an increase in brain tissue may not bring a proportional increase in docks and routes available to supply the tissue. Nor will the larger brain necessarily be supplied by an increased input of hormones, which come from several places, or electrical charges, to transport messages. Nor do we know whether the electric and chemical signals will carry on with their former speed or will move less rapidly, unless some other factor increases their speed (which may, for example, be an electromagnetic change in the state of the environment). That is, we cannot identify precisely the agents, nor the cause of their behavior. All that we can feel confident of, at this point, is that there are here the rudiments of an explanation for instinct delay, hence self-awareness, hence humanization.

The recent association of high or uncompensated adrenalin secretion with schizophrenic symptoms suggests offering this drug as a candidate for a humanizing auxiliary. One is inclined to distrust so simple a solution to so fundamental a problem, even after posting the usual warning signs: that the process is more complicated than it appears; that we know next to nothing about the circulation of adrenalin and other drugs with which it interacts in the process; and that historical proofs of such an evolution are probably impossible.

One might as well suppose, while offering the same type of warnings, that an electrical change has brought about human behavior. If the Earth has gained charge in recent millennia, the human body may be operating in a hyper-electrical mode relative to the environment in which it evolved. This would be the case with the biosphere generally; insects, birds, and mammals are all sensitive to electromagnetic fields and changes in them.

The hominid might then become the "nervous human" who turns

upon the not-quite-quantavoluted hominids and trains them to be human, meanwhile through adaptations and interbreeding creating a new race, whose members are quantitatively distributed about the genetic norm of the "nervous human."

As with every significant element in the quantavolutionary theory of *homo sapiens schizotypus,* the hypothesis of the physiological source of humanization is put forward to orient thought and method. The theory as a whole serves to show where we can go when deprived of the assumptions of a uniformitarian external force-field of evolution and of the free, long expanses of evolutionary time.

3.SIGNALLING HORMONES

A logical candidate for mutation and environmental transformation in the chaotic period is the endocrinal system. It is an anciently derived collection of glands, separate from but connected with the brain, the nervous system, blood pressure, metabolism, growth, sex, fear, and stress. It discharges numerous hormones that stimulate and regulate these systems. Its main components are the pituitary gland, the pancreas, and the adrenal cortex and medulla.

The endocrinal system, especially the adrenal cortex, is stimulated by stress and establishes counter-stresses in the organism. For example, rats bred in the laboratory have smaller adrenal glands and less resistance to stress, fatigue, and disease than wild rats. Their thyroid glands are less active and their sex glands develop earlier and permit greater fertility. They have smaller brains, are tamer, and are more tractable.

In humans, similar differences occur between people who are stressed by the environment and those who are not. New Yorkers usually have enlarged adrenal medullas, compared with the American population at large. Paranoid and obsessive traits, involving distortions of reality, are commonly observed among persons who suffer from an excess of adrenalin either as a result of great fear and anxiety or in consequence of inadequate suppressive and discharging chemicals and mechanisms.

Schizophrenia involves at least some separation of the "primary" self from a second self, which includes part of the self and engages in profuse identifications with the outer world. Frequently observed in mind-workers, it evidences heavy pituitary stimulation of the brain as well as insulin and adrenalin "excesses." The brain often becomes ungovernable owing to endocrinal disturbances.

Notable, too, is the association of fear, aggressiveness, and sexuality in variations of the endocrinal system. It is then reasonable to suppose, for instance, that sexuality is determined more by the stresses of the quantavolutionary period than by the aboriginal oedipal complex or simple sexual drives.

Other modes of mutation or transformation also point to the importance of the endocrinal system in developing humanness. Solar radiation stimulates the adrenal system, both directly and indirectly. Hence, abruptly changed levels of solar and other types of extraterrestrial radiation may have prompted humanizing behavior. The types of social imprinting imposed upon the first generations of mankind and all generations since then were, so far as we can tell, the same; delusory, symbolic, obsessional, and aggressive; these are typical products of endocrinal excesses.

Finally, the obsessive will to mutate, to change one's core-being down to the egg and sperm themselves, has been proposed by Freud as an evolutionary example of "the omnipotence of thought;" so strong a will would be more probably and capably generated in individuals who are endocrinally excited than by growth of the brain; therefore, the accelerated development and passover of hominid to human in a quantavolutionary period may be owed to the endocrinal system.

Man is supposed to be fetalized as compared with the apes since in the adult man the size of the head and the relative proportions of its parts resemble those in juvenile apes rather than those in adult apes. Bolk speculated that fetalization may have been caused by changes in the hormonal balance in the body, especially by a decrease in the production of the anterior pituitary hormone.

Dubrow has correlated growth and size of humans and many other life forms with changes in the intensity of the earth's magnetic field. We may wonder then whether an endocrinal change produced by a change in the GMF (geomagnetic field) might stimulate pituitaryism and expand australopithecus to modern human proportions. Since the left brain hemisphere is asymmetric with the right hemisphere, being larger occipitally, and this area is close to the calculating and speech centers, then a growth of the total cranium implies an important proportionate growth of this area and its special functions. That it may be more than proportionate is indicated by Dubrow's finding that the length of the skull geographically varies inversely with the intensity of the GMF. Thus humanization would accelerate.

The quantavolution that split man's mind and freed it to displace copiously upon the world may thus have been influenced by a declining GMF.

4. MUTATION

Let me consider now mutation, asking the ethologist and expert upon instinct, Tinbergen, to describe the situation:

> Present day theories of evolution consider mutations in the widest sense as the basis of all heritable change. The variability due to mutational change may show directiveness of various types, adaptive as well as nonadaptive. Adaptiveness is brought about by selection. Speciation, or the divergent evolution of populations originally belonging to one species, starts with geographical expansion of the species' range to such a degree that two or more populations of one species become reproductively isolated. The various populations thus isolated are usually slightly different in genetical make-up right from the beginning. This difference, together with the environmental differences leading to different selection pressure, account for divergent evolution of the populations which ultimately results, via the formation of geographical races, in the origin of new species, genera, and even families. Whether this 'micro- evolutionary' process is at the bottom of all evolutionary divergence, even of those often called macro- evolutionary, is a matter of disagreement. It is certain, however, that the causes of evolution can only be studied in micro-evolutionary processes.

A gene is a large molecule of deoxyribonucleic acid (DNA) wound on a double helix, along which are strung in fixed order some simple chemical structures called nucleotides. Only 4 types of nucleotides are ordinarily found in one and all chromosomes but their varied arrangements establish by code the behavior to be followed by any given gene. A gene mutates - that is, changes its code and hence its "building plan" - by a disarrangement or loss or destruction of one or more nucleotides of the helix.

This accident occurs when a foreign chemical or particle or charge or wave or organism enters the chromosome and its gene, with especial effects when the gene is in the process of duplicating itself. Once the gene is altered, it transmits new instructions and whatever aspect of the organism is under its command will accordingly change.

Mutations may also affect the organization of genes within the chromosome, rearranging them or even rearranging chromosomes. It can be estimated (following calculations by Wallace and Dobzhansky) that in the case of man, the number of "spontaneous" ("natural," "background") mutations that would occur for a world population of four billion people in 350 generations amounting to 10,000 years would be only around two hundred (200). Since practically all mutations are "cosmetic, " harmful, or lethal changes, it is embarrassing to place one's faith in mutation (at least as here construed) as the factor bringing about speciation from hominid to man.

If we are to use mutation theory at all, we must associate it with radionic turbulence of the most violent kind, extended over many centuries. And it is even more plausible if, to such catastrophic mutators, we add a permanent change in some atmospheric constants. And, then, too, we must continue to belabor mutation theory, for there is some deep mystery in it - a kind of genie in the bottle, something of Lamarck's environmentalism, of Freudian psychosomatism, perhaps even of the monads with miniature universes within them, which is to say, something of a Great Intelligence.

At this point in time, then, we still need mutation theory and catastrophe theory, with an open door to whatever other theory comes bearing fruit. We receive a hint that the merger of gene theory of mutation into macro- evolution or quantavolution is possible with studies showing that much DNA (like much brain tissue) is surplus, seemingly unnecessary. Is this material held in readiness for the sake of recombination? Is it potentiated for organizing a general response in the event of a mutation that would otherwise be too specialized to survive in the species? George Gaylord Simpson laid down several principles that are pertinent to humanization.

> It is now known that mutations, broadly speaking, an ultimate basis of variation, are discontinuous... The somatic effects of mutations vary from great to barely perceptible, or quite likely, to imperceptible by usual methods of observation... Despite the fact that a mutation is a discrete, discontinuous event at the cellular, chromosome, or gene level, its effects are modified by interactions in the whole

> genetic system of an individual (oddly enough, there is no generally accepted term for that important concept). They [mutations] are also modified by varying environmental factors. The results are that for many mutations, the somatic effects in different individuals vary in an essentially continuous manner.

The whole genetic system falls into line with the mutation, so to speak. This is certainly a hologenetic effect; one wonders why "no generally accepted term for that important concept" exists. A great many features of the organism (hence species) are systematically calibrated.

Simpson hesitantly comments on the likelihood of quantavolution of species:

> The instantaneous origin of a new species by a single genetic event can occur but is unusual. It is practically confined to cases of increase in individual chromosome numbers happening to produce a system both viable and capable of reproduction but not capable of backbreeding into the parental population. In usual... cases distinct evolutionary change involves the increase or decrease of proportions of genetic factors in whole populations, and this is a gradual process occurring in successions of generations. The prevailing modern theories of evolution are essentially, although not dogmatically, gradualistic .

No new species has been proven to form, through mutation, breeding, or otherwise, in human history. First mutations occur rarely; perhaps one in 25,000 spermatozoa or eggs possesses a gene that has been mutated. Still, with a large population over a long period of time the number of mutations will be high. Since women carry their eggs from birth, an egg mutated on one occasion may be represented in a birth as much as forty years later; male sperm is wasted and renewed, millions per ejaculation, so that a mutated sperm has very little chance of being partner to a conception.

5. MAJOR FEATURES OF EVOLUTION

An effort was made by the geneticist Richard Goldschmidt, in 1940 to provide a new material basis for evolution. He said that his lectures at Yale ought to have been called "the genetical and developmental potentialities of the organism which nature may use as materials with which to accomplish evolution." Evolution and natural selection, including the survival of the fittest, were accepted by him as facts. But, he said, selection and adaptation required "necessary hereditary variations" to work with.

So he strove to discover evidences of "macro-evolution." He showed how hereditary differences, that might have fateful consequences in appearance and behavior among species, might be attributable to certain mutations. He conceived the idea that "hopeful monsters" would be frequently generated, from among which some rare type might accomplish an evolutionary saltation. Although he could not demonstrate such directly, he conceived that novel patterning of chromosomes might instantly achieve the same effect as an accumulation of mutations, producing a new chemical system that would substantially alter an organism's appearance and behavior. So he could speak of "systemic mutation" as a complete change of the primary pattern or reaction system into a new system.

He might have added the term "hopeful scientist," to describe himself and others who were products of the "hopeful monster," homo schizo. The phrase: "To illustrate the presence and wisdom of God in the natural and moral world" meant to the naturalist, he declared, "the demonstration of law and order in his chosen field." This view is a common amnesiac sublimation of the characters of the gods Yahweh, Shiva, Zeus, and Jupiter, spreaders of chaos and lightning-like destroyers of the order of Mother Earth and Mother Nature.

Perhaps if he had investigated the character of his gods, he might have truly found the means by which nature accomplished changes -- by catastrophes multiplying infinitely the mutating forces and adaptative opportunities of the world. He then would agree even more with another authority whom he supports, O. H. Schindewolf, the German paleontologist, who not only surmised macroevolution but adjudged the causes to be catastrophic and extraterrestrial, in a set of studies published between 1936 and 1963.

The earliest men were in fact "hopeful monsters" who had to believe that the gods were responsible for their sorrows, as well as their welfare, but sublimated many of the sorrows.

6. INTELLIGENT MUTATIONS AND EVOLUTIONARY SALTATIONS

That genes instruct organisms via chemo-electric code is well known. That genes mutate occasionally has long been known. The mutation as an electro-chemical event with functional consequences is also appreciated. Puzzles remain: how, if at all, do mutating genes provide the non-random set of instructions needed to accommodate the rest of the organism to the new structure/ function of the changed part? The problem is made all the more poignant by the observation that nearly all mutations are relatively "meaningless" and mostly trivial; yet a given species is integrated functionally, and differs "significantly" from another species.

A new metaphor is therefore suggested. We assume that the mutation is a changed chemical message sent by one gene to all other genes as well as to the operations which itself commands. Every gene (hence chromosome) receives, upon mutation, not only a capability to provide a new instruction but also a capability for leadership. Every gene, like Napoleon's soldiers, "carries a marshal's baton in his knapsack." When it mutates, all other genes become dedicated followers. The gene, as befits the ideal field marshal, conveys to them instructions concerning the behavior newly expected of them. They do their limited best to conform to the new order.

Is it conceivable that the genes carry design accommodations for every successful macromutation? If so, where do they originate? Suppose that, in the beginning of life forms, each gene is possessed of designs that can cope with every form from an amoebae to a whale (this is, of course, not a new idea).Given a certain chemical stimulus, it will produce its part of the structure and function of any species known up to the present and many more. There is no logical reason why an individual gene capability of a bacterium of 2^{2000} combinations cannot foreshadow all life forms that have developed.

Mutations can be conceived to cause very little or very great changes in the structure and functions of a species. Further, mutations are considered statistical, that is, indeterminate increments contained in a limited number of commands. They are, of course, not models of the form-to-be. Their "intelligence" consists in their primordial ability to induce coordinated shifts of behavior in non-mutated genes. By implication, "important" changes occur by saltations, as quantavolutions.

In environments that provide mutational possibilities, radically different forms can emerge quickly, propagate abundantly, and branch quickly again. Long-time durations are of little importance; whether they occur or not is immaterial. The theory here is so simple that it may be merely a metaphor. It need not be justified by elaborate mathematical calculations. It preserves most of the general observations of Lamarck, C. Darwin, St. Hilaire, Mendel, Dobzhansky, Watson and many another geneticist, it can cope with paleontology and genetic engineering without strain. It suggests, among other things, that, in principle but against great odds, preexisting ancestral species can be recreated, and that the creation of future "major" life forms is within sight, perhaps at the level of probability of controlled nuclear fusion.

7. External Producers of Mutation

Schindewolf, Salop, and a number of other scholars, whether in the close fields of genetics, geology and paleontology, or in the general field of catastrophism, have brought forward volumes of material to support the likelihood of mutation-causing disasters. Not only mutations, but all other factors in speed-up of genetic change are provided by natural catastrophes -- isolation, adaptation, and extinction of competing species. Thus we hear Simpson say that "The chance of fixation of a favorable mutation may be considerably larger by accident of sampling in a small population than by selection in a large population..."

Catastrophes therefore simulate in quick time the supposed effects of natural selection. If man has been humanized within the past 100,000 years, or even within the past million years, actually at any age boundary, even granting the dubious long-time reckoning, he probably was humanized by catastrophe. Here the quantavolutionary model diverges from the evolutionary model most emphatically. In order to enhance the chances of a viable speciation by mutation, a heavy bombardment of particles is required such as has not been experienced in history; a radiation storm is called for.

Such storms must have existed on numerous occasions in recent prehistory, if the evidence assembled in my *Chaos and Creation* (1981) is accepted. They ionized thoroughly the environment by interrupting, deflecting, and reversing the electromagnetic field of the Earth; by mega-lightning electrical discharges upon the near encounter of bodies in space, upon the occurrence of great potential differences between space and Earth; and by removing cloud canopies and transforming the gaseous composition of the atmosphere. Meteoric pass-throughs collisions would have occurred. The Sun would be stimulated to hyper-activity. The electrical and atomic state of every organism and rock would be altered. The radiating effect of one meteor or comet of small size gliding through the atmosphere is heavier than that of a large cluster of hydrogen bombs because of its great heat - well over 2000° C - over a long trajectory; the wide distribution of fall-out; and its possible final explosion at a great speed of many kilometers per second.

A single such passage, of which there would have been many, should produce millions of mutations in the biosphere generally. A large explosion creates a catastrophic tube from the upper mantle into outer space, in and around which many millions of combinations of electrical, chemical, physical, material, thermal, and pressure events take place.

Yet, it is not only extinction that occurs, but also speciation. As soon as they will grant the occurrence of extraterrestrially caused disaster, paleontologists will arrive at a public agreement in favor of quantavolution. Essentially this would include first that the species have been created and exterminated in waves. The waves will probably be fixed chronologically at the passages between the conventionally named periods -- such as between the Pliocene and Pleistocene. Thirdly, they will probably settle upon radiation storms (or, for stretched-out changing, new atmospheric constants) as the principal force bringing in the great changes. They may well decide that these radiation storms are connected with cosmic explosions and encounters. Finally, they may, with greatest reluctance, turn to a shorter time-scale for measuring the succession of events in natural history. For radiation storms and geological disasters not only mutate and exterminate species; they also invalidate methods of dating that assume a constant chemical and geophysical environment.

8. VIRAL MUTATION

Quite recently, the role of viruses in genetic change has come to be recognized. Viral storms might accompany the large-scale penetration of the atmosphere by exploded material from extraterrestrial events. Various ancient myths report such occurrences. Apollo was the Greek god of plagues and arrows; he was a sky god (and not the Sun, as later writers supposed). Hoyle and Wickramasinghe have, in their book *Life Cloud,* proposed that early life forms were deposited on Earth by cometary fall-out. In *Disease From Space* (1979) they also claim space dust as the carrier of plagues to Earth. Deeply buried viruses might also be exposed and some of them mutated by large-scale earth upheavals. Large explosions can create drifting material that will disseminate both crystallized and already activated viruses in similar fall-outs, and with similar genetic results.

Hope-Simpson in 1978 reported that the last six peaks of sunspots coincided with pandemic influenza, possibly from increased cosmic radiation which mutated existing viruses, enabling them to evade human immunities. Not to be ignored, therefore, is the chain reaction of a virus, a viral mutation, and a human mutated by a virus. Again, the likelihood of successful mutation is small but the participating organisms are exceedingly numerous. We bear in mind the theory, advanced elsewhere in my studies, that several solar system bodies disintegrated during the past 14,000 years. One or more were probably carrying life forms. Viruses might persist for some years, possibly thousands, prior to their extinction, in a permanently hostile environment, and hence, while scarcely detectable today in meteorites or direct planetary sampling, would have been aboard their exploded vehicles in ancient strikes against the Earth.

9. Psychosomatic Genetics

Still another means for achieving humanization, and also mutational, would be the psychosomatic conversion of genes. For a time, the idea fascinated Freud and Ferenczi. They were influenced by Lamarck's theory of the inheritance of acquired characteristics. On October 5, 1917 Freud wrote to Karl Abraham to this effect, saying: "Its essential content is that the omnipotence of thoughts was once a reality." When Abraham responded that he had not heard of the idea, Freud wrote that it would complete the theory of psychoanalysis by providing a theory of change through an "entoplastic" adaptation of one's own body.

> Our intention is to place Lamarck entirely on our basis and to show that this 'need' which creates and transforms organs is nothing other than the power of unconscious ideas over the body, of which we see relics in Hysteria: in short, the 'omnipotence of thoughts.' Purpose and usefulness would then be explained psychoanalytically; it would be the completion of psychoanalysis. Two great principles of change or progress would emerge: one through one's own body, and a later (heteroplastic) one through transmuting the outer world .

Even before it was realized how minute was the probability of successful genetic mutation, Freud, like many another thoughtful person, like the theologians, like even Alfred Wallace and Lyell (until his old age), could not accept the piecemeal elaboration of *homo sapiens* according to the uniformitarian Darwinian model. With scientific catastrophism in disrepute and obloquy, they could not imagine an appropriate environmental stimulus to change.

The theory is not beyond discussion. Presumably the hominid bearer of sperms or eggs would be so drastically affected by environmental turbulence that it would will a chemical mutation upon them. Practically every tissue and organ of the body has been shown to be capable of physical change, usually deleterious, when an obsessed person focuses intense and prolonged attention upon the soma. The genetic material cannot logically be exempted from the obsessive influence; both point and systemic mutation could then occur.

The psychosomatic model has a low probability. Although the terrorized hominid woman may have had the most intense desire, conscious and unconscious, to change her off-springs, how could she have known how to target the eggs in her womb? Can terror act as a chemical bullet directed at the eggs? Psychosomatism, we can affirm, performs the seeming miraculous. But we prefer to believe here that psychosomatism is the cultural product of the already humanized homo schizo. Freud and his associates could not come to close grips with psychosomatic humanization; the chemistry, biology and neurology were not available then. They may not be now. Freud's reconstruction of the origin of conscience suffers from such basic flaws that one marvels at even the limited acceptance granted it.

10. An Atmospheric Transformation

External events can introduce continuous and to some extent permanent changes (operating as a new constant), if the events and the conditions they bring about persist. So long as heavy noise, air pollution, rapid movement, and other high-stress life conditions of New Yorkers are constant, New Yorkers will tend to have swollen adrenals. Or, so long as the proportion of oxygen in the air of the High Andes is relatively low, the people there will have unusually developed lungs.

A connection of the endocrinal system with megavitamin therapy has registered effects upon schizophrenia through facilitating the physiological discharge of adrenalin. A diminished oxygen supply or incompatibility of oxygen type in the atmosphere may introduce schizoid symptoms to some part of the population. The brain needs oxygen not only to survive but to energize neuro-transmissions throughout its domain. In schizophrenics the oxygen level in the brain is sharply lower than normal. Further, frontal lobe brain activity is low. Thought dissociation may be produced by oxygen deficiency in the frontal lobe. A radiation storm; a material fall-out; a sweep-out or in-take of atmosphere in transactions with extraterrestrial bodies; intense electrical storms; and the dropping of canopies (opening of skies) can drastically reform the atmosphere. They might change atmospheric constants abruptly or over a period of time. The new atmosphere forces upon the hominids a new "norm" of response.

In this case, Hominids "X" are presumed to have an already existing genetic capability of becoming human. They are genetically preadapted to quantavolution. The new norm is, at least among some individuals, within the range of genetic capability. The adaptible survivors behave according to the new norm, which is to say that they now behave as "humans." The reconstructed atmospheric constant may affect most importantly the fetal environment of the humans-to-be. This happens when the new chemicals in the air find their way into the hormonal food supply of the fetus. And/or the new constant presents its demands for changed physiology and behavior upon the infant after birth.

Man, and all life, lives off a radiation diet that is generally unperceived. Even today, delicate scientific instruments are required to detect radiation, and the symptomology of radiation poisoning is not very clear, or where clear, does not readily name its precise cause. The atmosphere of chaos was a mutator. The Sun of the later Solarian Age may not have been visible. Nevertheless, the finally settled atmosphere has played a role in humanization.

Legends around the world speak of a primordial cloudy sky. The opening of the skies would increase radioactive influences from perhaps still nearby and hot planetary bodies, and also and especially from the sun. Exposure to helio-radiation (including ultra-violet rays) generally increases physical resistance, relieves arthritic and muscular pain, lends a feeling of well- being, stimulates ergosterol and hence Vitamin D production, counteracts rickets and respiratory disease, and kills bacteria and fungi of the skin. It promotes the healing of wounds and athletic performance; it increases the rate of basal metabolism. All of these occur at the price of occasional skin cancers, and possibly of still unknown deep changes.

Although they would contribute to a higher general level of health and activity, they would not create the human. Larger events are required. The Earth's geomagnetic field has come under intense study in the past few years, because evidence now available points to reversals in the past. Whether the field has reversed quickly and often, as quantavolutionists believe, or gradually and rarely, as evolutionists think, a reversal of the North Magnetic Pole introduces an interval during which cosmic rays can descend upon the Earth unhindered and bring about mutations in great numbers. Some studies have indicated a coincidence of reversals with waves of biosphere extinction. Furthermore, the geomagnetic field (GMF) is declining slowly. If the decline has been exponential from some past peak, as I believe and am discussing with Earl R. Milton in *Solaria Binaria,* then the hominid was subjected to a sharply different paleomagnetic field.

So we must ask ourselves whether the relaxed grip of the electromagnetic field disorganized the hominid brain and in effect created homo schizo. For he would be presented with an intellectual freedom in the form of a bewildering number of options for action instead of the more closed system of stimulus-response accorded Hominid "X." The "constant" is still changing, but slowly, today. Still the frequency of heart attacks has been convincingly associated with internationally collected measurements of geomagnetic activity as registered by magnetometers.

11. Social Imprinting

A fifth means of transforming hominid into human nature might be by the social imprinting of shock upon the individual. The hominids again afford the basic genetic capability and a pre- adapted habitat. In this case, however, natural disasters inflict shocks upon the hominid beyond its "normal" tolerances of stimulation. The shocks in themselves are the grossly exaggerated homologues of the shocks of "normal" existence.

They take the form of a celestial scene inhabited by new symbolic references and other mind-openers; of terrorizing high-energy expressions including spectres and pandemonium; of crushing and effacing effects that are prolonged and of high intensity; of the raging of the natural elements.

Thus chants the chorus in Seneca's ancient tragic drama, *Thyestes,* of the shocking fiery passage of Phaeton in his solar chariot, when each and every constellation deviated:

This is the fear, the fear that knocks at the heart
That the whole world is now to fall in the ruin
Which Fate foretells; that Chaos will come again
To bury the world of gods and men; that Nature
A second time will wipe out all the lands.
That cover the earth and the seas that lie around them
And all the stars that scatter their bright lights Across the universe.

The shocks are so traumatic that the victims adopt response behaviors that become patterned as the essence of human nature. The traumatized catastrophical survivors retain the memories, but distort and use them in ways that are typically human. Most importantly, they devise in the very process of their own creation the social means of perpetuating their own changed mentalities and behavior. Human nature is then and thereby guaranteed by a collectivity of humans formed into a group or society. The memorial generations transmit and adapt new traumatic and "normal" tribulations to the fixated human nature.

In explaining the development of the human mind in relation to the catastrophes of Venus and Mars in the period 1453 to 687 B. C.

Velikovsky pushes beyond Nietzsche, Freud, Jung, and Eliade with the concepts of collective amnesia and aggression. Mankind is destructively aggressive as a result of suppressing its memories of natural disasters.

> "The inability to accept the catastrophic past is the source of man's aggression... Freud did not come to understand the true nature of the Great Trauma -- born in the Theogony or battle of the planetary gods with our Earth, brought more than once to the brink of destruction -- which was the fate of Mercury, Mars, and Moon. Freud died in exile from his home, when a crazed worshipper of Wotan was preparing another Götterdämmerung."

The view which I am setting forth embraces this criticism of Freud and the concepts of collective amnesia or repression concerning catastrophes. Also, aggression is to be correlated with this suppression, and the techniques of aggression are in a direct sense analogized unconsciously and consciously to events witnessed in the sky. Nevertheless, I perceive social imprinting as at best an auxiliary source of human nature, an intensifier, which itself needs to be intensified from time to time by fresh natural (or man-made) catastrophe. I argue that the reality recognized by the first human was catastrophic and his mind was as well. There was never -- and here I think we diverge from a common view of Velikovsky and a great many others, including conventional long-term evolutionists -- a clean minded, rational, evolved human whose mind was "blown" by catastrophic experiences: the recurrent disasters proved to homo schizo that his vision of the world was correct!

12. THE SUMMARY MECHANICS

It is perhaps apparent to the reader by now that I prefer, as a "holding position," a complicated mix of several means of humanization, altogether happening within a very short period of time. The mutation of an individual hominid is given prominence generally in the scenarios to come. But it is not difficult to switch from the one to the other, or to stress a combination. The changed atmospheric constant as the mode of humanization has the value of inherent continuity, and is as efficient as genetic mutation in explaining generational inheritance; also it permits humanization to occur simultaneously among many hominids at the same time, in the same month or over a few years. We might begin a search for humanizing mechanisms that are present in the modern atmosphere but would not have been present in an atmosphere in which hominids could thrive.

The branches of the human race have changed in some respects, mostly cosmetically, since the cosmic beginnings of homo schizo. But the basic ways of behaving as human were determined in the midst of great crises: the slowing down of the Earth's motions, the loss of electrical charge, the dropping of the immense cloud canopies in deluges, and the first openings of the sky. An allotment of a thousand years would have been sufficient for these tremendous experiences to bring about humanization.

Even while mutations were abundantly occurring among all species, a single group of hominids, largely potentiated as humans beforehand, in distress and in terror, would find amongst themselves individuals of flexible, if erratic, genetic constitution, who were capable of expanded symbolic behavior and signaling various interpretations of the new giant forces of the environment.

The same group would become capable of managing its newly installed communication system, and then lend its cooperative forces to the evolving interpretation of the universe, the aboriginal cosmology. The group would be driven to adopt the new system even before all of its members shared the mutant genes. In the endeavor to ease their pains and anticipate the sharing of the inheritable traits, it is possible that non-mutants actually mutated themselves by will power, adding a consistent but different emotional mechanism to the hereditary pool of the human-dominated group. Whereas the first mutants would operate by genetic instructions, the second kind of mutants would work out genetically a mode of hyper-excitation of the endocrinal system. This would lend the group an element of obsessive emotionality as soon as genetic miscegenation began.

The social imprinting of shock would come about not by itself alone but in the course of executing symbolic references of the first mutant type, in accepting the obsessive drive of the second mutant type, and in the development of followership among the erstwhile normal band, consisting of sophisticated crowd behavior already possessed by hominids.

All elements would be caught up in the atmospheric reformation. The mutations were consistent with it; they were in fact created by it and responsive to it so that, in a fundamental way, the correspondence of the new world with the new being was assured. Although it did not eradicate the old "normal beings," the radicalized atmosphere punished them and preferred those who responded readily to the new constants.

CHAPTER FOUR

THE GESTALT OF CREATION

The human creation happened all at once with a crackling and bursting of the hominidal dam. It was a "gestalt," a configuration of nearly simultaneous and transacting developments emerging from a central change.

THE GESTALT OF CREATION AND ITS AFTERMATH

(The Hologenesis of Homo schizo)

A.

Low-powered environmental forces operate, in a uniformitarian way.

B.

Hominid is not self-conscious. It has fully functioning instinctual reactions. It has an ape-like cranium, is bipedal, four feet tall, semi-human in appearance, and hairy.

C.
Individual concentrates its life energies upon physical well-being and sociability.

D.
Perception, cognition and affection are governed by a single coordinated instinctual being. Only rarely and temporarily are they "distorted;" no matter how bizarre or self-destructive its behavior (induced by disease or fright or chemicals) it does not ask "What am I doing?"

Postulate now a set of terrorizing natural disasters and distraught faunal populations. Problem now posed is: How could a human be created and survive?

A.
High-powered environmental forces are unleashed in sky and earth. All senses are bombarded. Radionic storms change the atmosphere and invade organisms. Physical well- being and sociability are everywhere damaged and threatened. A reign of natural terror.

B.
Instinctive behavior is generally frustrated by terror and strange stimuli. Microsecond delays in central nervous system and especially in brain transmissions occur.

C.
Schism of consciousness occurs in one or a few hominids with cranial enlargements. Proto-decisions are required for self-control. The "ego" begins as coordinating center for the fragments of the old conscious self; it is actually a poly-ego.

D.
Memories are intense. Memories are also suppressed in the struggle for self-control (ego versus alter-egos). Selective recall and forgetting spring into being.

E.
The alter-egos displace terror onto other people and the threatening natural forces. The primordial being does not know whether he is "talking to himself" or "talking to others." Self- punishment and self- mutilation are found to be ineffective but persist in efforts "to unite the soul."

F.

The ego begins to communicate with its selves by displacement and projection, and having begun the process, extends it to all subjects of displacement. Symbolism as internal language begins. Bilateral asymmetry (right handedness) is stressed to help centralize dominance in the left-brain hemisphere.

A second phase now occurs, organizing the world schizotypically.

A.

High-powered forces continue to impress senses with destruction, chaos, and threats of return. The poly-ego is fraught with ambivalence toward the forces, hence, by retrojection, also among its alter-egos.

B.

Perception, cognition, and affection are pliable (less instinctive and internally distorted "thought disorders") and mix up all kinds of phenomena of the triple-fear (fear of self, fear of others, fear of gods-nature) and triple control system of the person.

C.

Principal imprints upon perception of nature and affection are blocking (amnesia, catatonism); compulsive repetitiveness; and orgiasm (destructiveness, wild expressionism).These imprints of the new world order of the schizoid mind operate within the individual, between and among individuals, and between groups and divine (natural) forces.

Without time lapse, a third phase fashions the culture.

A.

Persons and groups, so as to control fears of self, others, and the object- world (animated),

B.

and to obtain subsistence, affection and the reduction of inner tensions,

C.

organize their perceptions, cognitions, affects, and energies,

D.

through the mechanisms of memory (amnesia and recall), displacements (associations and ultimately sublimations), compulsive repetition (rites, rituals, habits, rules and routines), orgiasm (aggression and nihilism), and communication (by behavior, signs and symbols),

E.

work upon materials and resources of selves, others and the object world,

F.

set up all behavior patterns ranging from informal to rigid, including (1) the regime of language, (2) religious rites and structures, (3) compulsory modes of coping with subsistence, sex, and conflict, all of which bear the stamp of the aforesaid needs, fears, and mechanisms but assume variegated culture-forms depending upon the "mix" of history, no matter how brief the history,

G.

and then exclude or punish "unaware," "sinful," or "sick" persons or groups who, in relation to a particular culture-mix are deviant (i. e., have too much or too little of the key ingredients),

H.

whereupon said deviants (e. g., officially labelled "schizophrenics") must fashion "mixes"of mechanisms and displacements, which, though great in number, represent and resemble in every case the peculiarity of the culture wherein they emerge.

1. A MIND SPLIT BY MINUTE DELAYS

The gestalt brought forth the prototype human instantly (which explains our use of the world "creation"). Whether by mutation or by trauma, the central event was a splitting of the mind in an essentially schizophrenic reaction. The split mind "recognized its other self." That is, it was forced into a basic, irreversible delusion that it had to deal with an inner person.

Self-awareness began. It was an awful feeling.

A permanent blockage (or suppression) was laid down before all instinctual behavior, creating a constant anxiety. The "anxious animal" could no longer act with instinctive ease although it could act more intelligently and with greater versatility. Now we have the answer to the questions: Why is human instinct so blunted in comparison with primates? How does it happen that all "animal instincts" in humans are within reach of psychosomatism? Instinct is the hair-trigger, set to go off without time for decision-making. Many critical human instincts are reachable by will and can be controlled; indeed they must be. They are set as slow triggers. This happened during the gestalt of creation.

Generalized delays of milliseconds in response time between the limbic and cortical systems and between the left and right brain hemispheres, experienced as environmental, electrical, and chemical impulses, introduce conflicts between the systems and the hemispheres. The delays occur not once, but repeatedly and continuously, because the external forces are not withdrawn immediately. The delays add up to a general depression of instinctive responses, which is sensed by the hominid as both crippling and frightful. Even with microseconds of delay, the organism senses piercing inner contradictions that call for proto-decisions by "itself vs. itself."

A host of proto-decisions fill the behavioral response-space left by the depression of instincts. But there is little experience to help understand, control, and guide the mass of proto-decisions. This new anarchy requires organization, but from what sources and how? A pure anguish, it might be suggested, should drive the hominid back into the archaic limbic system whence no self-awareness would ever emerge. But this cannot occur because the stimuli for the new order of mind have blocked the regression and thrown the bewilderment into the cortical arena.

It is too late to regret the passing of the animal. Either one or another of a pair of cortical referents will triumph by making a decision. Still, the several egos cannot contest indefinitely in a battle of all against all, else, like the warriors who sprang up from the teeth of the dragon that Cadmus slew, they will kill each other. The organism, to survive with its one stomach and conjoined limbs, must act as a whole.

The resolution comes from moving forward, not backwards. The organism widens the gap rather than closes it. What began as a set of millisecond delays becomes an alter ego. The alter ego grows through performance, habit and training into a *Weltanschauung* (a world view). The world order emerges, reconstructed by the human mind in a schizoid form. Drinking and eating, bowel movements, fear-flight-fight, copulation and many other behaviors are animal as well as human, but the human way of performing these operations encases them in schizotypicality. All the behavior that is authentically and ineradicably human is schizotypical.

Impulsiveness begins to become a vice, not a virtue, for the human. The organism comes to realize that at any moment it has the capacity to ask itself questions. As frightful as the experience is, the new human cannot resist the asking. The boundary of the brain hemispheres is the main locus for the sensing of the gap. The left hemisphere, losing slightly its near perfect coordination with the right hemisphere, accomplishes reflection. The reflection is fearful, but effective.

2. FRIGHT, RECALL AND AGGRESSION

Fright was all-pervading, both for what was happening inside the person and what was happening outside. Because of the terror and the split, a recollective memory leaped into being and with it instant amnesia. Recollective memory was a form of control, occasioned by the delay of instinct. Hominids might remember, but not recall. The voluntariness of recall summoned up the mechanism of the repression of recall.

Will is the spearhead of the drive to control oneself and the world. Now it was necessary to turn this weapon of will into a weapon of control. For the flood of terror demanded relief. A rapidly growing stream of symbols crossed the bridges between the two selves and flowed out to attach the symbols to the outer world and especially that part of the outer world that was threatening destruction, the turbulent skies and the effects they were producing on Earth.

Great fear was never to be eliminated from the human. It dominated his mind and set limits upon all of his behavior. It was the fear of his own schizoid character and fear of the outside world (and the gods). Of all unpleasantness, being two people is perhaps the most continuously unpleasant. Out of such fear comes the desire to control and somehow stabilize the situation, preferably by merging dissimilar selves into the original unity. Assuming some success in achieving stability, any increase in internal or external fears will excite the fear of loss of self-control.

In all of this a large role for human aggressiveness is prepared, for the world must be controlled if anxiety is to be relieved. Or one must delude oneself into believing that it is controlled and that one can take part in the control system to insure that it will work.

When the sensitive brute could not endure the intensity and scale of internal and external disasters that confronted him, he explored, besides flight and fight, other means of control to cope with "reality." And immediately upon seeking control he found it in the new exigencies of his constitution: in the ability to recall and forget, to perceive his individuality and duality, in "flights of fancy" and in the symbolization of his lines of communication within himself and between himself and the outer world.

When finally given respite from panic, these mechanisms could be used pragmatically, with brilliant and instant success, to organize and invent for all aspects of life. The human had become unconquerable, and lusted for conquest.

The very fact that the changed hominid could reflect upon itself meant that it was not itself, but a split self. So to primordial fear was added existential fear, the fear of one's own self- awareness, the distress of standing off from oneself, the basic schizophrenia of humankind, largely delusory from the standpoint of physiology since the same organs served the plural selves, but of course the schizophrenia was itself physiologically founded.

The origins of human nature were connected with the fearing components of hominid nature, and the subsequent history of human nature, as a result, has been mostly unhappy. The misery is generic, and therefore persists even when the rude clutch of disaster is released, as it was for periods of time, early and lately.

"What happened to me?" was the first question. Then came the gestalt of creation: it was composed of awareness, symbolizing, and projection. The proto-human strove to recollect himself amidst the turmoil of his kind and of nature. To exist and survive he had to discover himself amidst the disruptions of memory. His subconscious now existed in a way that it had not before, as a well of confusion, that overflowed with images that did not belong to the present, that offered uncoordinated seemingly unrelated elements that were taken care of by"unmechanistic ways," unfamiliar to animals. Before he could say "I am," or "I think, therefore I am," he had to come to terms with a new subconscious that distorted all perceptions of himself and others. His character was born of delusion.

The broken mind of the beast sought to restore itself, but could only do so under new terms. Restoration of the previous state was destructive to the organism. A consciousness had to be organized to seek materials to guide the organism in its disorganized condition. It had to pull what it required from the forgotten, which was not really forgotten, but which would no longer normally emerge in a flow of instinctive, directed, utilizable unconscious information that characterized the hominid. The tortures and triumphs of memory then began. The accommodation of an awareness to an uncontrollable but recognized history began. The conscious and the willful assembled together upon this small island in a sea of suppression.

Dominating the transition from a brutish to a human character was the psychological mechanism of projection which sprang from the creative gestalt. Projection is the imputation to another perceived existence, or being, of one's own motives and wishes. Once projection is achieved, a system of behavior in relation to the other is instituted, and a world of transactions, real and imaginary, is set up.

As his own self divided through self-awareness, man's gaze was fascinated by the sky. As he questioned himself, he questioned the intangible and uncontrollable world of the skies and all its mundane effects. The fact that the heavenly forces were abstract and impersonal became a matter of concern much later for a few generations of philosophers. Primeval man did not own a neuter gender. Everything in the world was alive. He did not have to acquire an illusion permitting him to reify or anthropomorphize. For he never had made and had now no reason to make a distinction between the living and the inorganic.

Projection to objects as living things was immediate. The gods came into being. What traits the gods came then to possess were the actual traits of a god as directly witnessed, the traits (later on) of remembered gods, the feelings and traits of mankind in chaos and birth, and such traits of life forms on earth as mankind perceived and found to be analogous to his own and those of the gods. What he saw in the sky confirmed and strengthened his projections and let them be retrojected into his own traits even more strongly. Each time this happened, there was a self-fulfilling prophecy, a growing obsessiveness, an enhanced belief that one was being threatened by sinister forces (paranoia).

3. Self-Consciousness

Self-consciousness, the poly-ego, was a village built upon piles driven into the sands of permanent existential anxiety. It was and is a patterned and integrated architecture accommodating to the neural blockages that deter instinctive solutions. The neural blockages are stabilized by socially elaborated mechanisms that take certain forms such as rituals, theology, and logic. Further, self-awareness involved the use of symbols, *first to institute an inner communication system,* and then to introduce transactions with others and the outer world so as to keep the far-flung egos fully operative.

Once achieved and begun, physio-chemically constituted and socially founded, self-consciousness is revived post-natally in each generation. The human poly-ego was both individual and social. Those possessed of it sensed themselves unique, and at the same time identical with their groups. It produced an anarchism at one extreme and a regimented discipline at the other that go far beyond the capabilities of the mammals. Herein lies the eternal cooperation and conflict between the individual and the group, which is the subject of so much philosophy, sociology and political science.

Self-consciousness in humans is not only awareness, as in a wakeful animal, but it is awareness of the (other) self or selves as entities. Further, the self looks at itself and at other people and objects with the same dynamic. Thus, in terms of psychopathology, the self - the poly-selves - is a form of delusional thought in the schizophrenic category of the split self.

This usage may have developed as a convenience for considering therapy; but in actuality the poly-ego is the only human self to exist and is a system of normal and sane delusions. *Since self-consciousness did not exist until the catastrophes began, the fundamental breakdown occurred only once, this in the days of "creation."* Repeatedly, in subsequent catastrophes, the mind might drift from its first moorings, but, with the help of culture, it would arrive at another anchorage in a new set of self-conscious delusions.

The mechanisms of the human conscious proved to be functional not only in obtaining relief from anxiety, but also at the same time in providing the goods of life. The proto-typical "madness" was "superior" for coping. So the mechanisms of the conscious found themselves to be generally released from their total service to emergency needs of disastrous times. They came to be used pragmatically for many other purposes, including the development of the useful arts and crafts and for social organization.

Ultimately there occurred an everyday dissociation of the emergency and pragmatic functions of the self-aware ego. The emergency functions, that are similar in effects to the superego, are more particularly the primordial functions, ordinarily engrossed by theology. When a new disaster occurred, when the poly-ego system again was deeply disturbed and dissolving, the old self displaced the new pragmatic self and recapitulated the mechanisms of defence as they were employed in the days of creation.

In times of great stress and fright, it is the primordial human self that takes command, not the unconscious nor the beast. A human organism will fall into a catatonic coma or die before releasing the self-consciousness it received upon creation. It will temporalize, symbolize, and control, up to the brink of eternity. Oblivion marks the surrender or death of self-awareness. The problems typical of the human species are in the regression of the ego-mechanisms to their primeval but human state, and not in the resurgence of the hominid (animal) traits.

The unconscious, when reviewed, is seen to be the reservoir of hominid instincts and the suppressed or forgotten materials of experience. However, it is commanded and transformed by the primeval ego, even though physiologically coordinated with the aboriginal instinctive animal.

4. Memory and Forgetting

The corridors of art and culture everywhere echo with the cries and gasps of remote recollections. R. G. Hoskins, in his essays on schizophrenia, writes how patients describe their mental illness:

> Very commonly it is as if the conscious self had descended to some lower region where it is no longer in control... The eyes are opened so that one seems to see back to the beginning of creation. One seems to have lived perhaps in many previous existences.

There was too much that was too bad to remember easily, and it was forgotten. Also too much was forgotten for even the unconscious sectors of the mind to bear. Recall may be regarded as the most obvious and "rational" function of the memory mechanism, mnemotechnology; it operates, however, only if the recollection does not destabilize the poly- ego. The "forgotten," that is, the memories that tend to destabilize the ego's confederational balance, provide essential subject-matter and forms to sublimate activity. They force their way into remembrance via the routes of theology, myth, literature, the practical arts and sciences, and social behavior.

The pride of man in a memory that is superior to that of the beasts is inordinate. Memory is a weak, self-imposed tool for displaying material to the conscious in a light that poorly reveals its sources. The special human memory, like everything else uniquely human, is a device that the beast may not need.

Accompanying the primary amnesia of events themselves, is a secondary amnesia that is associated with the forgetting of events. The amnesia of man came from the primal terrors and set up the mechanism of denial, which first insists that nothing was forgotten, and then persists in denying all sorts of traumatic memories, until we find him today the congenital liar, lying both consciously and unconsciously.

The memories were in the brain, memories of all except the most trivial and fleeting of events. They were diffused around the brain but could not be called up indiscriminately. The call had to come from one of the poly-selves and then would be subject to a veto from another self or from the central government of the selves. An endless complex track or network became probably an index of symbols, an inner language. When this language was developed as a political process, in communications with others, it formed an outer or public symbolism or language. But this is only a small fraction of the inner language that connects memories.

In the public language that ultimately developed were contained clusters of words that grew into creation stories, which purport to describe the days of creation of the world and of humanity. We expect the stories to be heavily veiled accounts of a true history, much like dreams that are internally distorted and censored but nevertheless lend themselves to scientific interpretation up to a degree. We see in the need for creator gods a determination to tell the truth in some way, to assert that the human was distinctly born.

Also stories were told of the environment before and during creation. In the new public language, the legends agree that there was a chaos, a formless, mindless order of the world, going nowhere until the divine intervention. We see two types of important "fact" in this chaos. First, there is a reality, an Earth with a dense translucent encircling high firmament of clouds. Second, the human is not there but is about to appear. Man appears as the canopy breaks and the gods appear.

Beyond this core of agreement, which I have not fully described here, the stories diverge. Running them together is like reciting a stream of dreams, all apparently referring to a single theme. This earliest extant public language is just what we would expect it to be, and what dreams are like, too, and what the world often appears like to persons suffering from mental illness. They hold a truth which can be deciphered.

5. THE STRUGGLE OF THE SELVES

The ancestor of homo schizo carried a bilateralized brain; two generalized and equally functioning hemispheres operated with a minimum of conflict. The cerebrum of Hominid "X" was large, perhaps too large already to escape conflict arising out of intra-brain and central nervous system dys-coordination.

Homo schizo inherited a larger brain, with immediate problems of electro-chemical and nutritional supply, and of neuro-transmission speeds. In homo schizo the brain conflict evades the earlier physiological compensation by moving out in all directions. So it gets less rest, is more continuously restless, awake and asleep. Rest often takes the form of diversion. As already pictured, specialization within the brain was sharply increased, and right-handedness developed. A general feeling of fear, inadequacy, and weakness was instituted that demanded obsessive attempts at self-control, which extended outwards as attempts to control the environment.

A great step of the suzerain ego is to consolidate its control by seizing and managing the right hand. In a typical neglect of transactional philosophy, it is conceived that right-handedness is "logical" inasmuch as the right side of the body is controlled by the "dominant" left brain hemisphere. May it not be more logical to conceive of the right hand as being developed by the left brain in order to strengthen the dominance of the left brain? Right-handedness is genetically predisposed, but only because the left cerebral hemisphere is genetically dominant. The left brain commissioned the right hand to be commanding officer in order to bolster its shaky regime.

6. Becoming Two-Legged

A number of physiological and anatomical changes accompany bipedalism, but perhaps all are ex post facto, such as the stretching of lungs and swelling of blood vessels to the head. Most likely, bipedalism is an adaptation for which an intense determination is required. There are no commonsense reasons for it.

We remind ourselves that the Indian feral child, Kamala, was totally adapted to quadruped motion to the age of perhaps eight years, and several years of training were required to get her to stand and walk voluntarily. Her muscles, tendons, hands, feet, knees, and probably her total body posture were quadrupedal. She walked on her tough palms with a full heel-to-toe motion. Kamala was comfortable on all fours and could run well. The human infant, of course, crawls for a year and more before being able to stand up and toddle. Only for sophisticated human activities is exclusive bipedalism superior, which presumes that humanism came first.

Bipedalism had some motive in the schizoid complex, in which aversiveness to others and ambivalence are prominent. Standing erect is a gesture of retreat and removal from others, which individuates beings. It is also a threatening and offensive posture, including the conspicuous chest-thumping that fiction writers overrate in gorillas. It goes along with (deliberately) smelling less, and with offering less in the way of hindquarters and front-features to nuzzle and smell. It encourages genital privacy because the hand and upper torso can exercise protective movements. The first homo schizo, one may conclude, would voluntarily seize upon bipedalism, if it were not an already confirmed behavior.

Bipedalism, therefore, matched the character of homo schizo and he is determined to master it. But what was this determination or voluntariness or will? Man was supposed to possess a will; philosophers and hoi polloi have thought so for thousands of years. Recently, however, the will has been removed by the philosophers of determinism, although retained by the masses. Did events occur during the gestalt of creation to give humans a will, yet permit it to be taken away under later rational analysis?

7. VOLUNTARISM

The "will" in hominid, we postulate, must be a "want," typical of animals, therefore an instinct - basically a will to feed, fornicate, flee or fight. In the disaster of creation, the new human achieved a new primary "want," to control himself and the world, to rid himself of fright. All of hominid's will - the aforesaid "Four F's" - is subordinated, rendered secondary, to the primary will to control.

Since the will to control is conveyed to a bewildering variety of human displacements and identifications, it acquires a new complex aesthetics that deludes humans as to its nature. People (philosophers and theologians among them) came to think that they were dealing with a qualitatively distinct mechanism, whereas it was a highly diffused aspect of all human activity, capable of exponentially more fixations than the simple "Four F's" of the beast. Some acts of obsession and compulsion came to be called "will," when they pertained to objectives of positive or negative value. These, however, if we ignore value preferences and their large variety, can be reduced to the great gestalt of instinct-delay, split self, existential fear, and consequent promiscuous and obsessive need to control.

The world is as will, then, just as Hegel said. It is a delusional creation of man's poly-ego confederation playing with its kaleidoscope. This game, with its dexterity and intensity, put all other animals to shame. And individual men came to be distinguished infinitely, in their applications of will, by the way their particular minds shook their kaleidoscopes. So that one man's iron will was to win a battle, another's to win a certain mate, another's to gather money, another's to die, another's to conquer will itself by willing nothingness. Much of this diversity probably occurred promptly after the time of the primeval gestalt. Its diversity elaborated into virtuosity, which doubtlessly played a part in intimidating all surrounding conscious animal forms, including our erstwhile hominid cousins.

8. Diffusion of the Gestalt

The hypothesis pursued here is that the gestalt of creation happened to one or two hominids, and diffused as a new dominant gene system. Were this proven untrue, we should proceed to a hypothesis of changed atmospheric constants. If this should be proven incorrect, we should retreat to a theory of psychosomatism, that combines psychosomatism, the "omnipotence of thought, " and potentiation of ever present lines of development of essential living matter. If this idea should be overthrown, we would put up a last-ditch defence with a purely cultural theory of catastrophic fright overturning the hominid mind.

All of these are conceived to have been quantavolutionary changes, occurring quickly and hologenetically, from the one Hominid "X" species to the present *homo sapiens schizotypus.* Further, it is likely that elements of all of three entered into the actual rise of homo schizo and his further development up to the gates of history.

In the quantavolution of homo schizo, what happened to the Hominid "X" ancestors, and to diverging strains such as *homo erectus* and Neanderthal? Many mutated, sickened, and died under the catastrophic conditions that were required to generate the new dominant gene system of mankind. Furthermore, the character of the new species was such as to intimidate the hominids and drive them into marginal living niches. Inasmuch as interbreeding was common, the human population would contain for some centuries or millennia hominid members and human members with hominid genes.

9. The Double Catastrophe

The necessity of natural catastrophe has been recognized, if a critical mutation of species is to be experienced. In other works and in later pages, I will have presented the theory and evidence for such catastrophes. Strictly speaking, this external catastrophism is distinct from the internal catastrophism of creation. Man is a catastrophized animal: both external catastrophes and the internal catastrophe of his genesis have awarded him this title.

Confusion between the two types of catastrophe can occur, as it did in some earlier passages that I have published. For, not only is there the outer chaos and the inner chaos but there is also the overlapping of the natural catastrophes with the earliest experience of homo schizo. He speaks the language of catastrophe out of experience.

Critics of quantavolutionary theory can turn this around and say that homo schizo, being what I have said him to be, naturally imagines all kinds of natural catastrophes to have occurred to which he was witness; that is, he would normally have hallucinated world-destroying catastrophes. For instance, Fenichel alludes in his *Psychoanalysis* to the manic's desire to control the world and Sebastian de Grazia in his *Political Community* to the ever-present ideology of the destruction and reconstruction of the world. Can I not keep the skies swept clean and in order, leaving catastrophes to occur solely in the mind?

To this, I would respond that homo schizo's stories of great disasters are too well supported, and too well detailed, to be either imaginary or highly exaggerated tales. It might be expected, too, that people who were genetically frightened, to appease their fears, would tell stories of a golden age and a gradual progressivism of mankind, which they do; these are partly there, but by their temporary historical framing they lend support to the disaster stories, so that both types of recollections must be accorded historicity, and thereupon further analyzed.

Because the terrors were sensible manifestations of high-energy forces, delusion and reality were forever commingled in the new species. The range of thought and sense material was great, including as it did the practically infinite combinations of sense data of the high-energy events and the immediately and infinitely symbolized associations of the events with the self and group. Not only are the earliest records loaded with catastrophic events and languages, but so also are Shakespeare's comedies and tragedies.

The great variety of detail in man's innumerable culture traits is an expectable and understandable resultant of all the psychological and real events attending the creation. For every controlled and uncontrolled construction that the mind emplaces upon events and objects, there are real events to fit into it. De Santillana and H. von Dechend explicitly commend a large, though unmeasurable, quantity of historicity in Ovid's work on metamorphoses. The palaetiology of the concept of metamorphosis may rest upon an abundance of mutated and damaged organisms accompanying atmospheric and radionic disasters. To hear it told, life was never dull *illo tempore.*

A portion of all religious expression and practice relates to such quantavolutions, among other things. These *spectra horribilem* then serve as religious lessons, teaching groups and individuals of the punishing power of the gods. The same events serve to connect the celestial with the mundane, inasmuch as sky images and stars are connected with the mostly terrible changes. The lack of control over mutants raises the level of terror. Therefore, what appear to non-quantavolutionists to be unconnected, inconsistent, and unexplainable varieties of the production of human minds here and there throughout the world, acquire under quantavolutionary theory a simple logic within a single framework of explanation.

10. Quantavolution and Hologenesis

The human probably was born from Hominid "X" in a brief incident that, for reasons given elsewhere, might be placed at 13,000 years ago, perhaps even a millennium or two later, but also perhaps within a 300,000 years period earlier. It would be well to fix the Holocene boundary at the point where the humans appeared. The aggregate of data on australopithecus and homo erectus promotes them to adjunct humans, also descended from Hominid "X. " Hence, in the preceding chapter, they were projected up the ladder of time.

The homeland of mankind cannot yet be ascertained, even though we agree with Washburn and Moore that man was born only once, at one time, in one place. We speculate that out of Hominid "X," whose behavior and appearance were distinctly different from those of the"hominids," "proto- humans," and modern humans of whom we know at present, there came a macroevoluted or quantavoluted type who intermingled with and dominated these families in short order and contributed to the destruction of many.

11. THE NEW HUMAN BEING

Upon a probable mutation the hominid was subjected to a general instinct-delay that left only "lower-level" and instinctive operations largely untouched (but not unreachable). The instinct-delay can be termed depersonalization, which was the first feeling of homo schizo, to be promptly succeeded by a splitting of the mind into multiple entities that ultimately became a typical human poly-ego. The depersonalization aroused the new creature to a high level of fear, a general anxiety, an existential fear, integral to its being, ineradicable.

The response to the fear was a grasping for control of the selves to reestablish the former hominid consciousness and its instinctive nature. This was also permanent. The human was marked by a mania for control. The control-mania could not stop with the selves, because the selves did not stay with the body. In the struggles among the personae, the whole world was embroiled; a splatter of displacements occurred. Streams of affect or identifications were ejected, with attachments ensuing, minimally at the level of attention.

Attention extended to habit and to obsession and to a sensing of property, all being mental strategies to fix upon objects to control, thence relief of anxiety. The "return on the investments" in real or sensed or illusory affect consummates the transactions, no matter whether with people, objects, or spirits; the transactions could be termed, also, projections and retrojections. The outcomes of this unceasing and uniquely human transactional process are numerous. They can be grouped into:

a.
Perception and attention with typical overlappings into perception disorders, hallucinations and illusions.

b.
Thought, logic and analogy, moving into rationalizations, delusions and thought disorders.

c.
Selective, recallable memory, often employing amnesia for fear- reduction.

d.

Emotional ambivalence respecting all persons and things, a mild anhedonia and general negativism, anxiety-freighted, as distinguished from the hedonic animal.

e.

Aversion or the non-acceptance of apparent prima facie resolutions of human relationships, including paranoia with its fearful denial of retrojected affect and the substitution of alternative hypotheses of threat.

f.

Psychosomatism, the stressing of the body to achieve higher control levels, often with healing and destructive effects.

g.

Guilt and punishment, ranging among all persons, objects and spirits to discipline and erase fear.

h.

Discipline and work as an outcome of attention, habit, and obsession.

i.

Drug addiction for anxiety-therapy and orgiasm.

j.

Anxiety, which continually presents problems for solution, and, when overabundant and impractical, engenders neuroses, neurasthenia, epilepsy, and depression.

k.

Internal speech, the coding of information bits and sets, in time and space, for quick retrieval by association or for computation, including speech disorders when pieces of code are compulsively expelled as speech.

l.

Language, public speech, to signal and control the outside world, real and delusional, using internal code elements that others agree upon.

m.

Sublimation, the elaboration of symbolic activity in a low-anxiety area of displacements.

n.

Basal activities closely parallelling earlier primate behavior, such as eating, sexuality, mother-love, aggression, and fear-flight resulting from immediate threats, except that these activities are continuously subject to uniquely human interventions.

In the outpouring of his new nature, the proto-human thus exhibited new methods of handling large portions of the range of animal behavior. He could think about, talk about, and do something about a world of problems of which his ancestors were unaware. He would give a new aspect to all the ordinary activities of the earlier hominid. However, if eternal "angst" be considered as a cost, the new person paid heavily for his virtuosity.

CHAPTER FIVE

CULTURAL REVOLUTION

In dreaming, the human brain works fast, conjuring plots and actions that would be not only physically impossible but also temporally prolonged. Persons who have narrowly escaped an abrupt death sometimes exclaim, "My whole life passed before me in an instant." Many creative artists and inventors, whether in the physical or social field, refer to their visualization or conceptualization of a total product in a moment of "intuition." Such occurrences point toward a theory of cultural hologenesis: if human, then holistic thought; if holistic thought, then holistic behavior; if holistic behavior, then collective instant culture, or at least a culture that develops as rapidly as the acting out of dream and thought sequences can be managed. A culture -- a group mode of mentation and behavior -- arose promptly with homo schizo. Just as man became psychologically holistic upon his origination, so did he become culturally holistic. Human culture was global from its beginnings. Culture was schizoid and remains so.

1. PROTO-CULTURE

The first culture was a set of wild moves in all directions guided by displaced instincts and an intense need to stabilize the psychic world.

Usually the search for culture begins with a search for tools, because tools can be hard and enduring, and because they exhibit a deliberate human effort to command materials to effect a purpose. We should acknowledge first, though, the inevitable and greatly convenient built-in tool kit of a human. The first human was a tool-user whose body was his portable tool-kit. The hands of the ape are not put to many of their human uses. The human made tools of his fingers, hands, arms, feet, back and shoulder muscles, tongue, spittle, voice-box - indeed of all his senses and organs that he could command.

Even today in a highly technical society where there is "a tool for every purpose," the built-in tool-kit is continuously in use in ways far exceeding the imagination and capabilities of the primates. One can indeed conceive of a culture without artifacts. But in reality man must go on to make other tools. He has no choice.

Like man's anatomical tools, his mechanical tools are projections of nature and analogies to it. They exhibit a sense of the future and represent the obsessiveness of humans. The tool is pragmatically rational if, in addition, it is functional (efficient) and conserves resources. A tool, then, is a socially transferable physical object believed by its user to confer a larger control over the world than he could otherwise achieve. A mechanical tool is a type of social tool, also, and there is some merit to defining a social tool as an organization of other people believed to add to the user's control of the world.

Who sports a club, supports a culture. He remembers to carry it, and foresees a use for it; so he has memory and foresight. The club is a versatile tool against living things and obstructions; it extends the arm and gives leverage. It has to be produced; a skill is involved, so we have *homo faber.* It is one's own, so we have a property right. It is a coercive threat; it is a sign of fight more than flight, so it communicates a sign of power and authority. As the "batons" of upper-paleolithic man evidence, the club converts readily to a work of art, employing symbolism of lines, geometry, living things, and carved depictions of the phallus. The club reaches into the sky to connect with shooting stars and comets, as in the snake-entwined rods of Hermes and Moses.

Thus the simplest tool, the club, represents the major areas of human interest: skill, subsistence, economics power, safety, authority, sexuality, religion, and aesthetics. It is required, however, that it be carried. If we knew when the club was first carried, we would have a sound basis for fixing the gestalt of creation in time. Alas there is no earliest club; wood rots quicker than bone; we have only the aforesaid early bone batons. We have practically nothing belonging to the earliest man, no ropes, skins, bamboo constructions, and so on; all subtle evidence is gone, leaving chipped stones and stone mounds.

Leroi-Gourhan has estimated the oldest cultures, the Acheulian, to have possessed 26 stone tools, the Mousterian Neanderthals to have 63, and the succeeding modern type to have 93. These kits do not include all of the tools by any means - not skins, vines, ropes, gut, shells, bamboo, leaves, clubs, wood levers, wood slides, bones, hair, fur, paints, glues and so forth. E. H. Man's survey of the isolated and simple-living Andaman Islander a century ago revealed no more tools of the stone type but more made of the material that would have been destroyed by time and nature. Such tools might raise the given numbers by a factor of five, giving 130, 315 and 465 which, averaging (for who can say what determined the ratio in each case), gives some 300 tools in earliest known human cultures. Then add the tool chest in the human body. We can take it for granted that the earliest human who used *any* tools, used *many*.

With such material uncovered from, or imputed to, paleolithic man, a world of intellectual, emotional, ritualistic and mundane variety can be contemplated. An engraved ox rib from Pech de l'Azé was called Acheulian and dated at 300,000 years by F. Bordes, and in 1982 Pietro Gaietto published in Genoa a treatise on pre-historic sculpture. There he moves the earliest artistic works of mankind "a million and a half years" back to the pebble culture of australopithecus and homo erectus.

He argues that the earliest busts and menhirs are as decipherable as the earliest utensils, and exposes abundant evidence of artistic mentation in material of a type hitherto disregarded and cast aside by paleo-anthropologist. Appearing first in what may be artificial modifications of naturally suggestive stones, they develop successively in pre-Neanderthal, Neanderthal, and homo sapiens excavations. Working independently, L. G. Freeman and R. G. Klein, University of Chicago anthropologists, announced a year later the discovery at El Juyo (Spain) of a sanctuary containing a probable altar, weapons, house-hold tools, animal relics, and a stone sculpture. The sanctuary was dated at 14,000 B. P. and the bust depicted a two-faced creature, half smiling man and half cat. It resembles a number of Gaietto's "sculptures."

2. LOST MILLIONS OF YEARS

Man should have reached a comfortable Neolithic level of culture within a thousand years of humanization, and stayed there unless general catastrophe intervened. The Neolithic is universally acknowledge to have been an across-the-board human culture with all basic practices, institutions and techniques invented and in use; it was certainly in being everywhere 8000 years ago.

3. TRIBES, CIVILIZATIONS, AND TIME

The great civilizations began to appear about seven thousand years ago with commerce and conquest. There have been perhaps fifty of them. They take about three hundred years to gestate and last for a millenium before handing themselves over to another civilization as with the Incas, and/or dissolving, as happened with the Roman Empire. During this time, and counting the component cultures from which they were amalgamated, there may have existed about 20,000 cultures in all.

It is difficult to "put a tribal culture back together again" once it has been absorbed into civilization. Sometimes a tribal culture will remember having been ruled distantly but not tightly or absorbingly. It shows almost no signs of having been included in a bygone civilization. Therefore it must have existed by itself since the beginning of human time, or since it split off from a tribal aggregate at some time in the past to form a related unit. The fission would have occurred because of natural catastrophe, flight from a growing civilization, internal disputes, or overpopulation and emigration.

It is unfortunate that all of these statements must be conjectural. Yet their thrust is unmistakable. There has not been enough time since the beginning of human culture for all tribes to have experienced participation in a major civilization -- except for the ecumenical proto-culture to which all peoples must originally have belonged. In this case, the demography of cultures would imply recent human origins and support the theory of cultural hologenesis of homo schizo.

If it were not for the accepted methods of reckoning time, scientists would probably have to agree that a hologenesis of both man and culture is logical and recent. To hallucinate further, if Solon of Athens had called on a panel of experts from Babylonia, Iran, China, India and Mexico, as well as from Greece and Egypt, in the sixth century B. C., all would have told him that man's history was short, at least since the last great catastrophe. But belief is firm in the tests that report long times for the early fossils and relics of man and life generally, and claim a long, slow ascent.

A century ago, when time reckoning was governed by our type of speculation, by the fossil record, and by the apparent ages of sedimentary rock strata, time measures were easier to assail. Today, radiochronometry lengthens human time and fixes it by elaborate chemical tests, the most critical of which are the radiocarbon dating and potassium-argon (K- A) dating.

Both tests are striving for validation in the crucial middle times between 10,000 and 100,000 years ago. It is expected and hoped that they will close this gap. Meanwhile the K-A test can be used to support very old ages for what appear to be human remains with artifacts; and the 14C test is keeping the Upper Paleolithic age far enough back to support impressions of a very gradual human cultural development. I have given elsewhere my reasons for disputing the validity of 14C beyond 2700 years, for regarding the K-A test as quite unreliable, and for questioning most other chronometry.

Here, I do not treat fully these tests, because the theory of human and cultural hologenesis is independent of the time-tests frame. Hologenesis could have occurred 13,000; 50,000; 200,000; 1.5 million or 5 million years ago, except that in all of these cases, an incredible amount of human history is missing.

For those dates that are beyond 50,000 years, one might postulate a limited jump in human and cultural evolution, leaving a final large jump for the Holocene boundary. That is, some hominid, perhaps not even a human ancestor, could have chipped a stone, with nothing else on his mind. Given my analysis of the club-wielder, I would not know how to explain this activity. It would not be modern man, but a different species. I find this solution easier to tolerate than a gap of millions of years between a true man, a chipped stone, and the Upper Paleolithic-Holocene periods.

4. ECUMENICAL CULTURE

There was in the beginning one human race, one language, one culture. The contrasting hypotheses seem to be losing vigor. That many cultures around the world originated independently implies that men scattered around the world and only then started up cultures from a delayed time-fuse in their brains.

Despite the tenacity with which this idea grips many people, it would appear absurd, unless one believed at the same time that humanization occurred immediately in consequence of an atmospheric change that affected the brain with some uniformity everywhere, an idea that I have not seen expressed except in these pages, and here is included as a partial and fluctuating cause of humanization. Since we cannot agree precisely when humans originated, for certainly most will accept Charles Darwin's series of insensible gradations in preference to my theory of holocene hologenesis, how can we fix a point for the beginning of culture? As I construe the conventional argument, it must assert that the ever extending ladder of evolution contains many rungs, some of which are physical gradations and other cultural.

There has until now been little chance of emerging from the source materials with even the beginnings of a division of culture traits as we conceive of them: elements that are assignable to times of common catastrophic experiences; independent inventions that came about owing to cultural peculiarities of given peoples with some parallels to be drawn from the independent inventions of other peoples; and innovations originating among one people and diffusing to others, whether in the wanderings after natural disasters and war, or in variously motivated migrations.

For instance, fire, which had been known to, and used by, hominids and other animals, would have been reinvented by mankind. "Fire was born when heaven and earth separated," says a Mongolian marital prayer. Fire -- in its modern sense of something to be used multifariously, made and remade -- was invented because the created human was terrorized by new intensities of fire, because the projected gods used fire in the skies and on earth, and because the new mind could remember its use and foresee its future utility. Credit for the invention was ascribed to a god and sometimes also to a god-hero who, partly man and partly god, could arrogate credit without displeasing the gods.

The earliest town plans were built according to a celestial model, and the planners were astronomer-architects. The conditions for planning were, again, an aware and awed human group, a sky religion, a skill in retrojecting and rationalizing a celestial scene, and then a science of measurement and construction. The orientation of the towns (Greek: polis) and temples followed first the North-South line of the Boreal Hole, a northern-most sky opening which happened in cloud-canopy times to represent the north geographical pole (from polis.)

In less cloudy and in bright times, measurements that were derived from the old monuments and improved by stargazing, permitted the practice to continue. The Egyptian and Mexican city and pyramid orientations were North-South. In protohistory, East-West orientations may have become prominent because the sky-path of Venus was East-West, and finally the Sun's regularities provided the lines of true orientation for planners.

A recognizably scientific astronomy is being sought farther and farther back in time. B. A. Frolov argues that an intellectual curiosity possessed early humans everywhere. The Russian counterparts of the Stonehenge monuments are at Lake Onega, and both are sky-directed religio-astronomical instruments. The Pleiades are called the "Seven Sisters" by aborigines of Australia, North America, Siberia, and other ancient cultures. Petroglyphs that appear to refer to astronomical constants and phenomena are found all over the world; it maybe mainly the prejudice on behalf of the "evolutionary ladder" that forbids the assignment of many such carvings to the earliest age of humanity; in some of such cases the glyphs are found among the earliest ruins of a people or are the only remains discernible. That is, the hologenesis of mentation and culture derives support from the increasingly early assignment of scientific works.

Two thousand years after humanization, a large number of humans possessed self-awareness, religion and rites, planned towns, armed forces, a full range of stone and soft material tools, special occupations, domesticated animals and plants, and complex language. They entertained a range of aspirations that followed their time sense into visions of improved life; they created the rudiments of the highest ideals of later times: freedom from fear through knowledge, individual autonomy, conquest of the environment, storage against future hunger, and social cooperation. But the high energy forces of the gods permeated history, life, and expectations. Destructions were frequent, and catastrophes, such as they already dreaded, were to recur.

Joseph Campbell puts our position here well:

> It has actually been from one great, variously inflected and developed literate world- heritage that all of the philosophies, theologies, mysticisms, and sciences now in conflict in our lives derive. These are in origin one: one also in their heritage of symbols; different, however, in their histories, interpretations and applications, emphases and local aims.

5. AMERICAN CULTURAL ORIGINS

Alexander von Wuthenau, in his book on *Unexpected Faces in Ancient America, 1500 B. C.- A. D.1500* scans the literature on Asiatic, African, Egyptian, Semitic, and European presences in cultures and races of Central America and presents his remarkable album of stone and ceramic countenances of the stated peoples. Despite conventional theory, there seems but little question that the Central Americans were a mixture of human types long before Columbus arrived.

But further, the American race had its own primeval forms. I argue that homo sapiens schizotypus was present in the Americas from his very first period, and despite repeated general catastrophes held on there in niches of survival, and was repeatedly reinforced across the Pacific and Atlantic oceans, with artifacts and cultural practices to remind us of these occasions.

Although this thesis is not central to the present book - because the theory of homo schizo can be argued on whichever grounds conventional theory chooses - it has important consequences for early American studies. As I foresee the emergent issue, it is not rampant diffusionism versus carriage across the Bering Straits, but rather how much of the similarity among races and cultures came from the ecumenical period of homo schizo and how much was transmitted via long distances thereafter.

The case for diffusionism is building up. Some of the material advanced before World War II regarding Asia-to-America diffusion is summarized in Lord Raglan's *How Came Civilization?* (Chap. XVII). He placed the world ecumenical culture of the first civilization in the region of the Persian Gulf. More recently Betty Ebers has marshaled the evidence for Japanese to Olmec (Meso-America) diffusion, by sea.

In another case, an authority on early Meso-America, Michael Coe, reports the "coincidence" from Needham's studies that "the Maya astronomers and those of the Han Chinese worked with an eclipse calendar of 11,960 days." The coincidence cannot be an accident, especially when one considers that the Mayans seem to have used "solar mansions," like the Chinese, rather than a zodiac, to mark the progression of constellations, and, further, indicated constellations, in the manner of Han China, by circles connected with straight lines, which was not seen in Europe until 1785.

Meanwhile, evidence of Phoenician, Egyptian, West African, Jewish, Arabic, Roman, Celtic, and Viking contacts ranged from New England to Middle Eastern America in the North and down to Brazil in the South. Indications of a European or Eurafrican presence in the centuries just before Columbus are not wanting. The idea that the Americas were a virgin to the Old World before Columbus deflowered them is an anti-historical myth.

As to the troublesome question concerning when these contacts took place, here we propose that the Americas have been in touch with the rest of the world throughout the history of mankind, except in the periods of great natural turbulence, with the contacts swelling in numbers whenever a few hundred years of technical development and cultural organization would occur. Whenever a catastrophe happened, which cut off peoples by splitting continental blocks, lifting mountains, creating great rivers, or interposing new climates between them, the isolated cultures developed very rapidly, requiring only a few centuries to exhibit different cultures, languages, and ways of life.

Let the editor of a recent collection of studies on trans-oceanic contacts summarize the situation for us:

> Clearly, the present status of our knowledge of American archeology does not allow us to attribute the origins of New World civilization to diffusion from the Old World with assurance. Equally, however, it does not demonstrate the independent origin of New World high culture. Just as the zero occurrence of artifacts originating in the Old World and found in America may be taken as a strong argument against the diffusionist explanation, so the early occurrence of a complex of Old World-like traits -- often very sophisticated -- in early levels of nuclear American civilization casts a strong reflection against the independent origins hypothesis.

This points to a very early heartland culture; then came divergence and sporadic exchanges.

I would suggest, concerning said passage and the same anthology of studies, that we should be looking for several periods of transference of traits; in *Chaos and Creation* I suggest six of them. Artifacts and usages can then be assigned by ages and the outcomes tested (for their logic and verisimilitude). Basic social forms, early ceramics, boat design, the lodestone compass, the pyramid, Semitic, Celtic and Roman relics, and many other kinds of evidence exist with which to clarify the periods of intercourse.

To summarize, a hypothesis of ecumenical world culture in the earliest times, attaining quickly the neolithic level, is supportable. Inventions require heavy motive power, both in the phase of mental gestation and of social adoption. The motive power must operate within and among individuals. Basic inventions came in a rush following the gestalt of creation. They flowed from the psychology of the new human species, originally a small group. They were tied immediately to astral gods and figures and to animals as well; this identification lent memorial power to the inventions and authority to the thrust of their diffusion.

Acting in the name of their gods and totems gave authority to the imposition of practices. The same aggressiveness that ultimately eliminated the hominids also foisted upon them the basic inventions. Those who grasped the meanings of the human culture, or at least could practice it, survived. The aggressors possessed ideology, skills, and zeal. No species could stand against them.

In this manner an ecumenical or universal culture was quickly created and diffused among a variety of human racial types. Potentiated genes were diffused and came to the fore quickly in adapting to a changing world. Culture traits were imposed under the most stringent conditions.

It was the greatest age of "evangelism" in the history of mankind. Within a thousand years of increasing natural terror, most basic skills would have been adapted from nature, developed, put into a framework of ideas and imprinted upon society.

CHAPTER SIX

SCHIZOID INSTITUTIONS

Totems and taboos are convenient ways of repeating and organizing obsessions. They are group elaborations of the schizophrenia of original humans. Both are found in all cultures and in varying degrees of weight. In large-scale cultures they are part of religion and bureaucracy.

Taboos are sacred prohibitions, whether received directly or indirectly from divine authority. The "Ten Commandments" include taboos. The name of Yahweh was taboo. At one time it might be pronounced only once a year. Violation of taboos is commonly supposed to have fatal results. Yahweh frequently concludes his injunctions with the phrase "... lest you die."

The totem, more strictly, is a symbolic identification of a human group with an animal or plant, which represents a divine force. Because animals (the owl, for instance) and plants(the sacred oak) were tangible, near at hand, and well-known, they could readily be fitted into the scheme of delusions; a communication system, largely imaginary, is set up between the life-form and its human patron.

In joining with a totem, a human group acquired a talisman and group representative. The totem life-form operated in the sky and on earth to the presumed overall benefit of its sponsors. Once the sun (earth) rotated too fast; the great rabbit, said some American Indians of the Great Plains, lassoed the sun and halted it, not releasing it until it promised to go slower (perhaps the rabbit was a cometary image.) But a totem also imposed limitations upon behavior by means of taboos, rites, and penalties.

A totem provides a complete schizotypical system: the injection of divinity into an animal denotes a cognitive disorder, a hallucination, a misplaced metaphor. The exclusiveness of the totem and its group towards other totem groups in its associated taboos reflects the schizoid aversiveness to others; the worshipping and cannibal sacrifice (sometimes) of the totem animal emerges from ambivalence; the numerous rituals and rules connected with the totem convey compulsive obsessiveness; and the secret and enduring aspects of the totem group's practices, going back to the totemic primal incident, display catatonism.

Enrico Garzilli writes of Faulkner, Joyce, Pirandello, and Gide in their searches for "the real self," notable names, to be sure, pursuing at the pinnacles of literature the primordial search for oneself within the poly-ego. He explains that the word is the self; becoming human is to become a word. Hence the importance of such ancient expressions as begin the Gospel of John: "In the beginning was Word; and Word suffused God; and God was Word." (My rendering.) We sense here the power and control exercised in the first naming of something and agreeing upon it with others. We should understand, too, here, that "Word" is "Logos" or "the enlightened life of mind."

1. Speech and Language

C. Levi- Strauss is of the opinion that "language was born all at once," thus supporting our position of hologenesis. He goes on to say that "whatever the moment and the circumstances of its appearing in the range of animal life, language has necessarily appeared all at once. Things cannot have begun to signify gradually. After a transformation the study of which has no relevance in the field of social sciences, but only in biology or psychology, a change has taken place, from a stage where nothing had meaning to a stage where everything had."

This is a surprising use of the word "relevance." Once we have understood what was happening biologically and psychologically, we comprehend what was happening socially. A quantavolution introducing language must concurrently involve a grasping for logic, for control over memory, and for the social consensus on meanings from which culture sprouts. The internal code of language was springing up and erupting here and there into public language.

Speech did not occur word by word, grammatical form by grammatical form, over millions of yeas of humanization. It probably sprang up in a mixture of counting, signs, and ejaculations. Counting has been connected (through Lord Raglan's *How Came Civilization?*) by Seidenberg with rituals, which fits the model of homo schizo well. "Counting was invented in a civilized center, in elaboration of the creation ritual, as a means of calling participants in ritual onto the ritual scene, once and only once and then diffused." Seidenberg explains that all people had religious numberings and taboos on certain kinds of counting. It is frequently imagined to be theft when one's name is counted. Today, a homologous paranoia underlies the hostility of many persons to the computer, which seems to steal one's name, carry one's number, and manipulate these and hence oneself.

A person is only created when named or announced and the creative word may have been the creative number. Marshack would seem to be moving along a similar path, with stress upon arithmetic and calendarizing (the catastrophized need to watch the skies for regularities that are hoped for, and irregularities that one must prepare for). He is also locating ever earlier symbolic forms.

2. GRAPHICS

Speech came promptly, but writing was not developed well until civilizations had poetry, art, religions, and social systems. A possible reason for this may also be supportive of our theory of language. It is logical that as speech is to the mouth and ear, writing is to the hand and eye. No one doubts that earliest man (or latest hominid) was as digitally adept as he was orally proficient. However, gestures, grimaces, and context could let the eye help the speaking process along.

But the hand and eye could not, like symbols, accomplish internal symbolizing or speech, which is probably what was occurring in the new creature to help him coordinate his several selves and their displacements in the outer world. That is, public speech was the extrusion of inner speech, like the small portion of the iceberg that floats above water.

Some people with complex languages do not write even today. Art of course takes the place of writing in respect to many messages from one's ancestors. A totem pole can take the place of much written history, depending upon the kind of history wanted. There is a clue here: a large society and an official class need explicit messages and records.

Until these criteria come into play, art can successfully block writing, somewhat as television blocks literacy. Art can say so much that, by comparison, the breaking down of pictures and symbols into writing may appear to be a meaningless and barren enterprise. Further, it may seem to be sacrilegious to openly admit that words are interchangeable tools. Hence writing was originally a holy profession, as in the Egyptian bureaucratic empire. It was carried over into government: "the needs of a centralized administration were a far greater impetus to the development of writing, among the Sumerians (cuneiform) as in Crete, than intellectual and spiritual needs." Earliest tablets speak mostly of rations and personnel in the palaces.

But, in maritime cultures, such as the Phoenician, the pragmatic value of messages finally broke the sacred grip. Words (orally spoken) had departed so far from their origins and symbols from art, that they might be used casually in practical affairs. The alphabet was invented out of numbers, phonetics, and calendars by people who were on the move, as in boats.

The invention of writing was an effective grasping for control of memory, behavior, and pragmatics. It delivered also a severe blow to the imagination; it caused massive disenchantment. It placed credit for works effectively upon the culture. No longer could one be taught by the gods, through subtle or at least mysterious parental and social transmission or from the depths of one's being, from inner springs. Besides memorization, one had exactness, repetition, a third party, an objectivity, a beginning of coolness and remoteness.

3. PRIMORDIAL LANGUAGE

Man spoke one tongue to begin with. As he diffused from his proto-patria, his speech had reason both to change and to remain the same. If there can be found a basic set of sounds and words that is common to all of mankind today, then one would have an original language, a proof of cultural hologenesis, and an indication of the recency of human origin. Searches for the first language have been modestly rewarding, enough so to justify a greater expenditure of time and resources, especially for computerized manipulation of data. R. Fester has proposed that "there is an original vocabulary of six archetypes common to all of humanity which still today comprises the basic of every language and which at the same time provides a clearly recognizable link between all languages." The root-words of "Pangean, " as we might call the tongue, would be BA, KALL, TAL, OS, ACQ, and TAG.

Fester claims to have discovered that in many languages, the syllable BA pertains to human relations and subsistence; KALL appears connected with the idea of concavity and the female womb; TAL refers to clefts, to the ground, to females; OS to thresholds; ACQ to water; and TAG to height, gods, erect humans. To Malcolm Lowery, who has kindly supplied me with his translated materials, the progression by which the words related by Fester to the roots are said to drift in space and among cultures is not intelligible.

Generally, the linguistic establishment has beaten back the numerous efforts to demonstrate speech affinities, regarding them as *prima facie* absurd. Such connections would be Gaelic with Algonkin, Chiapenec with Hebrew, Othomi with Chinese, Choctow with Ural-Altaic, these being Amerindian connections. The diffusionists have fared better in proposing Old World connections: Hamites with Semites; Sumerians with Magyars; Late Minoan with Greek; Egyptian with Hurrian; Etruscan with Lemnian; Berber with Basque, etc. Justus Greenberg says that the 750 indigenous languages of Africa were originally four families, and these were originally one, and possibly related to Hamitic, says Gilbert Davidowitz. Encouraged by the theory of hologenesis of culture, I would conclude that the search for the ultimate ecumenical Pangean language will not be in vain.

4. GROUP VS. INDIVIDUAL

Humans of the proto-age had immediately the problem of constituting themselves deliberately into a group. The psychology of the hominid band was gone. In its place was the fearful, distracted, individuated - even multividuated - person. He must belong, yet not belong, at the same time. The favorite topic of political philosophers and economists - the individual against society - took shape.

The bond between individual and collective psychology is tight. It is both genetic and adaptive. It is fully determined. It is unbreakable. Evidence of these statements gushes from history and anthropology on the one side and from many psychological schools on the other. Just as the brain can reach to the toe to express itself physiologically, it can reach to the stars to express itself psychologically. Where it happens to reach is a cultural affair.

Just as the human is a coordinated poly-ego, so a culture, and for that matter any group, is a mega-poly-ego, that typically selects a dominating ego-pattern as its design for the behavior of its members. A special concept of organization is required to grasp that organized behavior which is an extension of patterned behavior that is an extension of patterned mind-behaviors. The genesis of external organization is in the mind(s) of individuals and their groups.

There is no end to the process of "private-public" interaction from conception to death. That means also private-cultural. The individual and the group march along, side by side, from the dawn of mankind. Both society and the individual are schizoid in origins, structure, and functions. Their behavior and forms are not always congruent; the symptomology is varied. Then it is that deviance (medical schizophrenia) is defined. The individuals seek to evade the society or change its laws; the society seeks to make the individuals conform; else it treats them for mental illness or jails them on account of their menacing or destructive conduct.

The process will go on as long as human nature retains the form which it assumed in the days of creation. There are perhaps some non-schizoid culturally created humans, who have evaded hybridization with the schizoid, the fate of most hominids. Even if there were none at all, the idea of their existence should be retained for heuristic and theoretical purposes. They would be well-trained primates, although not discernible as such. The schizoids, and especially certain schizophrenes, are religiously and politically dominant. With their obsessions, suspicious hyper-awareness, penchant for symbolism, and their megalomania they control the world. That is, they try to control it; but the world is, by their own definition, uncontrollable. Homo sapiens schizotypus defines "control," and is insatiably anxious for control.

Human action moved back and forth along an axis of tension between the individual and the collective or social. Self-awareness was an inescapably individualist phenomenon. Never after creation could the sense of the self be exterminated. Never thereafter, then, could the collectivity perpetually and wholly dominate the individual soul. Incessant, forcible, and imaginative attempts to do so over all of history were doomed to fail and still are. The split self, a source of the greatest terror, could not permit its unification by the collectivity, even though the collective achieved its great resilient strength from its guarantees to the individual that it would assuage, diminish and even cure the terror of the split. There was no returning to the mammal.

Part Two

THE DIVINE SUCCESSION

Chapter One

THE GENESIS OF RELIGION

To the fresh mad eye of primeval man, the world was full of gods. The human mind worked so as immediately to create religion. It does so now and it did so at its beginning. This is a function common to all humans everywhere, at all times, intrinsic, inherited, irresistible. Religion is then naturally ecumenical; any two people anywhere can agree in general on what it is that they are talking about when talking about religion.

The mechanism is awesomely simple. To repeat: the human mind, as soon as it starts working, builds a multiple identity, a self-awareness. In the origins of the race, this trait is so pronounced as to set the creature apart from other forms of life. Self-awareness is the psychological manifestation of a physiology of the central nervous system, especially the cerebral cortex, which presents a person with the feeling of being at least two persons.

Since the body is one alone, it is "intended" for one mind, one spring of action, a single commanding organ. A person has the instinctive appreciation of and a nearly total apparatus for realization of unitary conduct. But this preparedness for the life of an ordinary mammal is rudely challenged by the sense of an inner conflict of selves, which can "change one's mind" and redirect one's energies at any time, with seemingly little possibility of control. It delays by microseconds the instinctive response that the mammalian physiology and neurology crave.

The result of the perceived conflict, that "I am I, but who am I that says 'I am I'" is fear. We can call this fear existential because it is the absolute quality of human existence.

Since the struggle of the selves is essentially psychological, it can be called supernatural. Then it is even more proper to call the projection and displacements of the self supernatural.

The operation of tying a world of external supernatural phenomena to the world of internal supernatural phenomena is invariably expressed in ritual practices, that is, repeated related performances.

"Fear made the first gods of the world," wrote the Roman Statius (c. 45 to 96 A. D.). In the long history of religion it is the only theory to come close to the truth. And man, in return, is theophobic, full of dread of god.

The simple mechanism of religion is then self-awareness, fear of the self, fears or anxieties displaced upon supernatural or tangible appearances of the world, and the development of practices to control and maintain transactions with the supernatural appearances. The drive to control oneself (oneselves) is paramount and moves man to wherever his rears alight. Again, Radin's anthropological surmise is acceptable: "man was in a state of fear, physically, economically, and psychically. Man thus postulated the supernatural in order primarily to validate his workaday reality." His aim was "the canalization of his fears and feelings and the validation of his compensation dreams."

Mircea Eliade has reported well the state of mind of the "religious man" through the ages.(He uses the term as, for instance, H. D. Lasswell uses the term "political man," as the "pure" or obsessed type of actor in history.) Where we employ the term "supernatural," Eliade uses the term "sacred."

"For religious man," he writes, "the world always presents a supernatural valence, that is, it reveals a modality of the sacred." Every bit of the cosmos has its sacrality. "In a distant past" (but why not include today?) "all of man's organs and physiological experiences, as well as all his acts, had a religious meaning," "Homo *religiosus* always believes that there is an absolute reality, the sacred, which transcends this world but manifests itself in this world, thereby sanctifying it and making it real." "For religious man, nature is never only "natural"; it is always fraught with a religious value."

Finally, "the sacred is equivalent to a power, and, in the last analysis, to reality. The sacred is saturated with being... Religious man deeply desires to be, to participate in reality, to be saturated with power. This rounds out an accord with our ideas of religious genesis. Man naturally sees the world supernaturally. Reality is supernatural. His heart and soul go into tying this reality to himself, to gain its powers. We should say that all of this grandiose ambition is to stabilize his mind, to let him live unanxiously, unfearfully, to be at peace with himself."

How good it is to be assured of this, too, as the Hebrew Elohim assures man, that he shall "fill the earth and subdue it; and have dominion over the fish of the sea and over the birds of the air and over every living thing that moves upon the earth," and furthermore has given him "every plant yielding seed which is upon the face of all the earth and every tree with seed in its fruit..." Elohim is thinking and working like any ideal reasonable man would think and work. All is divinely created, by hard labor. All is sacred, therefore.

Yet, granted that humans are bent upon creating the supernatural and tying it into themselves, why should they dispose of the credit to gods? Why should they not be frankly proud of the world that they create and control, whether it be supernatural of tangible? First there is the fearful fact that they do not control it. Second there is the fear that disregards fact. They fear that they may not control anything; man is born with an inferiority complex from not controlling himself. Third, there is the appearance of purposeful control of the world by non-humans, an appearance, one may insist, that has both invisible and perceptible substantiation.

Chapter Two

THE SUCCESSION OF GODS

The first god who was, remains in the latest god who is. The gods have been of the same descent, always, everywhere. I mean this not in the sense of many theologians, that, "Yes, God has been eternally Himself but we have gradually learned more about His nature," nor in the sense of many sectarians that, "Yes, people have forever worshipped false gods but gradually we are coming to see my God," but rather I mean it to say that the gods were discovered once, in the earliest times, and that there had been a direct descent of the same divinities down to the present. By "discovered" is meant that the first humans perceived gods in the world; they perceived the supernatural, and they took immediate steps to control it.

The history of religion permits this statement. Leroi-Gourhan believes that the Upper Paleolithic hunters were probably religious. An engraved reindeer bone from Langerie Basse in Southwestern France, which I reproduced in "Chaos and Creation" probably represents the mating of Heaven as a bison and Earth in the form of a woman.

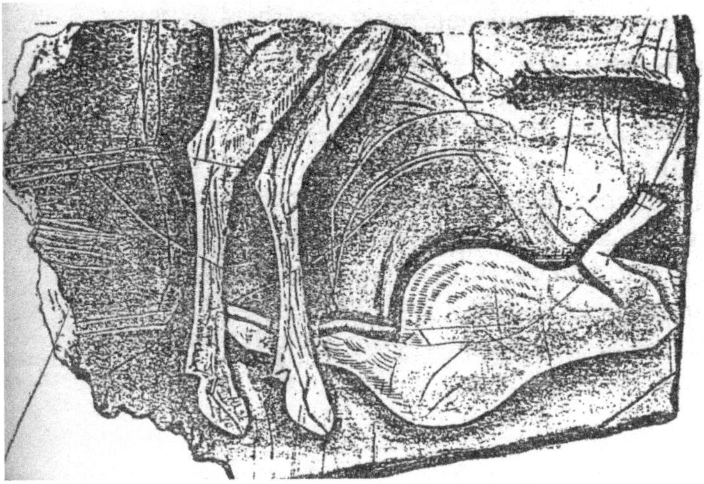

Engraved Raindeer Bone of Bull and Pregnant Female, Langerie Basse, Southwest France.

The hind quarters and phallus of a bull hover over a pregnant women facing up. Rendering is by Piette, pictured in Marshak. Dated conventionally to Middle Magdelenian (14000 B.P.?). Marshak asks: "Does this composition depict in relation to a horned animal which may be a sky symbol. (p. 320)

Much earlier practices respecting burials and the mounting of bear skull accord to Neanderthal man also basic religious ideas. Leroi-Gourhan (in *Religions de la Préhistoire*) produces a scenario of a large primordial religion from an "insignificant" incised tablet. What is revealed by relics must be only a token of full-scale rites of religion.

A Soviet excavation found religious incisions on animal skulls hundreds of thousands of years ago; for that matter, Pietro Gaietto, whom I have mentioned earlier, attributes sculptures to "hominids" of 1.5 million years of age; but, as I have argued in other works, the measurement of time is in a sorry state of disrepair.

Earliest humans gave preeminence to sky gods, as soon as one or more might be discerned through the thinning canopy of clouds. Ouranos and his counterparts in other cultures were, first Heaven, then God, corresponding to the canopy and the appearance of a great sun-like object (among many others) in the new skies. However, since we believe this tumultuous set of natural events took a part in creating the human race itself, we would maintain that man was never human before he was religious.

The belief in sky-gods is attested to both by the most ancient sources of religious practice and by the studies of modern so-called primitive peoples (whom we prefer to call "tribal"). All of the "great" religions begin their stories in the skies: The Judaic complex, the Greco- Roman complex, the Egyptian, the old Chinese religion of Heaven, the Meso-American complex, the Teutonic, the Persian, the Hindu. "The Chinese T'ien means at once the sky and the god of the sky." So far as one can tell, all primitive religions have important celestial referents, and we may quote cases from Eliade again:

> "The supreme divinity of the Maori is named Iho; iho means elevated, high up. Uwoluwu, the supreme god of the Akposo Negroes, signifies what is on high, the upper regions. Among the Selk'nam of Tierra del Fuego God is called Dweller in the Sky or He Who is in the Sky. Puluga, the supreme being of the Andaman Islanders, dwells in the sky... The Sky God of the Yoruba of the Slave Coast in named Olorun, literally Owner of the Sky.
>
> The Samoyad worship Num, a god who dwells in the highest sky and whose name means sky. Among the Koryak, the supreme divinity is called the One on High, the Master of the High, He Who Exists. The Ainu know him as the Divine Chief of the Sky, the Sky God, the Divine Creator of the Worlds, but also as Kamui, that is, Sky. The list could be easily extended."

Why is the sky the seat of the gods and even the gods themselves? From his unmatched scholarship, Eliade fetches a proposition which we believe to be incorrect: "Simple contemplation of the celestial vault already provokes a religious experience. The sky shows itself to be infinite, transcendent... For the sky, by its own mode of being, reveals transcendence, force, eternity. It exists absolutely because it is high, infinite, powerful." This speculation stands without supporting evidence. It seems to say, "since heaven is divine, and the gods are celestial, there must be a reason; the reason is a) since the gods are there, the sky must have impressed man and b) the sky is impressive (for the gods are there)." The logic is confusing and borrows, though not with conscious purpose, the propaganda technique of showering agreeable statements upon the reader.

One must return to the beginning. Granting that the sky-gods and sky-religions are primordial, how is man prompted to perceive the supernatural there, place preeminent divine activities there, and make the sky the centerpiece of religion? If humans existed long before religion was invented, then it should be embarrassing to argue that the skies might exist for millions of years before the idea of religion popped into the minds of people everywhere (and very much the same idea of religion, that is, sky-religion without aeons of animism, pantheism, ghosts, totemism, and such other forms of religion).

Eliade does not explain how early religions would move from sky-gods to demonism, totemism and animism, and sometimes back, for modern tribes of this kind meet no insuperable problem in adopting a sky religion such as Islam or Christianity. We offer two explanations. First, these religious practices were originally, have been, and are always with us, and are not at all embarrassed at co-existing with sky-gods.

The second explanation is consistent with the first. The sky-gods seem to have disappeared from many minds of our "high" civilization in favor of the worship of technology, cinema and political heroes, and a number of psychopathological quirks. "Primitive tribes," since explorers and anthropologists began their profuse reports, seem to have lost their sky- gods, too, or never to have had any, or to possess *dei otiosi* (removed, inactive gods). May not these tribe people be acting like civilized people in focusing upon the sky- gods when the gods are active, or when the memory of them, consciously or unconsciously, is acute, tending to dismiss, forget, and deny them when they are not causing great trouble? The skies became peaceful and the world stopped shaking; people turned to the supernatural manifestations of their closer environment. In this case, we may surmise also that the sterner the institutions of memory (records, graphics, priesthoods, bureaucratic churches, holidays) the longer the sky- gods will persist in a culture.

In his book of *Timaeus,* Plato accepts and rationalizes in its early pages the existence of "everything visible, and which was not in a state of rest, but moving with confusion and disorder" prior to the work of the Divinity or Demiurge which in its plenitude of intelligence and power "reduced it from this wild inordination into order."

Here is the first revolution; a Chaos, worked upon by a Demiurge (God) produces Order. This is a common ancient myth but we recall that Timaeus is a highly sophisticated Pythagorean and thinker. I conclude that the first of all great events remembered by man and emplaceable in primevalogy is the separation of Heaven from Earth.

The Divinity, according to Plato-Timaeus, using earth, fire, water, and air from the universe formed (generated) it into a figure, an animal containing all figures and animals and gave it the "most becoming"... "spherical shape, in which all the radii from the middle are equally distant from the bounding extremities." So says Taylor in his great commentary on Timaeus. This universe moves in a circular revolution.

Taylor concludes that to early and late Greek philosophers, the boundless, the universe before god, was composed of thick cloud or mist. Fire made it visible and that is why it became the first of the elements.

There is a major dilemma in *Timaeus,* faced by all philosophers and theologians who explain creation. Was God always around but disinclined to do anything about the Chaos? Then finally did he act and make order, i. e., the universe as man knows it?

My interpretation is as follows:

The Cosmologist is Man.

Man senses ancient experiences.

He asks when did experience begin. In fact, he is asking "when did I begin?" i. e. *my* inquiring mind.

He thinks everything always was, because this is a logical thought.

He recollects, however, a time before the time he recalls, and remembers such time as chaos or disorder (or thick fog).

This time of the ordering of chaos must be either a memory of when man first got his head straight, i. e. could reason and ask basic questions, or an actual revolution of his nature or environment (a catastrophic set of events involving perhaps the lifting of a low canopy from Earth) which he recalls because he was already *homo sapiens* in all or part; but he cannot recall any specific catastrophic events before this time; therefore it becomes his creation moment, his gestalt of creation.

The primordial heaven and god do not endure forever. And at this point, Eliade recalls the famous ancient concept of the *deus otiosus,* the distant, removed, hence disoccupied god. Having created the world, the first gods generally retire. "Celestially structured supreme beings tend to disappear from the practice of religion, from cult; they depart from among men, withdraw to the sky, and become remote, inactive gods (dei otiosi);"

Eliade presents relevant cases. "Everywhere in these primitive religions the celestial supreme being appears to have lost religious currency. . . Yet he is remembered and entreated as the last resort. . ." Eliade does not bring out the most striking fact about the retired god. His is often a forced retirement, following a bloody, world-shaking revolution. The Greek Ouranos was castrated by his son Kronos in a terrible revolt, and moved into exile, with no intimations of a return to power. A new great age begins.

Moreover, the new great gods are also celestial. They are not household familiars, woods sprites, or volcano ghosts. The Greek pantheon is well-known, but there are others as well. All of the great Greek gods are sky gods, though they may keep house on Earth as well, Hephaistos on Lemnos, Hades in the nether regions, and so on. The great ones are identified with the moon and planets: Aphrodite, Kronos, Zeus, Hermes, Athena, Ares, and possibly Apollo, Uranus, and Poseidon. (We do not refer, of course, to contemporary nomenclature.)

When these gods are entered upon the historical record, dim though this time be, a period of greatest power can be assigned to each; this project I undertook in *Chaos and Creation.*

Then the sequence goes: Ouranos, Aphrodite as Moon, Kronos, Zeus (Hera), Apollo, Hermes, Athene and Hephaistos as Venus, and Ares. And there is substantial reason (not commonsensical) that these gods achieved power, fame, and worship because they were identified with great sky bodies, such as the planets, upon the occasion of great natural catastrophes befalling the Earth.

Scanning Samuel Kramer's collection of *Mythologies of the Ancient World,* we find persistent outcroppings of the procession of gods and ages despite his complete disregard of events in the heavens that might differ from the behavior of the sun, moon, planets, comets, and stars today. We find dual splitting creation gods, of the type of Earth and Ouranos; we identify Saturn, Zeus, Venus, and Mars, and also stories of cataclysms of the raising of the sky, and of world ages.

In the *Epinomis,* Plato accomplishes a significant trick of theology. Discarding the mythology that places the gods on Mount Olympus, he replaces them upon the planets where, he says, they belong, hoping to reform their bawdy characters thereby. He says we must get rid of any notion of the strife of the gods. They move always in order. (Elsewhere, Plato would have any disbelievers in orderly skies punishable.) The astral gods are the real ones, he insists, and he gives them their names. (He anthropomorphizes the vault of Heaven, Kosmos.) Their names, he suggests, should be coordinated with Syrian and Egyptian observations, which are much older and "tested by vast periods of time." To us it occurs that bringing the gods down to Olympus had been psychologically an effort on the part of Greek myth-makers to control the gods; they became human and tied to human fortunes directly. Now Plato, feeling no threat of planetary disorders, wants to send them back to their former homes, which are once again safe. Dei otiosi, the removed gods, will be doubly safe, safe for themselves and safe for mankind.

We note that the Greek and many other cultures regard their sky gods as blood- related. To the Greeks - to us, for that matter this could only mean that their history was intertwined, overlapping, of the same order of celestial experiences.

We note further that the greatest Greek philosophers and scientists did not argue against the succession of gods. They did not challenge the succession because somehow it was real to them. Somehow they were experientially or psychologically inhibited from claiming that the gods were born together.

Thus, there has been a succession of gods and goddesses in human history. Yet human nature is obsessive, that is, faithful; further, it was a great sacrilege to forget god, and severe punishment and expiation not only followed forgetting but were performed as prophylaxis. The compelling reason to change gods is to be found in reality. The reality is that the gods have changed, and, despite all his efforts to be loyal, man has been forced to worship new gods over the ages.

The ambivalence of the gods caused mankind from the beginning to exert itself strenuously to control them. A continuous redefinition occurred. Yet never has the nuclear complex of a god been put aside without great external pressures, the most excruciating of which has been the advent of an apparently more flexible and potent deity. In these cases, people have, as often as they could, tried to merge the new and the old; any evidence of continuity and any confusion of identities, whether physically or psychologically produced, have been seized upon to establish that the worship of the new is faithful to the worship of the old.

Therefore it happens, consciously or not, that all gods have an unbroken line of ancestors going back to primordial chaos; there the gods are made from the abstract elements such as air and water, or the world begins out of nothing. We should bear in mind that when Egyptian history opened, with the Pyramid texts, Osiris (Saturn) was already dead, deus otiosus, and Horus (Zeus) reigned.

The god Nun of Egypt, first god of the first recorded cosmogony, bears in his hieroglyphic name that he is of the primordial wastes of water in the sky, and Egyptian legends state this to be the case. Mother Earth, Terra Mater, the Universal Genitrix, Gaia, is the most durable of the gods, and found practically everywhere. In Hesiod's *Theogony,* she gives birth to

Ouranos who is "a being equal to herself, able to cover her completely." It is clear, however, that Earth (who may even be conceived of as masculine sometimes) reacts to the changing gods of change. This Nun or Heaven is "father of the gods" and father of Atum or Re. He or it is the demiurge of the boundless, featureless darkness, from which evolved the first hills or eminences. There appeared in early Egypt four different cult centers with special creation myths, all of which were essentially the same.

In Sumer in the 5th millennium before the present, as legend has it, Nammu, whose ideogram carries the meaning of "sea," was called the mother of heaven and earth who also bore the gods. Fluids and gases are favored elements of chaos and materials of creation. There is more than a semblance of logic alone in this accord of legends; the idea that gases go with chaos is attractive but is more than ex post facto explaining of legendary fiction. Fluids and gases must indeed have enveloped primordial man and attended the birth of the gods. Ouranos emerged out of the watery and turbulent wastes of the sky cloaked in robes of clouds.

Philo Byblius anciently reported from earlier sources that the first Phoenician god was Elium or Hypsistos ("the Highest") and was succeeded by Ouranos who was succeeded by El or Kronos. But I would interpret this primordial god as the first stage of Ouranos, the adamantine condition of the sky prior to its breaking open to reveal the great light of Ouranos.

The Babylonians, successors to Sumer, in the early third millennium

B. C. worshipped Marduk-Bel (Baal) as patron god and world creator, exalted over the old Mesopotamian pantheon just as Jupiter came to be exalted over Saturn in the Roman Greek pantheon. Poseidon (brother of Zeus and son of Chronos) remained in heaven after his father fell and only later, upon agreement with Zeus, descended to rule the seas. He also flooded the land so doing and was known as the "land-encroacher." Thus the descent of Poseidon (Neptune) is to be identified with a great deluge, perhaps a name for, if not a later part of, the same great deluge that is connected with the crippling and binding of Kronos (Saturn) and is the same as the flood of Noah brought down by Elohim in Hebrew Genesis.

An important line of attack may be levelled against our assertion that he succession of gods reflects a series of natural catastrophes upon Earth. Religions have continued to acquire new gods without actual catastrophes and have spread widely without catastrophes to help them do so. Some of these religions have been militarily aggressive, others peaceful. Thus Islam conquered large areas at first by the sword, as is well known, but in recent years has converted peoples readily with little bloodshed and compulsion, as in central Africa. Farther back in time, as Wheatley *(The Pivot of the Four Quarters)* asserts, the Hindu pantheon moved into Southeast Asia along with its social institutions. Along with the religion went peaceful commerce. Many shrines were erected, around which there grew up cities. So enthusiastic were people for the peacefully inculcated religion that sometimes the near totality of a state's economy was given over to oblations to the pantheon.

Yet, the 2600 years since the probable last great natural catastrophes have not been distinguished by peacefulness. War and slaughter have been conducted in the name of a warlike religion (or interpretation thereof), or of a peaceful religion, or in the name of no religion but the state or tribe. We are led, then, to conjecture that homo sapiens himself, though relieved of direct models of destructive behavior in the skies, continues to carry out deeply rooted impulses to destruction, whether through unconscious memory or because he is constructed genetically to do so. That both are in fact the case is a main thesis of my Homo Schizo.

So long as the skies were disturbed, and the Earth with it, the character of religion reflected clearly natural events and imposed models of conduct upon man. But religion itself was born in the creation of man and, if he were other than true to his nature or were of another nature, he would not have a peaceful religion and behave peacefully in all probability. Religion is a dependent variable of human nature. It is a dependent variable of natural events. We shall have to inquire, as we proceed, whether, in some other sense, in another kind of reality, religion may be an independent variable, owing its existence to conditions freed of human nature and ancient natural disasters.

To speak of religion as a variable reminds us of how vague and intangible are the materials of the history of religion and even of religious behavior today.

To speak properly about the religion of a person, a standard intensive interview at the least is required. "What precisely are your perceptions of the supernatural?" "What practices, life pattern, or habits do you possess that are related to these perceptions?"

Were a representative sample of the people on Earth thus interviewed, and the results properly classified, tabulated, and analysed, we should be able better to generalize about the relation of present religion and gods to the historical religions and gods - provided, we should add, that we have assembled and ordered all that might be known about historical religions back to their origins in the origins of man; this, however, we should probably be incapable of doing unless we were to adopt as the guiding hypotheses those already suggested in these first chapters, namely:

The earliest human cultures were simultaneously religious.

The earliest and most important supernatural objects everywhere were celestial.

The Ouranian complex of Heaven and Gods was the first list of Dramatis Personae of religion everywhere.

The Ouranian complex was overthrown by nature and simultaneously by man.

All successive gods everywhere have descended from and relate to the Ouranian complex.

Man believed himself forced to change gods from time to time by evidence in nature.

Man, as he changed gods, accomplished the transition with as few variations as possible in previously assigned powers, traits, names, vestments, rites and religious conceptions.

In these transitions, man became adept (to his way of thinking, which was and is delusory) at reconciling and controlling his gods through his religion, whence, by controlling the gods, at controlling the world, all with the ultimate and impossible goal of obtaining self-control and peace of mind.

CHAPTER THREE

THE HEAVENLY HOST

The animation of nature is an instinctual interpretation, primordial with humankind. It occurs with humans today, more obviously among the young. To exorcize it takes training.

The earliest gods took shape as the Sky and Earth. Logically, and in accord with most evidence of what was manifest, the primordial welkin was densely packed, without brilliant separated lights, until the sky was broken up and these appeared. There developed next a more definitely formed solar god of the Sky. A change in nature was responsible for the change in divine forms. If the sky had been penetrable, the great god would have come first in this solar (or planetary) form.

We believe that until nearly 2700 years ago the skies were periodically invested with changing forms, and much of this turbulence was impacting upon the Earth physically, as well as upon the minds of humans. The scene was conducive to polytheism. Divine presences of all types might be discerned. Yet there was usually a great god, a father of gods, an Ouranos, a Kronos or a Zeus.

We infer from this fact that such beings were at some time most impressive features of the sky and, when they were not, were scalding memories, which had so dominated the human setting that no successor, no matter how prominently active could match what its "ancestor" or "father" had achieved.

Some cultures, such as the Roman, Greek, and Hindu, did not conceal the succession of fathers, and assigned family roles to junior actors, while the Hebrews over a period of time accepted the Mosaic rationalization which fitted several great gods into a unity. This did not come without ideological and political struggles of great intensity and long duration, some of which are recounted, in expurgated form, in the Old Testament.

"Varro had the diligence to collect thirty thousand names of gods - for the Greeks counted that many. These were related to as many needs of the physical, moral, economic, or civil life of the earliest times." He found 40 Hercules alone. So writes G. Vico. The sacred book of the Mahàbàrata (1: 39) claims 33,333 Hindu deities, and later sources say that there were a thousand times as many. The Nordic Grimnismal gives over 50 names to Odin. The Babylonian Emunia Elis culminates in a recital of 50 names of Marduk. In the history of symbolism and language, words may actually have begun as god-names. Words might have been more sacred than pragmatic, until an advanced state of collective amnesia and sublimation had been achieved.

If the Greeks had 30,000 god-names, and the Hindus even more, then all the world's cultures must have had hundreds of thousands. The great numbers, however, reduces to a comprehensible order when a proper theory is applied to them. The total of this heavenly host includes, first, a few great gods, whose real existence in the sky lent structure to the ages. Second occur the thousands of names of the great gods, most of which have yet to be identified with their referents. Many of these names are concealed references; others are what foreign cultures call another culture's gods; some names isolate a quality of the gods; some names are used to marry the gods of one culture to those of another.

Appellations may thus be congruent and complementary, that is, logical and harmonious qualities that a single personality may possess. Or they may appear nonconforming, leading nonparticipating observers (enemies or scientists) to question the nature of the god. However, as with great contradictions "God vs. Devil" - so with lesser contradictions - "god of arts vs. god of war" - the contradiction might be only apparent, the same supernatural being having apparently produced a variety of effects during his primary effective manifestations in nature.

Another major source of divine names (besides the attributions) is the outcome of processes of memory and forgetting. To forget the disasters that characterized the appearance of the gods was an urgent demand of the bruised mind; but any lapse of memory would be accompanied by fear that god will not permit himself to be forgotten and will punish forgetfulness. The mind then works to define and characterize god so that his image will be tolerable upon the conscious level. It further adds new words to its vocabulary of the divine, discovering that a god called by another name is less threatening. Still further, by the logic of delusion, a god whose name is mysterious or hidden will respect the awe and fear bound up in the secrecy and at the same time will restrict himself to activities that do not threaten the very core of terror that crouches in the human soul. A plethora of non-names, secret and cult-names, and common partial names comes forth.

Effects of many kinds are produced, the least of which is the confusion of names that confronts the outside observer; the selective remembering is tolerable; occult elites can dominate societies; the language and concepts of a people are enriched as the naming of gods flows through the symbolic world by association, analogy, and implication.

CHAPTER FOUR

LEGENDS & SCRIPTURE

The biggest difference between myth or legends and sacred scriptures is that the latter are selected legends, called "divinely inspired or spoken" by their believers, which have been carefully guarded and edited to pursue the continuous but also continually changing religious goals of their custodians. Myth and legends, not so regarded, or whose line of custodians died out, were left like abandoned children to wander through time as casual history and unconstrained imagination, until caught up by scientific mythological studies.

Giambattista Vico was the first modern scholar to perceive this process when, two centuries ago, he wrote:

> *"The fables in their origins were true and severe narrations, whence mythos, fable, was defined as vera narratio (a true account).. But because they were originally for the most part gross, they gradually lost their original meanings, were then altered, subsequently became improbable, after that obscure, then scandalous, and finally incredible. . . These are the seven sources of the difficulties of the fables..."*

One of many debts that we owe to Plato is his respect for myth and legend. He, too, fulminated at those who dismissed or, worse, corrupted history by their misuse of legends. In my skeptically minded exploration of the story of the destruction of Atlantis, the attitude of Plato mitigated my doubts. Plato goes out of his way to insist that the story be taken seriously, despite its prehistoric origins. Critias, his protagonist, is given to claim repeatedly that he heard and learned the story from his grandfather as a true and exact account. Significantly to a modern mnemologist, Critias declared that although he had forgotten much of what he had heard of the previous day's discussions, he had forgotten none of what he had learned as a child about Atlantis.

The Atlantis story is generally disbelieved, yet if an educated unbeliever were to compare it with the story of the Deluge of Noah in the Bible, it would appear to be just as (im)plausible. It is no less specific. The "author" of one is Plato, of the other, Moses; who is more reliable? True, Atlantis is no longer to be found, above or below the sea, and therefore presumed not to have sunk; but the flood that climbed to great heights all over the Near East has vanished, too. Objectively, one would have to be as skeptical (and no more so) about the one account as about the other. The difference is that a great many millions of people believe in the Noachian Deluge because they believe in its sacred format, while the Atlanteans are long dead and the moral of their story - that Zeus destroyed them because he found their squabbling and vices intolerable - no longer lives in people's minds.

A legend is history which has been largely unconstrained by realism and objectivity since the happenings that it describes.

Myth may be defined as a religious and aesthetic interpretation or story based upon legend and history. Its goal is to serve essentially non-historic functions while reminding its audience of a significant historical happening. Myth is closely related to rituals and sacrifices, which have the same goal, but, like sacred scriptures, are under severe theocratic constraints.

Myth is often indistinguishable from legend, but this occurs in part because the original culture to which a myth and legend belonged no longer exists to explain to us the difference between the two; myths and legends intermingle in a flow through time which we experience much later and find indistinctly composed of both. The famous myth of Phaeton, who drives the Sun's chariot, burns up the Earth, and is destroyed by a thunderbolt of Zeus, is by common standards today an entertaining myth, but appears upon investigation more as a legend, supporting an historical intrusion of a cometary body upon the Earth's atmosphere.

Sacred scriptures will always contain a high proportion of vague, indecipherable, incomprehensible, contradictory, and substantially untestable material. They will also have lost much, as historiographic methodology increasingly shows, owing to the alteration and accidents of their form of transmission, through cultural miscegenation, by reconciliation of older history with later history, by imposition of patterns of integration and new styles, by the collective amnesia that seeks both to forget actually and recall symbolically the traumas provoked in terrible ancient catastrophes, and by other changes in referents to accommodate ancient to present conditions, as a comet becoming a star, or as invisible electrical discharges which are now referred to as purely symbolic manifestations. Therefore there are limits to the scientificity that can be granted to the *Rig Vedas*, *Bible*, *Eddas*, *Book of the Dead*, *I Ching*, *Popul Vuh*, and other scriptures.

Nor can it benefit the credibility and influence of believers in sacred scriptures to be relegated by general consent, including their own, to the nonsensical remnants of the works. For example, many Biblical scholars refuse to employ or give credence to Talmudic commentaries and ancient legends of the Jews, when these documents will often testify to the authenticity of Biblical statements and elaborate them in a way that enhances their credibility.

Ominous conclusions emerge from these several pages. There is much history in myth, legend and scripture everywhere in the world. In a sense, all religions are desperately honest in their fundamental statements. Yet it is appreciated that, in a memory choice between a delusion and a historical fact, a religion will prefer the delusion. An attempt to "clean up" an historical religion by eliminating historical and empirical errors cannot succeed. Meanwhile we affirm that a religion cannot subsist on delusions alone: it must make historical and empirical statements. Are we to believe then that historical religion must be abandoned? We are not yet ready to answer this question.

CHAPTER FIVE

MAN'S DIVINE MIRROR

That gods are often snares and delusions must be admitted. Yet the occurrence of the delusions takes on patterns evocative of actual events and of common mechanisms of the analyzed human mind. Natural expressions of high energy occur in cometary approaches to Earth, deluges of water and other material from the skies, anomalous intensifications of heat and cold by conflagration or sudden icing on a large scale, simultaneous large scale volcanism, and otherwise. Much evidence goes to show more of such catastrophes in ancient and prehistoric times than over the past 2500 years.

We say that the more frequent these occurrences and the greater their intensity, the more that gods appear and the more religious humanity becomes. If these be called gods insofar as they are apparitions and because of their enormous effects, then there is a real historical reason why mankind once was much more religious than now. Geology and archaeology can demonstrate (with much more research than they are inclined to provide) the actual basis for enhanced early religion. Psychology and the history of religion can show how the religious mind has expectedly peaked in these actual stress periods and subsided when the strains relaxed.

If one wishes to animate the ancient apparitions (metaphorically or delusionally) and assign the fantastically great natural events to interventions of the gods, defining gods as "whatsoever can produce such effects," and further goes on to distinguish and assign gods to the different effects of, say, air, fire, water, and earth, there can be no logical objection. So long as one does not proceed beyond the evidence to impute motives, make misleading classification, and imagine an organization of the cosmos, none of which can be even partly demonstrated, the gods of nature can be said to exist as truly as "democracy" or an "infinite regression series."

Here is where mankind gets into trouble with the scientific authorities of anthropology and psychology: it assigns a great many undemonstrable qualities to the gods and spirits. Then, hardly pausing, it fashions such qualities into a mirror of man. The mirror lies.

We can make two principal statements and several dependent propositions about the Divine Mirror of Man: first, all human qualities are found among the gods; second, divine organization portrays a reorganization of the human mind.

To demonstrate that every human quality has been sometime, somewhere, and even frequently, a divine quality requires hardly more than a list of references on the history of religion and anthropology.

Is the god assembled anthropomorphically? The implication, even when not stated explicitly in sacred scriptures and legend, is that all of the traits of the divine do amount to a creature not unlike man. That Elohim created man in his or their image is, of course, a direct statement of the Hebrew Genesis, and if one were to compose a physiological mosaic from all references to Yahweh, the mosaic would evolve to look like Moses and act like him, including how Moses would like to have acted.

The Divine Mirror, it seems, is more perfect than the gazer. For it contains all of his qualities and all of his dreams and desires. Sometimes these are contradictory, but the mirror finds a solution. It may show a god with devilish features, or a god who is both female and male. Does it ever show a god who is both brave and fearful? Often, despite the fact that fear creates gods who are afraid of other gods, afraid of themselves, or mistrustful of their worshipper, this last being a kind of fear that drives gods (as it does men) to excesses of all kinds. So, indeed did the Lord behave toward Job, when the Devil drove him to be suspicious of his devoted and good worshipper.

In an early work, C. J. Jung wrote an *Answer to Job* where brilliantly but in a fundamentally naive form, he hints that man is too clever for God. "It were better," however, "not to wax too conscious of this slight moral superiority over the more unconscious God." One notes the marvellous schizoid behavior of the human, Job, when he is trying to control God. The making of the ambivalent god and then controlling him becomes the greatest work of man.

God suspects and is jealous of the game that man is playing, a contradiction-in-contradiction, mirror in a mirror in a mirror, contra-contra-contradiction, which the schizoid can continue indefinitely, always one step ahead of God. In the story of Job, one finds the full range of schizophrenic conduct, including the creation of the Lord as the preferred instrument for working out human delusions. I trace the schizotypical character of the human race in the preceding chapters.

Significantly, Jung separately focuses his research upon Job and then upon schizophrenia. In the story of Job and God we even locate a tendency of humans to make of gods what they would make of themselves if they could, a kind of unreflective healthy instinctive animal, rid of the curse of self-awareness -- though this same self-awareness is the only true mark of the human and the source of god as mirror of man.

The experience of all peoples has been generally the same, intense ecological stresses anciently operating upon a divided, fearful mind. To say therefore that gods are "good" and men are "evil" makes anthropological history impossible, theoretically or as fact. The obverse to "how the gods could be believed to do evil to people" is, "how the gods could be believed to do good." The efforts of humans to justify the evils visited upon themselves are extraordinary, considering the gravity of those evils. Some profound reason must prevent them from declaring that gods and devils are one and the same - a disaster. Why do humans not recognize the animated high-energy forces of the world as the open enemies of the human race? Indeed, this did become finally the feeling of a great many people in modern times, whose change of attitude coincided with a de-animation of the forces of nature.

Primeval man and his successors found good in the gods because in the first place the ideal of the good god itself performed useful functions.

The gods created man, and man appeared to be superior to the mammals whom he resembled and lived among. Therefore, gods should be loved for their creative deeds.

Still, gratitude is a refined subliminatory trait that would hardly result from this syllogism. There had to arise a satisfying powerful identity out of the gestalt of creation: the creative god was built into the mind of the creature; it was his first projective delusion. His first great relief from fear was placing the responsibility for his creation, not upon himself (an idea that must promptly have occurred) but upon "some himself not himself," ergo a god. Who denied god, denied himself; who denied himself would not survive. The madness of great delusions was the condition for survival.

There remained only the elaboration of the madness into human norms. A quick transfer of traits occurred - man gave to god all of his abilities and took them back as blessed gifts, down to the rudiments of stone age technology, the very fashioning of a club. Because of the obvious powerfulness of the gods, the gifts acquired power in the human mind, and man would step forward to control the world with an obsessive confidence, a false confidence, very often, yet with enough successes to accredit the transfer. At the same time, man could deny his personal responsibility for all that he was creating.

Further, by imitating the gods, invention was promoted. More and more objects and procedures for controlling himself and others were imagined to descend from the gods and more and more were created under divine inspiration. This despite the interference of the gods thenceforth in inventions of all kinds, wherein nothing could be invented and applied unless it had come from the gods or was blessed by the gods. The psychological mechanism had its drawbacks; in the most peaceful and pragmatic periods, the wellsprings of invention were overlooked, while the subservience of practical innovation and social reforms to religious dogmas and rituals was condemned.

The mechanism for projecting and retrojecting gifts of power and techniques was in itself adequate to explain why a punitive god could be assigned benevolent and beneficent qualities. Yet it was not the only source of the idea of the good god. The first mutant humans came into being in the midst of chaos and destruction. That they had survived while all around them lay a biosphere of death and destruction, including what had been their own kind, was a miracle; their minds were now equipped to reflect upon it.

Mourning was a trait already possessed; mammals and primates mourn. Beyond mourning, however, or if human mourning were to be distinguished, was a new consciousness of the self, an individuation from the group, that could see what had happened to others, see what oneself had escaped, and assign to the escape a selective feature, a blessedness, a sense of being chosen for survival.

Thus arises the quality of personal satisfaction and joy amidst ruin, that interjects itself into the most grandiose human tragedies, and causes people to dance, laugh, and sing when the world shakes and burns around them. It was a primordial human acquisition, directly connected with the animated forces of destruction. Sailors, returning aboard a ship off of Krakatoa in 1883, who watched the desolation of their families on the shore from volcanic explosion and tsunamis, laughed and jumped with joy that they had been spared. Hysterical conduct, to be sure, in awful fear, but such is the nature of hysteria, and laughter often is a fringe around hysteria.

The divine identification and imitation justified and provided morale for survivors to revive and conquer. A newly-acquired super-mammalian aggression abetted the profits of survival. Those who survived could move out, reinforced by grace of the gods, and in imitation of the gods, readily loot, kill, or enslave whoever remained alive and strange. The material gains of aggression were thenceforth regarded in the category of gifts of the gods, and regularly some portion of them was returned to the gods by means of sacrifices. From old Mexico Brundage gives us a song composed by the Emperor Axayacatl: "The flower death (for sacrifice and cannibalism) came down to Earth. It came here. It had been created in Tlapallan (Heaven)."

Outstanding in human behavior is the voluntary and unconsciously motivated repetition of actions in every sphere of life. In individuals, instinct serves for habit, the distinction generally being that instinct is untrained. Habit and custom are inculcated by training or imitation. Not only is habit pervasive of normal activities of individuals and groups. It is also characteristic of many psychopathologies, where it is called obsession.

The origin of habit and custom lay in the primeval fears of the self-aware human, and the discipline that such fears sub-consciously and later consciously impressed upon him. First came schizophrenic obsession. The more intense a blow or trauma to the body (mind), the more intensely and frequently it is auto-inflicted neurologically afterwards. An obsession is an auto-inflicted reiteration of some or all of the initial reaction to a trauma.

An obsession discharges quantas of the stored force of the trauma, which originally could be tolerated short of death only by its redistribution (i. e., memorizing) in successively less related circuitries contacting the affected area. Some effect of a trauma also are discharged through interfering circuitries, some of which were developed in primeval man in analogously obsessive and some in non-analogous behavior, especially symbolic manifestations and erratic uncontrolled seizures.

These forms of dissipating the impactive force of the trauma are founded upon analogous primate behavior. They establish themselves as quasi-voluntary and voluntary activities of the split self, which more or less observes its own reactions and discharges. They are seen by men as voluntary because the self views the action as a decision of two or more compromising internal selves.

CHAPTER SIX

INDISPENSABLE GODS

Historical religions conserve the memory of a certain time when the world was created and humans came into being. None says that mankind always existed, or that it evolved mechanically by random association of particles. A purposeful act took place at a certain time. Most religions say that mankind was subsequently destroyed and recreated. Almost always the extermination of humanity stops short at a surviving couple or the equivalent. The subsequent homologue of the first chaos is a subsequent set of catastrophes by flood, fire, wind, and earth movements.

To preserve the memory of the first time of creation is a function of rituals, liturgy, anniversaries, and sacrifices. Many religions have strenuously sought to reproduce, short of deliberately re-annihilating themselves, the exact circumstances of chaos and creation. They have obsessively kept forms, practices, and words that go back to the beginnings of all religion and the first experience with the gods.

All historical religions are therefore highly conservative and weaken their foundations as soon as they admit deviations. The function of inescapable and exactly repetitive practices and symbols is to relieve the massive anxiety stored from the earliest times by confessing what happened in those times and reliving them successfully.

What appears to be radical in religious history is reactionary. Practitioners of the religion, wrought up beyond sufferance, find even the rigid rites of their church insufficient to recapture the moments of chaos and creation. Prophets, apostates, evangelists, and orgiasts arise. So do whirling dervishes and *hashishin* (assassins) and berserkers. They are chiliast or millennialists. They proclaim the end of the world while demanding that everyone acknowledge the full and immediate meaning of the creation of the world. They prepare to die and be saved in the recapitulation of the original catastrophic times.

All historical religions are based upon punitive gods, are self-punitive and are punitive towards others. Gods are adjudged good to the degree to which they refrain from destroying their creatures. Humans exist by divine tolerance. A common word for a good person in most religions in "god- fearing". Personal merit through skills, altruism, and dogmatic belief and practice is sometimes, but more often not, a guarantee to a greater of lesser extent of the gods' benevolence; never is merit a perfect or universal guarantee. This belief in the denial to merit of its due is not, therefore, as some think, a connivance of religion with the envious mob.

Sacrifices are forms of punishment of the self and others to forestall, and therefore to control, a punishment from Heaven. The concept of representation effectively lets a partial sacrifice stand for a full sacrifice and a sacrifice of others stand for a sacrifice of oneself. Sacrifices are said to be gifts freely given; yet it is acknowledged that withholding sacrifices will be followed by divine retribution. The more valuable the sacrifice, and the more strict the rules under which sacrifice and all other kinds of punishment occur, the more pleasing to the gods.

Guilt is self-punishment. It is the refusal of pleasure to some negative degree. We often know it in its late and rather pragmatic sense: guilt is what makes a fickle creature responsible; without guilt, personal and social discipline would be impossible. To get relief from guilt, one follows religious directives or some secularized substitute such as warring for one's country or pursuing "the work ethic."

But primeval guilt originated from the terror of "the other self," the terror produced out of the minute systemic delay of instinctive impulses. At the same time, the heavens were turbulent and terrifying. To control one's unbalanced self, one signaled the gods to arbitrate; and the gods responded, saying, "Your soul is a struggle of good and evil. We, with your cooperation, will take care that the good dominates you. You are not sick. Be hopeful. Help is on its way." This formula, although it can be called delusion, was a great invention. Granted the essential incurability of human schizotypicality, it alone could lead to a manageable psychic world.

Important anniversaries or holy days are celebrations of divine destruction and near-escapes from destruction. Every truly religious anniversary celebration is therefore ambivalently tragic and joyful. Anniversary excesses and orgies, at both extremes of somberness and exuberance, are nevertheless occasions for the relief of tragic memory, more or less deeply suppressed. Anniversaries cluster around the great cycles of the ages, which give evidence of having been common to most of the world's cultures.

Calendar diversions, not psychological changes, have driven apart the anniversaries of different cultures; they are farther apart in days than they are in mind. The end of the year inspires saturnalia in many cultures. Roman Catholic and Greek churches mark a different Easter holiday for unessential reasons. Anniversaries sometimes are pulled together in a given culture by their original proximity during a cycle such as a solar year and by their psychological resemblance. Thus, Venus (perhaps at -3437 B. P., where Before Present =1984 A. D.) and Mars catastrophes (perhaps in -2671) occurred around March 23, close to the Spring equinox; the holidays were merged ultimately, and are submerged at Easter time in Christendom and comparable holidays in other cultures.

Sublimation, like ritual, is universal in religion; it pacifies, dissembles, represents, and rationalizes the strict conditions of the fatal times. Sublimation becomes more secular and pragmatic with the evaporation of stored anxiety over long periods of prosperity and peace. Disaster, deprivation, and frustration raise anxiety levels; they tend to cause reactions against secular sublimation occurring such as happen to occur in the artistic, social, political and religious spheres; these activities are then attacked as irrelevant and blasphemous.

Furthermore, all religions incorporate directives for every aspect of life -- work, sex, property, power, relations, health, and knowledge. Humanity was created and made deistic at the same time; the human mind is not logical, but it is wholly occupied by a way of looking at the world as a supernatural creation. The question of separating special values and calling these "the province of religion" has no meaning to a mind that was originally formed with every value at stake.

One god has been replaced by another on various occasions. Almost always, the replacements were successful because of unconscious techniques of cross-identification and rationalization. Sometimes men sought to replace gods by deliberate choice, with or without the help of events such as cultural amalgamation; invariably then, compulsion and heavy propaganda were employed. Such occurred when Hinduism moved over Southeast Asia, when Christianity came to dominate the Roman World, and when Islam moved across Asia and Africa.

The replacement of all gods by materialistic and atheistic ideology is a special case, discoverable, in non-catastrophic times, among philosophical schools such as the ancient cynics, among scientists and humanists of the post-enlightenment, and among communists.

Invariably secular replacers have argued the lack of empirical proof of the existence of gods; they have also stressed the contradictions of ruthlessness and mercy in the concepts of god; and they have attacked the behavior of religious establishments. As alternative behavior they have recommended principles of brotherly love, cooperation, and mental health, among humans, or principles of an ideally organized state that provides enough goods to satisfy people's needs without recourse to supernatural agents.

The major proof that such ideologies might succeed is based upon the waning of the gods when societies possess a pragmatically optimistic morale and are materially prosperous or believed to be potentially so, as recently. Then the gods have seemed remote and unneeded; considerations of logic and efficiency would appear to dictate their abandonment, removal, and forgetting.

Even under optimal conditions of prosperity, secular morale, compulsion and propaganda, the replacement has proceeded slowly and painfully. At the peak of their success, the ungodly ideologies have been undermined by new gods (e. g., Christianity in the Roman Empire), resisted successfully by the masses (e. g., communist Poland, 1945-- 1983 A. D.), transformed into secular religions of temporary duration (e. g., Roman Emperor worship, der Fuehrer Hitler, Comrade Lenin), or transformed into pseudo-scientific therapeutic or philosophical sects employing substitute semi-divine agents (e. g., gurus, anthroposophists).

The fundamental obstacle to ungodliness has been the construction of the human mind. Inasmuch as the events of creation that split the hominid character introduced the splitters as gods, humans become god- seekers as part of becoming human. The particular manner in which the universe was seen for the first time implied perforce the instrumentality of divinity. Self-awareness formed a nature which was unceasingly prone to discover gods.

Far from being an afterthought, the gods were a first thought. To excise this thought, after thousands of years of experience with it, was not only most difficult pragmatically; it was structurally impossible, at least as long as the origins, function, and mental structure of religion were not understood.

To forget the gods is impossible; the memory deck can only be reshuffled. To retain self-awareness without schizotypicality is a contradiction in terms. Human creation involved a basic reconstruction of mammalian mind; to extinguish this essential schizotypicality would restore man as an instinctive mammal, but is in any event now physiologically and psychologically impossible. Symbolism, as the effect of the split self, flows naturally and cannot be obliterated. By the same logic and dynamics, treating symbolically with both the "other self" and the "outsider-others" must inevitably result in projectional thought, that is, treating the "outside other" with the same mechanism and feeling that the self utilizes in dealing with its "own other."

All of this process is transactional and the transaction is of the essence of human being. Therefore a group mode of projection, a group communication, is inherent in the individual-social complex. Thence, naturally, whatever is unanswered and questionable becomes a matter for resort to authority -- that is, a prevailing, preponderant group opinion. Since the group is forever under historical and existential stress, it is forever seeking authority and incapable of receiving satisfactory answers to its questions without a symbolic, abstract and animate referent that provides solution. Thus it happens that, if humans exist, god exists.

God is the closing of the circle -- both question and answer. But so inextricable are the question and answer that only logical artifice can distinguish and designate the two.

Man's need to control the terrible and the terror causes him to invent gods. Nowadays, if one were asked how to control or stop an advancing comet, he would dismiss the possibility, and say that we must await it.

He is not prepared to undertake all the actions that ancient man had ready just for such approaching catastrophes - propitiation, sacrifice, ritual, saturnalia, "going on the warpath." Nevertheless, as catastrophe approaches, at first slowly and then rapidly, and then hysterically, the modern human will act like his ancestors, including the excesses of guilt for not having foreseen the deserved end of all folly. He will draw upon the dwindling and remaining reserves of the "old time religion."

If the fossil voices telling us of the nature of the gods and of the rules for man's behavior respecting the gods are distorted and incorrect, and though they are not valid and reliable guides, yet these voices have told us things of positive value. They have given us foundations of history. They have recounted the basic facts of existence repeatedly. They have conjured existences differing from ours. They have in effect performed innumerable experiments with the allegedly divine from which we can learn what not to do religiously, and to a lesser extent what to do.

Perhaps the greatest lesson they have taught us (by negative inference) is that the religion of today and tomorrow should not be sought in the religion of the past: that humans, until they reach some certain level of perfection cannot be trusted to have known and arrived at the nature of the gods. Whenever historical man has said "Let us change our religion," (even if he does so in the name of preserving the old religion) he is saying "We were wrong about god and religion and it is up to us now to find a new way to god and a new religion."

The gods have retired into new forms, but they still operate through the busy humans whom the poet Rilke called "the bees of the invisible." The gods are still everywhere and are not as remote as our scientific texts would have us believe. They are in astrology, in magic, in fortune-telling; they fly to the scenes of disaster; they augment the forces of authority; they heal and console; they scare; they make anxious; they set the rituals for a multitude as they have done since the times of Ouranos. They assume their own negation: for they argue with themselves in Natural Law, Bureaucracy, in Dogmatic Materialism, in Reified Words, in Mummified Heroes, in Times and Worlds without End. They let themselves be moulded into One and the One obliges his necessities by becoming Many. Beyond all else, they stand at ease waiting for Armageddon and the Day of Judgment. Then they will don their armor and gather their hosts.

Although they have retired it still takes rare courage to contemplate all of their continuing manifestations and to resist the invention of new negations. There is yet nowhere else to go and few who would follow.

By skimming along on the thin ice of the cerebral cortex or by mathematical astrophysics or metaphysics or another such exercise, the gods can be sublimated. Dumb bestiality may be equally functional. We think that of all ways of facing them, the best is to look at them everywhere, contemplate their every manifestation, anticipate their reappearance, but do no more. If there is any question of human madness, it is erased when one pretends to be divine. Our human destiny is an open question. We deny our humanity if we try to close it. We belittle ourselves if we plead with the gods to answer the question at any cost.

Whenever gods and religious practices have been abandoned, put aside, forgotten, changed consciously or unconsciously, those who made such changes are saying to us their descendants: "Do not think that our ancestors, or us, or even you, will have the answer. There will be New Testaments without end." We the present generation are told that we are not the first, nor the last, but a truth-seeking figure in the series of forevers until the day when somehow, somewhere, we shall be perfect. At which time, we might, if we dared, claim omniscience, omnipotence, and the fullness of virtue.

In hastening to accuse traditional religion of claiming falsely absolute truth and morality, we often fail to see in the seeming absoluteness its inherent self-confessed contradiction. Just as psychiatry has proven that excesses of anger, self-destructiveness and aggression have ordinarily come out of self-doubt and self-hatred, so we can see in the madness and excesses of historical religious behavior the same psychological sources of self-doubt and self-hatred transformed into dogma, authority, bigotry, punition and guilt in the name of absolute achievement and arrival at the nirvana of perfection.

Yet even while civilizations and peoples are being destroyed in the name of absolute truth, newly arrived at, a class of readers or priests of the absolute are contradicting the behavior in gushes of explanations and interpretations of the ways of the gods. As the Hindu Brahmin calculates, the warrior slays. As Anselm seeks proof, King Richard crusades.

The most important question of religion is not how to eradicate gods, but to establish gods at one with humanity and the human soul. For there can be no logical or moral objection to the concept of and belief in gods in themselves; again the human being, insofar as he knows any happiness, has known it in activities of a sublime sort that are inextricable from the divine. A formula and model is required, which is physically possible, and which will forego conflicts of the self, among humans, and between devils and gods. Specifications are:

a) sufficient relief from fearful stress to permit the search for a new formula;

b) a search for physicochemical change agents (whether mutational or continuously operative) that would eliminate terroristic memories, with all that subtends from such in the way of self-destructiveness and other-destructiveness without damaging, and optimally while promoting, the affectional and inventive facilities of humans;

c) and, while the search goes on, and anticipating that the search may be unsuccessful, the invention of social strategies (therapies and institutions) that will hold the conflicts in abeyance indefinitely.

Secularism is a negative counterattack against religion, justifiable as a restraint against malpractices known to everyone. Generally, however, humankind is not in a state to abandon religion and the gods. At best it is capable of achieving a concerted view of an overall divinity and the sacredness of existence. It can borrow from and encroach upon science. Great good would ensue, provided that the concerted belief could work its way into the aims and practices of myriad rituals of human lives.

THEOTROPY

Whatever its failings, past and present, mankind continues to pursue religion. Even when most intent upon relieving himself of its falsehoods, constraints, and burdens, he exudes the divine and the supernatural, and these coagulate into habits and practices definably religious.

There are three causes of this incessant and probably inevitable theotropy. One lies in the delusional structure of the human mind, which must exist in the supernatural no matter how it may transform its perception into operations of abstractions and logic.

A second cause of religion rests with the potent limitations of mankind, who has learned that much more than himself and his activities occupy the universe, a something that may be unmeasurably greater in its goals and influence.

The third cause of theotropy is the human's need to expand his sphere of inquiry. One cannot rest where one is; one must forever seek to expand -- in effect, then, to divinize oneself. No ending in defeat is allowable, no surrender to entropy. Theotropy, as much or more than entropy, is the rule of the universe.

In this inescapable commitment to religion there is little which connects to the problems of ordinary ethics. Goods and bads are the bickerings of passengers on a train rushing through the night. And the consolations of religion are the resolutions of these bickerings in all the compartments of all the classes of all the cars. Religion here is peaceful accommodation which rarely succeeds, and never fully succeeds.

The real religion, or call it theotropy, does less for one than one longs for, but ever more for one and one's kind towards an indeterminate destination. We do not know where we are going, but we are making better speed.

CHAPTER SEVEN

SACRAL VS. SECULAR MAN

Any old religion is likely to have a complete life-program, guaranteed to give satisfaction. It will include answers to all problems that arise, with a counselling service from birth to death. This is no mean achievement, but rather a work of unceasing genius characterizing all ages and all cultures, and therefore thousands of designs and operative systems. Our admiration of the astronomical universe pales in the light of universal religiousness.

Indeed, if one is hungry for proofs of the existence of ultimate design and intelligent gods, here is fertile ground to plow.

But why, out of all this experience has there not occurred one religion of all times and places for all people such that a model human being would lead a happy life? Why should not one formula have been discovered? Why all the changes, conflicts, misery? In replacing the instinctive existence of other creatures, why could not man rapidly invent just that proper set of behaviors that would satisfy the respective and combined needs of his human mechanisms and culminate in expressions of satisfactory existence? Is there some practical impossibility, the fault of the external world? Or is there some inherent contradiction of the mechanisms of human nature?

Regarding these issues as a whole, one large risk seems to confront model religious citizens. The near impossibility of a general religious system being all things to all people all the time causes universal individual problems within the religion. It also causes divisions into priesthood and parishioners, mystics and ritualists, managers and managed, and so on, which aggravate the insecurities of all affected by the divisions, that is, of all believers. Ritual resembles instinctive behavior and may cover most aspects of life except revelation.

No religion exists without a place for mystic revelation. Yet revelation is the opposite of ritual. Somehow every church must give birth to and nurture this hero (or assassin).

In addition, every religion exists within at least a partially secularized society; even in the most simple tribal society, where all seems to be definitively sacralized, there is an everyday need to confront and exploit nature, to use tools variously, to deal with outsiders.

Conditions change; religion is conditioned; religions change. Every ritual change is a slap in the face of the religion, and face-saving tactics are numerous.

The modern secular man was emerging in the Renaissance. Machiavelli was living at the same time as Saint Ignatius of Loyola (1491-- 1556), founder of the Jesuit order. Loyola, unlike the author of the *Prince,* who moved fully and confidently into the modern disintegrated secular society, was seized by the need to keep the total image of Jesus under control and in mind, and to capture and reintegrate any escaping impressions and thoughts. Roland Barthes has grasped the essence of Loyola's mission and procedures, as spelled out in Loyola's book of *Spiritual Exercises.*

> "The obsessional character of the Exercises blazes forth in the accounting passion transmitted to the exercitant: as soon as an object, intellectual or imaginary, appears, it is broken up, divided, numbered. The accountancy is obsessional not only because it is infinite, but above all because it engenders its own errors... [Every failure induces, requires, more accounting.] Everything is immediately divided, sub-divided, classified, numbered off in annotations, meditations, weeks, points, exercises, mysteries, etc. [That is,] the Exercises can be conceived as a desperate struggle against the dispersal of images which psychologically, they say, marks mental experience and over which - every religion agrees - only an extremely rigorous method can triumph."

The whole aim and process is a totalitarian domination of the mind for the purpose of putting oneself into a position to ask God questions and to receive passively the answers. All vagaries were returned to the Source. There is no denying the social impact of the Jesuit method and practice. Allowing that traditional Catholicism continued inertially, Jesuitry became a great active sword that held much of secularism at bay while causing it to involute.

Evidence abounds that secular man is actually a form of sacral man with Jesuitical control. What is sacred possesses for its experiencer an aura of the holy, of awe, of fear, of divine arbitrariness, of supernatural animation. Sacral man in his extreme expression sees the cosmos and all its details as sacred; there are few of such men, of course. The extremely secular man sees everything as void of the supernatural and fully accessible to the senses; there are very few of such men, too.

Let us provide some categories of behavior that might be regarded as sacred or at least non-sensible, to which most so-called secular men adhere. For one thing, they believe in many myths, myths of their descent and families, of their country, of the history of their locale, of wars and voyages. More, and now we make a few specific allusions applying to some, by way of illustration, they hold myths about GM, GE, Microsoft, their President and political leaders, Albert Einstein, Hollywood, the Mafia, the flag ("Old Glory"), Harvard University, the "Spirit of Saint Louis," the Philadelphia Eagles Football Team, Bellevue Hospital, the "Monopolies," "Justice," "free will," "reason," "truth," "nature," snakes, elephants, diets, and so on and on.

What do we mean by associating such first with myth, then with the supernatural, hence with the sacred? The myth has in common with the sacred a non-empirical aura of "emotion" or feeling attaching itself to a non-existent or otherwise psychologically incomplete perception such that, whatever it is, it would not recognizably exist unless it were mythified. International Business Machines (IBM) does not exist as entity, but only as hundreds of millions of mental and physical operations of people, partly related to machines.

But "it" is "mighty," "global," "venerable," "rich," "progressive," "losing money this year," "in need of revitalization," and so on. One is "loyal" to it, "depends" upon it, "accepts its policies," "questions it sincerity," "sues it," tries to "break it up," "ignores its complaints" and so on. Lawyers hop around on "its giant body like fleas on an elephant," "defending it," "justifying it," and of course "living off of it."

A great many people derive a feeling of the supernatural and sacred form when functioning in the corporate ambiance. The Chief Executive of the Schlumberger multinational enterprise declared that a corporation nowadays must learn from the Japanese that "we have the responsibility that religion used to have."

Are these behaviors and beliefs any less religious, say, than the behavior of believers in a volcano religion? The typical secularist worships a dozen such volcanoes; he is polytheistic; he believes in the supernatural and practices rites in regard to it. I do not argue here the consequences: this mythicized aggregate produces millions of hard objects for people; what does the religious aggregate produce but "useless objects" such as church buildings and a superabundant "software?"

Inseparable from myth in practice are symbols and fictions. Language is but the greatest set of all fictions. That it is magical is provable in the behavior of humans in regard to it from their beginnings up to the present. Words lead a life of their own, in the world of words, distinct in part from the objects to which they ordinarily refer. Modern secularists use words freely; a candy is "divine;" every accident is a "catastrophe." No matter; that the world turns with an energy of 10^{37} ergs of energy does not deny to a leaf wafting down from a tree its own erg. What we have in secularism is a disintegration of the sacred cosmos into infinite particularistic ergs of the supernatural, but at the same time a denial of the cosmic supernatural.

Words merge into symbols, which may be words, pictures, displays, but also contain the impact of sets of words, without integration with the grammar of the language. A symbol contains a stimulus to arrive at an attitude or predisposition of mind or behavior. The symbol of the cross has been found throughout the world from the time of the earliest gods up to the present, denoting the chief god or a reference and extension of the god. Wherever a cross occurs, the supernatural does as well; in the ancient world, stones of Hermes were put up at crossroads. Many symbols are likewise ancient. Some of them, like the cross, find their way into the secular crests of noble families, secular institutions, international organizations (the Red Cross) the trademarks of modern corporations, and the escutcheons of government agencies.

Such modern references are very weak, it is said; this is true, and art designers and public relations experts will invent trademarks and other symbols for a price, using scientific techniques for determining how readily the public will recognize and accept the symbol. Still, unauthorized use of the trade-mark can incite a law-suit for millions of dollars; something sacred must be conveyed. It contains more than a single erg of the supernatural.

So it is with fictions, which are of several kinds, including the words, myths and symbols referred to already. We need only to mention that others remain, and also contain qualities of the supernatural, and they are continuously and necessarily employed by the secular mind. The "average" is one of the most useful concepts of science, but it does not exist. Very often sought, like the Golden Fleece, once found, it leads to marvelous gains. That "everyone knows the law" is a fiction treated as fact in a court of law; "ignorance of the law is no excuse for an offense."

Science, law, literature, drama, and music constitute a veritable fictional world that no amount of secularism can eradicate. Secular man can only claim that these are all piecemeal tools, that he "uses" them, that they do not make him a believer in the supernatural, and that he can understand me when I tell him that these are unreal. But this must be a very special secular man, not an ordinary one, for the ordinary one does not see the dizzying use of hundreds of tools; he is used by them, attaches all kinds of fleeting supernatural associations to them, and does not understand well at all when I speak of them as unreal.

So the ideal, extreme, purely secular man will try to squeeze out of life all that is fictional, we suppose, if it ever ended in anything but the most mad hermeticism, with various rituals for exorcising fictions, in a direct confrontation of the real. Pure secularism would be a life of instinctive stimulus-response: wordless, thoughtless, myopic, and solitary. Wrung out of existence would be the arts, politics, law, the market-place, love, human relations, and science itself, including both the conception of all these and all of their ritual accompaniments.

Since he must himself employ the supernatural and its rituals, secular man, we see, does not so much want to destroy religion as he does to particularize it, to make it pantheistic and kaleidoscopic. He wants to keep all his options. He wants full freedom to pick up and lay down any iota of the supernatural or any practice connected with it. He is like the sophisticated Roman of 2000 years ago who also wanted to pick up and lay down any god or rite as he pleased. He does not wish to be part of an all-embracing and integrated cosmic religious system, not even to be reminded that everything in the world and in culture is tied to everything else, even secularly, if not sacraly.

Religion as such threatens his options. He wants to freely disperse his affects and attentions. He wants to be free to change them. He admires the composer who builds idiosyncratic tonal works or the sculptor whose "Composition in plastic, number 18" pretends to communicate with nothing or nobody. Just so, he wants individually to compose and recompose the vignettes of his life.

There is accordingly a strong trend toward the disintegration of morality. Morality, too, is piecemeal in secularism. Each item is judged right or wrong by itself. We note this in pragmatism where the consequences of an act determine its morality. We note it in American law where social consequences tend to be the measure of a crime and its punishment. We note it in the press, where instantaneity and shock push aside moral priorities. We note it in democratic politics, where the politicians must, and willingly do, fix the plight of whoever is complaining most, generally ignoring the "good of the whole," scales of values, or long-term considerations: "The wheel that squeaks gets the grease."

Still, the supernatural of everyday life in modern society is not enough religion for a great many secularists and they solicit new religions, inventing them, so they think, actually "reinventing the wheel" time and time again. These are by no means to be dismissed; they are heroic endeavors to join science and traditional religion, to worship the Divine and the Good without reference to the succession of gods, to build peaceful humanistic communities, to make contact with presumably intelligent beings in outer space, to achieve sacred communities with new rituals that dignify rather than abase their members, and to build a satisfying non-materialistic life around ideals. To ridicule them is by implication to ridicule ourselves. (To ridicule ourselves, on the other hand, is not far from our minds, as we mistake one turn of the road after another; we feel always on the brink of absurdity, that the whole enterprise of penetrating and ordering religion is surreal.)

We hear of physical therapy communities, where diet, exercise, and love build new souls, and of group therapy communities where, in one case, one learns to love oneself and, in another case, to give up selfish love of oneself to love others. We learn of astrological networks of believers who adjust their lives to the elaborated meaning of planetary motions and conjunctions.

There are communities and networks of haute couture, work, skills, fraternity, rock'n roll, sexual practices, diet, outer space communications, sports, and many other special areas that go far beyond occasional meetings and informational exchanges into the dense supernatural and ritual affairs of religious cults. They are voluntary. Participation may be brief and intense; it is for that period sacred, supernatural and ritualistic.

We begin to see an overall pattern of the people of a secular society; they live amidst many intense but sporadic religious episodes, where their minds are fully occupied in recapitulating birth, baptism, initiation, marriage, priesthood and death in brief compass, and in between these episodes, they float and paddle in a swirling world of secular symbols, legends, myths, and fictions. Are they happy? Have they found Truth and Morality? Once again, I would warn against a hasty denial. What is "happy"? Who is happy in this world? "Happy" may be a little thing, quite evasive, quite accidental and lucky, though subjectively grand in its effects.

As for "moral", that, too, may be the accident of a soul that is bumped and tossed about like flotsam, until finally jettisoned onto the shores of goodness. Modern secular man may have never existed, except as a disintegrated sacral man.

CHAPTER EIGHT

ETHICS & THE SUPERNATURAL

There used to be a major area of study called "the moral sciences." It is defunct. In turn, every field of the natural sciences, social sciences and humanities has tried to extricate itself from moral responsibility and qualify for the name of science. Even practical schools of business, medicine, dentistry, law, agriculture, engineering, architecture, nursing, social welfare, etc., claim to provide an objective education; they have achieved the logically impossible feat of inculcating in their students an abundance of the best ways of doing things, while pretending not to consider good from bad, right from wrong.

We know this to be nonsense. All applied science must exhibit preferences for lines of conduct. Scientific method is itself a moral system. And just think of the vast proportion of alumni of schools who confess, with a quaver in their voices, to all that they know and owe to their alma mater. Somebody is teaching somebody something in the way of morals! What is happening?

Is this hypocrisy? Are the schools and students, the society and its people, claiming one thing and practicing another? Yes. They are using a technique that places upon an unreachable, untouchable level certain problems such as god, religion, and the supernatural, along with the associated problem of the ultimate sources of morality and their justification; they take up all other problems as only of instrumental importance, as problems of means, not ends, as problems whose solutions can be taught to burghers, brigands, and beggars alike.

Whereupon a society becomes secular, segmentalized and instrumental (hence exploitative) in its behavior as well as its morals. From many a segment are cast many grappling hooks for the larger morality, some of which catch hold and from here and there spring the many varieties of religious practices characteristic of the secularized society.

Where there is not a grappling for religion there is often a contradictory pair of behaviors: the one a specialized nose-to-the ground empiricism, the other a hopelessly dispersed attention. Religion generally focuses attention onto a few, high-priority objects of value; secularism dissipates attention.

Attention is itself a value imposed on whatever is attended to. It is a preference for its object, selected out of all potential substitutes as objects of attention. Attention is instinctively determined in non-human creatures and modified by parental and group training in many species; the ambient force impinging on the creature also helps to determine the objects of its attention. As with other creatures, man's attention in part is a valuing of the object, elementary, without training, without justification.

So, straightaway with birth, we fix the infant, if he had a mind to wander, upon the right, proper, goods things -- the nipple, the nurse, the movements of the nurse, her voice, his bowels moving, his eyes lightening, his muscles flexing, all following after the not so good things -- his wonderment at himself, a loss of his boundaries, a panicky feeling of loss of his warm pool, stunned dissolution exposed into infinite space.

Suppose his family to be church-goers. He is habituated to church as soon as he can be counted upon to be quiet most of the time there. Time passes, and one day, when he hears, "We are getting ready for church," he displays a mind of his own. "Why?" "Because..." "Because of what?" "Because it's Sunday." "Why do we go to church on Sunday?" "To worship God." And so on. It is almost entirely a morality of means, that carries him from one step to the next, not "really explaining."

Sometimes this begins, or he is catechized, even if he asks no questions. "Why should I worship God?" "God gives us our blessings in life." "Like ice-cream?" "Yes." And like your mother, and father, and bed to sleep in, and food to eat. To train him properly, the trainer is usually clever enough to number only things which the trainee likes. But there is small pay-off for the trainer unless he slips into the list of blessings things that he, the trainer likes. So they go to church to assure the blessings that each wants. They already have different religions, in a sense.

What can be said of morality in this simplistic story? There is a great deal of moral training and moral response. The church and its religion are part of, and will always be part of the child's life. Unless he undergoes heavy secularization he will possess hundreds of ethical views that are connected directly and indirectly with his religion. Almost none of them has come about through autonomous action, reasonable analysis, a survey of cases. The morals collect upon him like fuzz upon a rubbed glass rod.

I am saying merely what dozens of writers have said before me. With regard to practically all those who have practised religion throughout history and today, the whole of religion may be regarded as a generally effective machine to structure a collection of behaviors and bring about their enforcement. The key to the ramshackle edifice is the reduction of cosmic, existential self- fear.

For all that religion has dominated the human world from its beginnings, its ethical results have been paltry. The one thing that is supposed to justify religion is precisely the thing that religion does worst, making the human a satisfactory ethical creature.

But it must be said that religion has forever assumed the most difficult of all tasks: supplying human existence with an objective morality. The problem is multiplex: how to deal with oneself, one's inner relations; how to deal with others; how to treat with the animate and inanimate world of nature. In the end, one is supposed to be able to say "ought" confidently, to live according to the same "ought," and to be happy. In all of this, one's morality ought to be consonant with the real world and its operating principles, science, that is. Hence, morality is the governance of behavior by rules for preferring and achieving certain human and natural relations and states of being.

Unfortunately the simplest, most general rules crack under the stress of psychology and anthropology. "Don't drive while drunk" is a reasonable rule. It should readily illustrate what Emmanuel Kant meant when he propounded his famous dictum: "Act only on that maxim through which you can at the same time will that it should become a universal law." Yet Kant's rule, though it might work to his personal satisfaction, might bring about continual disasters if it were allowed to justify others, such as many suicidal and dying persons who would be pleased to have the whole world die with them. Even the drunk may a) deny that he cannot drive safely, b) suggest that everyone should enjoy a drunken drive from time to time, or c) suggest that drunken driving is a good way to play the necessary game of half-wishing self-destruction. If he does not express such ideas, it may be because he realizes that the police make no distinction between common drunks and drunk philosophers. But now we speak of authority, not Kantian rationalism.

If we ask what functions are performed by an ethical judgment, we get a more lively sense of this feeling. Feeling ethical, one praises or reprimands, one rewards and punishes another. This sometimes changes the behavior of the targets of such feelings in the direction desired by the moralist. More broadly, then, one exhibits a preference in order to arouse enthusiasm or indignation, to rally support. One raises an ethical feeling in order to determine a policy, and to get on with affairs in an orderly organized way. None of this would be done without our or someone's expression of value.

Subjectively, too, the very power to make an ethical judgment is a satisfaction in itself, which often is sufficient unto itself, regardless of consequences. To express one's feelings is in fact synonymous with giving vent to ethical judgments.

Thousands of unsuccessful moral philosophers attest to the frustrations abounding in the pursuit of morals. Voyaging to the Moon is less difficult than the problems of morally justifying the effort involved in the accomplishment. Nonetheless, all humans behave morally and always have. By moral behavior we mean acting one way rather than another because, among other reasons, one feels that it is right and good, and that not acting that way would be wrong and bad. This "feeling" is a "real" thing, physiologically compelling, with physical disturbance and mental states called frustration, indignation, anger, humiliation, and anxiety if the moral act is not performed and euphoria, satisfaction, and physical and mental relaxation if it is performed.

The easiest way to "solve" the moral question is to deny it, that is, to assert that people feel moral or immoral, right or wrong, in consequence of a heap of experience, commands, forces, and natural traits. There would be found in this heap no specific independent moral quality. Morality then is no more than what is in the definition above, "among other reasons."

The only fault that I can find with this idea is that I do not like the way people behave, and I feel that I am not alone in this regard, so I wish to change people. But how do I extricate a moral principle from the heap? Why should anyone else care for what I like or what I do not like, unless I had power to force compliance with my morals and they would do well to obey my rules, or else - "lest you die ..." as Yahweh might say.

In *Civilization and Its Discontents* Sigmund Freud points out the commonly known problem of ethics:

> "that ill- luck -- that is, external frustration -- so greatly enhances the power of the conscience in the super-ego. As long as things go well with a man, his conscience is lenient and lets the ego do all sorts of things; but when misfortune befalls him, he searches his soul, acknowledges his sinfulness, heightens the demands of his conscience, imposes abstinences on himself and punishes himself with penances. Whole peoples have behaved in this way, and still do."

He calls this an "original infantile state of conscience." "Fate is regarded as a substitute for the parental agency. If a man is unfortunate, it means that he is no longer loved by this highest power; and, threatened by such a loss of love, he once more bows to the parental representative in his super-ego - a representation whom, in his days of good fortune, he was ready to neglect."

Fate is looked upon as an expression of Divine Will. Fatalism is very strong in early religions and ethics. Why? The authorities and experts say: because primitive man was at the mercy of savage natural forces. Still, if man were to be of the same ideological cast today, he would also be fatalistic because obviously, when one think of it, very little real control has been exercised over the immense and infinite area of difficulties besetting us. Rather, the change of attitude has come about as a result of changed ideology, *Weltanschauung,* and this has changed because of a fairly long calm condition of the Earth and the skies, and the development of a progressive, free-will, uniformitarian (self-contradictory) philosophy.

Perhaps the distinction between traditional sacral and modern secular man is that the former has not forgotten his primeval scenarios, whereas the latter has suppressed them very deeply and become overtly pragmatic.

The thousands of cultures existing in historical time and space have given us a fair sample of the ideal and practical ethical capabilities of religion. The experience on the whole has been unimpressive to one looking for a happy human way of life. The more one trusts to religion, it seems, the less good one can obtain from science and politics. On the other hand, science in itself (that is, science which is entirely positive and empirical) is quite helpless to address the moral perplexities of man.

Politics, moreover, has, if anything, a poorer record than religion, speaking now of politics as a secular approach to human issues; for politics tends by itself to depend upon sheer physical force to order a population, and systematic violence is hardly an improvement upon whatever chicanery and delusions historical religions employ to rule a people. Would one have preferred to be governed by the barons or by the monks of the European Middle Ages, by the warlords or by the Shinto and Buddhist priests of Japan, by Aaron or by Joshua?

The questions are difficult, enormously complicated, and perhaps biased. Still they are worth considering if only as a means of suggesting that ethical progress in a society is not to be identified with its secularization. The key to good governance is an ethical system beyond facile contrivance. Neither religion nor secularism, as such, promises success.

Yet occasionally the human becomes ashamed of living a lie and hates himself and hates his religion and gods for having created his dependency upon delusions. He admires the "honesty" of the bear, the trout, the dog; they are not of two minds and forked tongue. Why cannot his morality be so straightforward?

Blame part of it upon his obsession with history, his compulsion to repeat his worst experiences. He demands that his morality today be that of five thousand years ago. He demands that it be of the highest order: We know what that means; it must come from Heaven. Further he demands that all people share in an ecumenical morality. The logical and sociological impossibility of both demands will not deter him. He is implacable. He will not pluck his morals from a garbage heap.

Besides providing people with morality, it is said, religion puts them directly in touch with the supernatural realm. For the mass of people this is untrue, just as it is that their religion gives them some special ethical competence. A few practitioners must enjoy the facilities for communion with the spiritual universe which churches and temples provide. The mass media (motion pictures especially) and drugs, as T. Leary has eloquently argued along with others, and, too, many gurus, seances, and non-church rites provide this type of communion.

We seem to be at an impasse, owing to my downgrading of the creative moral and spiritual functions of historical religion. Supernaturalism appears to be all manner of anti-scientific folly. Morality exists concerning countless particulars in human activities, even while neither religion nor secularism can justify its source, hence their application.

We see no easy solution, perhaps none at all. Later on, we may offer some grounds to justify a "relatively absolute" morality, meaning by this verbal barbarism some unchanging moral propositions that are themselves changing. If one might conceive of a religion that is an integrated whole, accommodates change easily, and that does not fundamentally and continuously violate the controls and benefits supplied by science, then this religion may not only be superior but also popular.

Does this mean that morality is human and mundane, part of an endless process going on in millions of transactions every day everywhere? Yes. Does it mean that the supernatural, the divine, the gods are not the source of morality, that ethics exists without religion? Yes. Does it mean that mankind is morally sui generis and autonomous? Yes.

Does it mean that humans are "immoral" and "wicked," with no means of setting ethical standards? No. Does it mean that the supernatural, all that is divine and sacred, has no effect upon ethical behavior? No.

The supernatural, as non-knowledge, is knowledge of a sort. Those who transact or seek to transact with the supernatural in order to think upon the divine, engage in an ideational relation with the divine, and are affected by the knowledge which we possess of the divine. They will behave differently than those who deny the supernatural and avoid it.

Religion, to put it in commonplace language, can make people better. It should be the "right" kind of religion, and, of course, this would be the form we are here advocating: self-aware, open, relativistic, non-historical, connected with the sciences of natural and socio-psychological processes, non-anthropomorphic morphologically, anthropomorphic structurally. Let us see what science is doing that is religiously relevant and can be adapted to religion.

CHAPTER NINE

RELIGIOUS ELEMENTS IN SCIENCE

Out of religion came politics and then science, each reacting upon the others while going its own way. Science is a set of interests that is religiously, socio-politically, and autonomously determined. Science struggles to conform to a scientific method in whatever it does. The struggle lends it its distinction, providing it with its social character. Without the method, it is useless to speak of science. The method is applied to whatsoever extension of the senses is of interest and controls such extension; both operations sometimes fail but also often succeed in our day.

A scientific procedure typically puts forth a hypothesis about what is measurably expected to occur under certain conditions, and, by finding or producing the conditions, finds or produces the event. Wherever conditions permit, these are produced under controls; wherever they occur naturally, they are overseen as strictly as possible. No place is allowed in theory for supernatural conditions or supernatural effects, that is, for the intervention so factors that are undefinable in material terms, or of an external ungovernable will.

As Alexander Hamilton quipped, when Benjamin Franklin suggested prayer at an impasse while composing the American Constitution in 1787, we should not call upon the help of a foreign power. Hamilton, intending for politics what Franklin had already practiced in electrical experiments, had in mind a republic whose behavior might be predictable when certain regular operating conditions were established by its structure.

The incident reminds us that science includes social as well as natural science. Humans are a material factor in the one, if not in the other; they are a contaminating factor in both. The human factor has so continually disturbed the scientific method in its application to natural phenomena that, in a sense, all science becomes social science, especially as the material conditions of study become more difficult and less amenable to continuous ordinary sense observation. We cannot go here into the progressive discoveries of the intervention of anthropo-sociology and especially psychology in the workings of natural science, citing the works of P. W. Bridgman and others, but we can, without fear of rebuttal, warn of the inevitable effects upon experimenters and researchers of their psychological as well as physical presence amidst the supposedly materially and logically observer-proof conditions of scientific work.

Willy-nilly all sciences, in their healthy vigor, are wrestling with the supernatural and contributing to its expansion thereby. In this sense, all sciences are addressing the foundations of religion and theology. The more scientific work that is performed, the more areas of uncontrollability and contradiction come upon the stage. Science itself is the biggest factory of the supernatural. It tears holes in the fabrics extending reality. It works all the while surrounded by amateurs of the supernatural and theologians, pelted by derision. Perhaps one might forecast the most esteemed and influential religion of the future by locating the contemporary cult that is closest to the anomalies and radical new interests of science.

I have examined human mental structure and operations in other works, so am permitted to relate here only the central relation of religion and science, and of this most clues are already familiar to the reader. Science emerges from the limited but most significant ability of the human mind to capture pragmatically, that is, to control, the connections between the person and an immense world of identifications and displacements. From its very beginnings, mankind has identified and sought to control the heavens and the gods, the mountains and oceans, the plants and animals. No other being on Earth is so ambitious; all others are confined to such rational activity as instinct requires for the purpose of survival and propagation.

The human mind, disordered by genesis and at birth, has the immense problem of extending pseudo-instinctive (that is, voluntary) controls over connections with existence that have very little to do with survival and propagation. The human, for instance, will sacrifice (both in the functional and symbolic senses) everything - food, family, sex, lesser powers, safety - in his efforts to command the skies.

Furthermore, besides the skies, there is many another realm of being that he is compelled by his mind to deal with, an infinite set of realms it seems, even though his mind, we must remember, is assisted by only moderately competent sensory organs, so that he is encumbered in his ingesting, questing, and adjusting.

So the need to order one's head requires that the cosmos be set in order, and it is natural for one to apply the pragmatic(scientific) techniques that substitute for instinct in the obtaining of both very close necessities and the most faraway necessities, and hence the elaboration of science out of immediate pragmatism occurs on both the intimate material level and the cosmic level.

Science is a human activity and therefore can be characterized as such, no less than religion is a human activity. It has a history, a sociology, sub-cultures, a psychology. It exhibits struggle, cooperation, ambition, failure, success, inducements, payoffs and penalties, a total range of material subjects to study, just as religion is subject of study, by the scientific method. It has religious and political aspects. It deals in authority, fictions, myths, claims, anomalies, rituals, and hypotheses, all of which are perilously reminiscent of religion and the supernatural. And, of course, to distinguish it especially from all other social activities, it is obsessed with the secular ritual of scientific method, and tends to extend the practice to all spheres of life.

The basic rite of scientific method is similar everywhere. But there come into being elaborations, embellishments, and variations of the basic rite. Some scientists like to think of the changes in naming, conceptualization, procedures, research interests, and so on as "progress" or at least "different ways of looking at the same thing." Other scientists know that they are in the grip of fashion and fads, whether in astronomy or geology, psychology or sociology. Magic, cultism, and other overtones, usually sounded and noticed in religious practice, can be heard in any science.

Every science must have a supernatural auxiliary. I would call it a suprascience, if such a term would not offend. I mean that the science itself consists of a stripped-down method and its findings, and that there must form around it not only a halo or encrustation of fictions, hypotheses, and non-empirically derived speculations, but also an attitudinal complex, rather like a system of illusions and delusions, or like a ruling formula (a term which Gaetano Mosca applied to the field of political science).

This auxiliary suprascience functions as a propaganda machine to make the science appear to its practitioners and public as continuously worthwhile, to tie it non-empirically into various problem areas of life, to act as a lightning rod (I will not argue whether lightning rods really are effective against lightning) to dissipate attacks gathering against the field, to give the field a history (much of it pseudo-history) and a future (much of the genre of science fiction).

As political science is impossible to consider without its ruling formulas (elites, democracy, kingship, laissez-faire, militarism, etc.), so astronomy cannot exist unaccompanied by schools of astrology, or geology without forms of uniformitarianism, or economics without models of "economic man," or literary analysis without fads and fashions, or medicine without magic and homeopathy, or chemistry without suprasciences, one or more for each of its numerous subdivisions such as diets alongside food chemistry, drug cultures alongside drugs, aesthetics alongside plastics, and so on.

There is no fakery here; there is strict necessity; man lives in the skies as well as in his hovel; culture marches along all paths and all paths are psychically connected, even when, especially in a scientific and pragmatic age, they may be, by an effort of will, separated for specialized solutions.

Under these circumstances, man lives throughout the cosmos, effectively. He lives pragmatically in the cosmos that he can experience and command through sensory manipulation. He lives mentally (and, by ritual, pragmatically) in the cosmos that is beyond experiencing but which he can imagine and bring into order. We may fancy that Jesus of Nazareth would speak this parable:

> A woman of the mountains saved her money to buy a rain cloak, for she was often wetted by the rains there, and when she had sufficient she ventured to

> Jerusalem to buy a cloak. But the cloak was so beautiful, that she would not wear it, so as to preserve it, and all her clothing became wet and damaged. Now I say unto you, wear your beautiful cloak of religion and all of your other clothing will be saved, and your Father in Heaven will replace your rain cloak with the raiment of angels.

Reason, many theologians and secularists pray, will serve religion, and show a person what is good and bad in religion. So, if there is bad religion, it is because men do not use their reason to find the good, or they exercise their free will to choose to do bad with religion.

Rationalism is thus used in two ways to damage religion. First, it becomes secular and refutes most or all religious pretension, as with Voltaire and Marx. Second, and more important here because it is a lesser known argument, rationalism erodes religion because it claims that mankind, possessed of the gift of telling what is "true" religion from what is "false" religion, only needs to be educated to distinguish "truth" in order to pursue true religion.

Thus the problems of religion can be said to be solved by the independent pursuit of the principles of reason with regard to supernatural beings and rituals. In this second situation, the rationalist theologians, counting here Saint Thomas Aquinas insofar as he is Aristotelian and rationalistic, lend themselves to the continuation of evil in the name of religion; for evil becomes the result of ignorance and neglect of reason.

Reason, as conceived in traditional and conventional philosophy and theology, presumes "free will." Free will is considered as the endowment of human nature with the capacity to choose one out of two or more alternative options as the basis for action upon an issue. Thus, employing reason, a choice of good over evil is imposed by free will, and an opposite decision becomes a free choice of evil. It is this "free will" which has been used in many cultures to explain the harsh effects of religion. Man is wicked and is therefore punished by his gods; by no means can the wickedness be blamed on the gods.

This argument would appear to constitute an imposing defense of traditional religion and may even explain why all other life activities are dealt with by the principles of rationalism and free will (rather than the other way around). If so, it is one more important indication of the extent to which the religious sphere permeates and dominates the structure and operations of the other seemingly separated spheres of life.

Actually, the belief in free will can be viewed as a primary obstacle to the improvement of religion. Not only does it make of man in his own eyes a wicked sinner, much more fearful of the gods, the authorities, and the people around him than he would otherwise be, hence aggravating his natural paranoia, ambivalence, and hostility to others. But it also makes it impossible for man to govern himself; for he believes that he has within him, quite divorced from the really essential set of mechanisms according to which he behaves, the ability at any time to change himself from good to bad and from bad to good.

Furthermore, the "bad" and "good" are themselves applied in the religious sphere often quite apart from any connections which they might have with the other spheres of life. "Free will," and rationalism as well, are fantastically individualistic fictions. They permit the dissociation of an individual decision from all that in fact determines, and should determine, the decision. Neither a balky donkey nor the gods themselves can prevent man's exercising his will upon them to turn along his way.

By contrast, the theory of homo schizo holds that man derives his religion from the same set of mechanisms whence he derives all his other interests and activities. One cannot allow the concepts of free will and rationalism to enter. All of human behavior considered as a mind transacting within himself and throughout the medium of his culture is of one piece, holistic.

Free will is no longer, if it ever was, a useful idea. The known and experienced deviations or range of choice available to us is large enough, whether determined or free, to allow for extremely diverse decisions.

Now see what this theory of homo schizo does to the status of the supernatural and of religion. It elevates their status, rather than depressing it. But, more than that, it makes sacred and religious man impregnable to separatistic assaults upon his religion. For he can say and he can prove, or others can do this for him, that even if his religious aspects are suppressed, he will be different only in those particulars where a transference occurs, from the prohibited areas of religion, to the permitted secular areas.

Religious man can further declare that the elimination of religion does not eliminate evil, but merely introduces more evil to other quarters of human behavior. And he can heap up evidence showing that secularized societies and secularized man have shown no noticeable improvement in conduct denominated as good.

Until we decide who we are and what we want to be, we are at fault in what we are and want to do. Unless we shut the doors against all unwanted conduct from all spheres of life, shutting the door against religion in the hope of stopping all unwanted conduct is futile; it will enter by the other doors. As well as saying that religion cannot be suppressed, and as well as stating that much of religious behavior is true both in itself and in reconciliation with science, we are now prepared to say that the suppression of religion will not consign evil beyond man's ken. For that great task, a reconstruction of human nature is required.

Such a reconstruction may well be impossible. We do not know enough yet to define the terms of reform. What we can do at this stage of our study is to argue for the incorporation into religion of our findings, both to prepare the ground for the possible coming reconstruction and to maintain the best possible, the least damaging, of religious as well as of all other systems.

The problem of absolute morality - of the standards of good conduct and the means to practice it - must go unsolved here. Absolute morality may be forever beyond human abilities to demonstrate. Short of this, we resort to what many philosophers before us have advocated, a natural law of human behavior: How people have always behaved and seem compelled to behave is restructured so that the consequences which people seem always to have wanted - even when acting in contradiction - will ensue.

Since we do not appeal to gods, reason, or secular authorities, nor to charism, faith, and revelation, it would appear best to label our natural law as hypothetical, tentative, and only so good as its consequences are acceptable to most people, whether educated or not, in all cultures. This might be called a natural moral consensus.

To summarize from suggestions offered in various passages of our work, we perceive four essential and general human demands: for **freedom from fear**, for **material subsistence,** for **new experiences,** and for a **disinterested arbitration of human conflicts.** Fearlessness; subsistence; experiencing; and justice: these words may be used also.

All of these require controls over the self (selves), others, and nature. Control requires skills (considering even brute force as a kind of skill at leverage, if nothing else), and mankind is obsessively driven to elaborate his internal and external control system to a stage where he has obtained what he can regard as minimal and sufficient guarantees of his several needs.

The overall problem of a culture is, unconsciously or consciously, to provide a network of practices that will supply its people with excellent chances of obtaining these guarantees. And so we proceed to human relations, technology, politics, religion -- family government, world government, cosmic government -- and science, which acts to supply better ways for cultures to fulfill these needs.

CHAPTER TEN

NEW PROOFS OF GOD

Ideas of quantavolution - of sudden great changes - attract the attention of historians of religion in especially two regards: religion was, and must remain in whatsoever guise, the companion of the newly born, traumatized, self-aware human mind; and the earliest religious voices, still speaking through the sacred documents of ancient times, were telling many truths, even literal truths about natural events, human nature, and human institutions - truths that bespeak quantavolutions.

Yet quantavolution also presents a distressing problem to those who believe that, if once they could approach closer to the earliest days, when "the gods walked on earth," they would be inspired, ennobled, and reinforced in their faith. This is not the case and they may well become downhearted and skeptical.

They should realize that even before quantavolution was assembled as a body of theories, those theologians, mystics and millennialists who ventured into the great first days of creation had to feel the terror and suffering that comes with "looking upon the face of god."

Hence, indeed, most religions have calibrated the approaches to the sacred, so that only the well-prepared and thoroughly-warned would attempt the journey. That is, long before quantavolutionism, it was fully known that only the hardiest of souls could cope with the revelations of the first ages, could endure the historicity of the apocalypse.

It is possible to absorb the theories of quantavolution solely in the form of science, eschewing all contact with the religious experience as truth, while pursuing every avenue to religious experience as sociological and psychological fact. That is, quantavolution would only ask of its students that they exercise its hypotheses and evidence according to the current general methodology of science. This would be the more comfortable and easier choice.

Alternatively, however, one may confront the issue of the religious truth contained in this body of revolutionary theories, probing, inquiring whether perchance there is inherent in them something that those seeking a truth that is religious will recognize as valuable. The road is hard, and lined with the philosophical tombstones of many catastrophists and uniformitarians who have gone before, trying each in turn to transport a body of science into the realms of religious truth. Nevertheless I shall aim at the same goal.

I shall try to reach my goal in several steps. A metaphor of a box is used throughout. The human mind is inextricably contained within a physiologically limited box of perceptive possibilities and cyclical redundant logic.

The truths of religion are not within this box. Yet the existence of the box is proof of the supernatural, this being what is outside the box, which, we must admit, has an outside.

It is well that religious truths are not within this box, for the box limits and shapes its contents, and therefore disciplines the fields of knowledge that it holds.

The box nevertheless is indefinitely expansible. Its resources and limits have not been fully tested or strained. (This differs from the expansion mentioned above, which was only communicative expansion.)

In whatever direction the box expands, it is not likely to be limited. Hence the supernatural is not bounded by the accomplished full testing of any of its facets.

The simplicity and complexity of things are subjectively perceived or operationally invented. Things in themselves cannot be defined as absolutely simple or complex. The same is true of the concepts of space (size), time, past, and future.

The same is true of "life" or "animism." It is subjective perception or operational invention, not defined other than by the human mind.

Now, if god is whatever is beyond the box (i. e. limitless), god must be also what is in the box, inasmuch as what is in the box is speculating upon what is outside of it, as we here.

Therefore, all that we sense and think in ourselves and our perceptible and thinkable world is part of the supernatural. If the supernatural and god are joined, we are pantheists.

Putting god into an animistic metaphor, god is our judge; god has already judged us. We already are composed and function according to god's cosmic spirit, by intelligence, and necessity.

The very nature of our ignorance, then, cribbed and confined in our cosmic box, constitutes a proof of the existence of gods. The agreement of extended ignorance from the crypto-blind bio-box probably stands up better under modern scrutiny than the traditional arguments for the proof of god. Furthermore, a second modern proof may tend to confirm the existence of gods, and support the proof from the cosmic box.

The universe is presently thought to be some billions of years old, probably finite, although the boundaries are not clear, and populated by many billions of stars. Many stars, if not most of them, are believed to have spawned planets. Planets will have had ample occasion to acquire atmospheres and "the building blocks of life," as we like to say from inside our cosmic box. We are also forced to admit that life as we know it is defined from inside our box, and "intelligence," as we might conceive it without actually knowing it, may be a product of other means of manufacture and assembly.

Even if we have to conjecture the birth of gods from the elements of the atomic table, the combinations and permutations of this, plus the practically unlimited conditions of time, space, temperature, and pressure can provide the substance and form of a god whom even scientific materialists must recognize as authentic and in being.

In the creation of all things, we must contend with the principle of entropy, a term invented by Clausius in 1865 to refer to the state in which thermal energy is no longer available for mechanical work. Later, and especially with Norbert Wiener, the term was broadened to describe "the running down of the universe." Clausius himself had written "the entropy of the universe tends to a maximum." (Thus the idea merely subsisted until a century of history changed the optimistic mid-nineteenth to the pessimistic mid-twentieth century intellectual climate.)

Given even a time of short duration, or of a conventional dozen billion solar years, the number of occasions for "phenomena" or "phenomenal intelligence" to appear in the universe is extremely large. Cases of "negative entropy," that is, of existence moving toward creation rather than desuetude, must be very numerous.

There is no reason to use life on earth as the archetype of the universe. As the *Encyclopedia Britannica* reports, "all the organisms of the earth are extremely closely related, despite superficial differences. The fundamental ground pattern, both in form and flesh, of all life on earth is essentially identical." Nor ought we take mankind as the measure of the product of potential habitats of intelligence. In some proportion of them, something much more than homo sapiens schizotypus must have emerged. There should exist planets or complexes where beings of much greater intelligence and competence than ourselves exist. There must be a range of such superior intelligences from superman to gods.

Whether individuals, conglomerates, complexes, spirits, physiological aggregations unknown to us, or even creatures suggesting ourselves, these will all have many times our abilities.

Perhaps some will have supernatural capacities (for we cannot understand them) a billion times our own. Perhaps one of the beings may have generated power to move the universe itself; for, as the second law of thermodynamics maintains that the universe and all matter within it is running down, but an exception is made in the case of life which is negatively entropic, so there is an excellent chance that somewhere in the universe is an intelligent being, of which we can conceive but which we cannot become, whose powers are such that it is in control of the universe moving in the direction of intelligence and progress as we conceive of it: this being certainly must be called god. It would then be for all practical purposes omniscient and omnipotent. That it would be all-caring, omni-benevolent, may also be presumed, for to take care of itself it would have to take care of the universe in some part, as a case of "enlightened selfishness," in our limited human terminology.

Thus the traditional concept of god is exercised with a new proof involving the probability of supreme negative entropy. God is created by the universe, working in opposition to the principle of entropy with the equally universal principle of creation. The creative principle, arising like the phoenix from its ashes of entropy, must naturally turn to controlling the universe.

If this god is not already a fact, still, in the aeons of time to come, it must become a certainty. As the local gods of the solar system were born and died in succession, there may have been many temporary or quasi-omnipotent gods in times and spaces beyond all solar system experience. The universe offers billions of chances for a supreme god to arise in the future. Sooner or later, the universe will create its supreme master, just as the earth, this indescribably minute place, has created its locally supreme master, the human. Whereupon truly the universe would be intelligently ordered, as contrasted with the present chaos, and the far-flung parts, including our own, would be irresistibly induced to cooperate.

Let us proceed to discuss this theory of divine actual or potential existence at greater length.

To establish a new religion on solid grounds requires that the history of religion as the history of the true god be rejected. If one relies upon the scientific history of religion, one would be led to the conclusion that gods do not exist.

Luckily for those who yearn for gods, one can go beyond the history of religion, to psychology and philosophy. There they will learn that the human mind is basically limited. Its perceptions and condition are structurally bounded. To exceed this structure they must rely only upon corollaries of the cosmic proof: 1) extension of some hitherto neglected remote recesses of the structure of mind and body and 2) a type of reasoning that proceeds on an "if.... then" basis which says: This is desirable; the question is open; the desirable is therefore not foreclosed. If god exists or gods, and is/are as we think god ought to be, then we are happier and can seek progress. Since the "if" cannot be foreclosed by any known means, the "then" is always possible.

What is greater than the self can only be known anthropomorphically, that is, by extensions of the self as it is known to one. Hence, if the universe has dimensions that are quite divorced from human traits (or their extensions), we can never know them. But the premise that more exists, which we cannot possibly know, is itself a proof of the existence of gods, even though we cannot know them in any other way than in this paltry manner. Moreover, it is possible that dimensions of the universe hitherto unknowable to us will make themselves known, whether because they change so as to be comprehensible ("God makes himself known,") or we change ourselves structurally by genetic accident or manipulation.

If the universe has only those qualities which we now possess or may in the future possess, or if the universe changes its qualities, then we can come to a knowledge of the gods that we, in our limited way, know must be there.

Are these possibilities of knowledge additive? Can we say that our full knowing potential plus the potential of the unknown gives us virtual certainty that gods exist? Like the lost sailor, we know that land lies in every direction. Also we know that land may be far away if we go in some directions. Can we determine in what direction the divine land lies?

A wee mouse, five centimeters long, is in many ways superior to the human. In proportion to size, he can run 20 times as fast, jump fifty times as high, scale walls, swim naturally well, has senses superior to those of men, and trains readily for reactive tasks. His brain and his organs are marvels of miniaturization, relative to ourselves. The outstanding difference is that homo is schizotypical, that is, self-aware and all that flows from this fact.

We know nothing about any species that has the equivalent of schizotypicality and what this affords us. We can conjecture how many species in all the universe might be schizotypical or have other systems capable of performing operations that we designate as being along the parameter of the human-as-divine up to the exceedingly divine, that is, the full god.

Moses and his followers claimed that Yahweh could see and punish malefactors and delinquents. The Christian religion says that God can know the minds of all persons. Paranoids will sometimes say that they can tell what all minds in a crowd are thinking and single out individual minds, too. A body containing 10^{20} cells can pass a signal to most or perhaps all cells in a brief time so that they are all reacting consonantly. The coordination of "nature" exceeds that of gods, so to say, in some respects, and goes far beyond the most paranoid human mind. (Indeed, many humans are content to control one other person, such as a spouse or child.) Coordination means two things: communication and control.

Thus far, the shocking modern revelation of the numberless stars and vast extent of the universe has been converted into constructive thought regarding the possibility of there being other intelligent beings in the universe, with whom we might possibly communicate. Inevitably the thought has been elaborated into contentions that at some time in the past, astronauts have settled upon our planet, assimilating biologically with lesser breeds, or constituting the human race itself. The thought has also moved, theoretically, to the contention that more intelligent or hostile or flagrantly incompatible beings might be confronted, to our embarrassment, should we be successful in communicating with exoterrestrials.

These discussions employ formulas not essentially different from what we employ here. We take up estimates of 10^{11} galaxies of 10^{11} stars each, without counting dark stars or clouds, reaching thus 10^{22} stars. We count 10^{22} dark stars and dark clouds as having theogonic possibilities ("darkness" is our problem). Gods take time to develop, but we may assume that the average body has had enough of such time, billions of years. Whether, for instance, the Earth has subsisted for 4×10^{9} or 10^{6} years, it has had at least a $1/10^{44}$ possibility of generating a god. Of the total source bodies, some 10^{11} (plus dark stars and clouds) would exist in our galaxy alone. We are not counting the separate planets or comets, that would multiply these several stellar figures by 2, 5, 500, 1000 or some other multiple not known, but depending on the average number of planets per star.

Suppose that science on Earth expands its capabilities ten times, a figure not in excess of many predictions from various fields. Suppose the human achieves an IQ of 160, lives to be 200, and can travel to the neighboring star cluster of Arcturus. Suppose the human is even morally set upon acting as god. The human will probably not be a god, but he will show that gods are possible somewhere. That is, it does not take too much more than man can be in order to define a god or demigod.

If there also are and have been 5×10^{11} centers for realizing divine beings in the galaxy and this over a period of time - in fact, why not infinity? - then the chance that one or more gods have developed is certain. Probabilistically, at least one is certain; let us say five are highly probable; 5000 are likely; 5 million are at least 50% probable; and some 50 million gods are possible, this in our galaxy alone. In the universe as a whole, these figures would be multiplied by 10^{22} . At 5000 per galaxy, the gods would number 5×10^{25} , too many by far to crowd into Valhalla.

If gods should die (speaking of the Götterdammerung), that is, lose some or all of their capabilities, one would halve the number of gods in the confines of the stipulated universe.

Among all of the probable godships, should not many have evolved to a multi-galactic god, and at least one to supreme god of the universe? An interesting feature of the results here is that there appear to have been more gods than the conventional formulas claim there to be planets with intelligent life forms. This paradox occurs because one does not constrain estimates by looking for something close to man, to technical civilization, or intelligent life as we know it. Further the requirements of an environment similar to man's can be waived; the gods need not be limited by humanly severe temperatures, or the presence of a long string of prior primitive life forms called for by non-quantavolutionary evolution.

Thus, as soon as less conservative considerations than are customary are set for intelligent forms in the universe, the number quickly exceeds the number of gods estimated here. Ordinary calculations of life spans are irrelevant, too; the occurrence of gods presumptively reduces time constraints; the possibility of divine viability stretching over much, most or all of the age of the universe adds to the probability that gods are active now.

In sum, of the terms of the formula used in many discussions of communication with extraterrestrial intelligence (CETI) only the gross number of celestial bodies is usable in estimating the likelihood of the existence of gods. To find the number of extant technical civilizations in the galaxy, by the "Green Bank formula" of F. D. Drake, *N,* one multiplies R* (the average rate of star formation over the lifetime of the galaxy), by *fp* (the fraction of stars with planetary systems), by *no* (the mean number of planets per star that are ecologically suitable for the origin of life as we know it), by *fe* (the fraction of such planets on which life in fact has arisen), by *fi* (the fraction of such planets on which intelligent life has evolved), by *fc* (the fraction of such planets on which a technical civilization such as our own has developed) and by L (the average life of a technical civilization). That is, *N*= R* *(fp)(no)(fe)(fi)(fc)* *(L)*. [Product of these factors.]

Results, depending upon the estimates fed into the formula, have ranged from one to millions of technical civilizations in the galaxy. Our own technical civilization capable of interstellar radio communication is only a single generation old. The 1000-foot-diameter telescope at Arecibo Observatory in Puerto Rico, existing transmitters and receivers, and a presumption of the same type of equipment on another planet would provide a communication medium of 1000 light-year diameter, providing 10^6 stars. Space travel and laser transmission are technically near availability to extend somewhat the range. There is therefore some chance of a communication exchange now.

But how have we defined a god that gods should be so numerous? By god is meant a coordinated divine activity such that *1)* it can endure or reproduce or replicate itself indefinitely under highly varying ambient conditions, that *2)* it can act so as to expand communication pathways and thus its influence at an exponentially increasing rate, that *3)* its proven scope and domain of intervention is extensive within a galaxy or is multigalactic, and contains no inherent limits, that *4)* it provably (in human terms) acts so as to increase the aptitude and appropriate behaviors of the most promising existences (including humans) with the end in mind of reducing entropy and establishing theotropy as the dominating principle of the universe.

How do these qualities, if applied to the human condition, reduce fear, war, and famine, while increasing love and knowledge? I have mentioned but a few of such moral connections in these pages. Their deduction from the principles of godship do not appear to present problems in excess of those traditionally and successfully solved by theologians such as Saint Thomas Aquinas when deducing human moral behavior from the qualities of gods.

Theotropy can be considered from the standpoint of gods and of humans. Regarding gods, the achievement of influence is by means and in terms that we understand or cannot understand. So far as we can understand, gods must extend themselves either immediately or by a succession of moves.

Insofar as our world is governed by no intelligent divine influence - at least no sufficiently powerful and satisfactory influence - then no "great" god has even in our short-time view extended itself over us, whereupon we can probably more correctly imagine that any god occupying itself with humans is proceeding by a succession of moves, that is, by growth. Both may be occurring, a onetime immediate assumption of our world and a succession of moves to change us.

A reader who has pursued our works on quantavolution knows how we believe man to have acquired his nature and how the world as we know it has come about. Thereupon he may ask: "Why does man need a god, considering all the troubles gods have appeared to cause?" Worse, "What legitimate reason has man for seeking god?" Worse of all, "What can any god do for man that is good for man?"

First of all, none of these questions can destroy the gods if they do indeed exist. No more than one can dispose of tax collectors who are troublesome, unwanted, and useless. From a homocentric point of view, however, we can be more cooperative in responding.

We need all the help we can get, plainly and simply. We are inadequate to our dearest wishes for the universe: that it be controlled and beneficent to ourselves and the posterity with which we identify.

We need help of a quality that is beyond the ability of everything whose qualities we know directly. Our faithful dog, Shep, is not up to the task. Nor are even our most trusted friends and allies. If there is a god, we need him.

The third question gives us pause. If the price exacted by seeking, finding and cooperating with god is our most priceless gifts, we may prefer our troubles, death, entropy, and oblivion.

This leads abruptly to the question which we seem to confront at every turn of the way. What does, what ought, the human wish to be? If he wishes to be like the gods, and the gods are likely to be so indulgent, then all is well and good, and we should search eternally, if necessary, for the gods.

One's purpose on Earth will then be answered from the divine point of view: the human is created for the divine task of helping to save the universe. He, and all developing and positive matter, are assigned this overall function. The universe has bred the human as a way to its own survival, as a challenge to its death, as an antibody against the death and dissolution foretold by the second law of thermodynamics.

Elsewhere we have written of man's basic needs, to fearlessly subsist, experience and live justly. If the gods are theotropic, we have nothing to fear from them except the loss of that element in us which is self-destructive and entropic.

What might this element be? Let us call it the diabolic, because it will turn out to be that often highly attractive mixture of uncertainty, fear, hatred, spite, lies, greed and egotism that goes into some of the most wonderful human creations. Will not the gods take from man the taste of evil for which he slavers? Or will the gods, like certain historical gods, allow man the gift of diabolism with all that it does for his music, dance, art, inventions, and politics?

This is one question; another question, equally important, is related: will the gods take away self-government, self-rule, decentralization of decisions, whether large or small? There is indeed an argument, posed as, "Let every man go to hell in his own way." The felt uniqueness, the exultation, the happiness of determining one's way are not to be given over, even to the gods, one senses - and we can hear the most stupid as well as the most brilliant of humans saying so.

Perhaps the gods will be indifferent to such trivialities, perhaps they work sloppily, letting as much as we know of life pursue itself along their general guidelines. Or perhaps it will happen that in their intense pursuit of godliness, humans will get their fill of risks, conflicts, imagination, and autonomy. In any event, this crisis is far down the line of theotropy, whereas man's decline and destruction are always close at hand. We prefer to think therefore that, in the pursuit of the divine, humanity will have all that it will want of symbolism, diversity, and excitement.

From the human standpoint, gods are to be awaited and solicited. If they are awaited, the presumption is that the gods are interested in expansion for its own sake. Any part to the universe will do. This is probably an unsafe assumption because it implies a certain kind of god. But god is more than a mere "land-grabber," we reason. He is interested in his own development; he is maximizing his opportunities of theotropy and not interested in entropic refuse. Therefore, gods are to be invited. For some lucky mystics, gods may indeed already have been entertained. I cannot understand the means, hence cannot confirm the encounters.

But what are the occasions for conflict among potential and actual gods in the galaxy and universe prior to the universal achievement of a single supreme god? Will there not occur what even mankind has experienced on its low level of achievement, a set of squabbling barons, a battle of the gods? Then the gods themselves will do what it is now widely believed that man will do - destroy themselves and contribute to the entropy of the universe?

As they move out to order and exalt the universe, what will determine their jurisdictions and, as implied in their aims, will merge them into one? Let us look once again at the traits of the divine bodies. They excel in expansiveness, in sensitivity to domains of potential theotropic existence, and in promoting theotropism (countering entropy). It is this last that determines outcomes. The theotropism or divinity that competes most effectively to eliminate entropy will merge with other divinities to the degree that they operate in the same way. It is to their interest to behave in this way. In the end it will be the constructive principle of the universe that will influence and absorb all potential theotropy in the universe. Creation will triumph over destruction. This is the aim of the universe, the greatest of natural laws, and is the ultimate good.

A second objection occurs. If, as has been asserted in this work, man is not a rational animal in any usual sense of the term "reason," and if sublimation is employed to move him from his great fear of himself and the world into large intellectual, imaginative and real worlds far beyond himself, then why is this proof of the existence of divinity not another sublimatory consideration? Is this all "mere" sublimation?

The answer is that sublimation is not unreal, even though it may refuse to treat directly with its origins in human nature. Its rationalizations are testable by rules of reality, logic, consensus, pragmatism, and evidence; this, too, we have said earlier and will discuss later as well. Granted this, the theotropic proof must contend with all other assertions about divinity on the basis of which ones best fit the state of the world as we barely know it and of whatever provides the best consequences for the human condition. Homo sapiens schizotypus is released from his fearful bind and contradictions by this view of the supernatural and is directed to employ his energies constructively - theotropically rather than entropically.

If the principle of entropy exists - and we think that this is so out of our material perceptions - then its opposite principle may exist because, first, the world is not fully entropic, next, there is an anti-entropism observed, and, third, entropism must originate from something that decays. In this last case, the something that decays must have been non-entropic, possibly anti-entropic, that is, theotropic.

The entropy and theotropy can co-exist: they do so under our eyes. It may appear that the theotropic is declining, but this may be false. Our narrow perspective may be giving false measures, and we are better conditioned to detect entropy than theotropy. Especially with our present confidence in materialism, that is, our indifference to theotropy and our desire to emulate the ideal instinctive animal, we may be today underestimating theotropy.

But is not the theotropic also material? It can properly be conceived as such, but only if we realize that most of what we call material is the refuse of theotropic materialism. As to what composes theotropic processes, we submit that theotropy is composed of what is tangibly material, of some extremes of the knowably material (particles, waves, light, etc.) of material potentially known to us but not yet known, and of material unknowable to us. I only call it material for fear of erecting barriers between the "material" and "immaterial."

Under the regime of theotropy, it appears that mankind is to be more of an observer, thinker and admirer of the abstract than the active being who is acted upon. How can he behave religiously otherwise, and how can his morals connect with this religion?

First, one who possesses this religion will be occupied with the future, as historical man has first sought a heavenly salvation and lately has sought salvation in the future also, but in a more scientific and technological way.

Second he will be more objectively self-searching and theological than historically he has been. He is looking for a different kind of divinity; this affects the quality of the search.

Thirdly, he has to consider the question: Do I wish to attract gods? Do I wish to be adopted by gods, lightning-struck so to speak? Do I wish to become chosen by the gods? Do I wish to be embraced by a larger theotropy than I have means of becoming in myself??

Surprisingly the answer to all of these questions will be a strong affirmative. (I say surprisingly for I feel personally that we have no right to expect such definite answers to questions that we have formulated with such difficulty and hesitation.)

Behind the banners of entropism stands a sad, scientoid diabolism. We do want to live in theotropy, in the future, in the realms of the gods.

Then the question becomes : How do we attract the gods? Do we do so with signals, search parties on vehicles, sending care packages of our little technical tricks into outer space?

Or do we go seeking the gods with a message that we think will have meaning for them? What could such a message be ?

Our best message, our invitation to the gods, is our ability to take care of our own world and its surroundings. It stands to reason that the gods, if they have already reached us actually or potentially, or if they were to come upon us in their expanding operations in the universe, would either embrace us or dismiss us by indifference or destruction.

What would achieve their embrace? Obviously, they would embrace theotropy, for that is their essence. What are the signs of theotropy, which in our older language we might call blessedness? They would have to be signs of which we are capable. These signs are not negligible; they are signs of godliness.

Theotropy is the trend of existence to achieve divine influence. Inasmuch as humans may be capable of it, it calls for an expansion of the influence of life over death and of mind over matter. Thus, it appears that the very principles that we have ascribed to the theotropy of the gods are principles that reverberate down the corridors of human time and thought.

If these principles go unattended or are unsuccessfully pursued by mankind, the gods will not punish us; the gods have more important matters on their more universal "minds." They will ignore us, and let us continue in the predictable shortness of our forever to suffer both from our own behavior and being godforsaken, which must mean the loss of our hopes, of our development, and of our future.

There are a great many people who believe that god may exist but always has reason to punish people, so much so that it is useless to attempt even a decent peaceful and material subsistence for mankind. Famine, plague, flood or war are seen to be inevitable divine visitations. Such apathy and fatalism go along with the succession of gods who could hardly allow mankind to recover from one catastrophe before bringing down another upon it.

At the other extreme of materialism stands the vanguard of the technical achievers. So flushed are they with the successes of empirical science, that they predict a never-ending invigoration of life and conquest of vast reaches of outer space. Among these are the ones who would fill capsules with gimcracks to fire into far space.

The fatal flaw in their vision and plans is a misperception of human limits. The human race stands at a crisis of will and belief, of world disintegration and warfare, even as vehicles hurtle into outer space. Humans have not solved their basic issues of life over death, and mind over matter. They may be incapable of doing so without the help of a greater theotropy than they can neurologically possess. The most that they may be capable of, and it would be indeed a great achievement, is to invite the gods for help on matters of life and mind. This the technocrats and military operators of the political economy and outer space are of no mind to do, whether by acting in the name of mosaism or atheism. In their arrogance, they see no need to invite the gods to their feast. Or they try to beckon to them by exercises paralleling the long history of sacrificed beings and the destruction of nations.

We conclude that gods - or god, if you will - exist. They do not exist in fear; fear is human alone. They exist in our mind as the mind tests the limits of reality and invents, while integrating these limits, a special kind of reality in the supernatural -- the area of the divine. Gods exist outside the mind with as much probability as the universe that we contemplate is real. In these two senses, god is reality.

Granted reality, the divine must be our most important reality. This may seem to be skating on the thin ice of scholasticism. "Tell it to a starving man." But it is a statistical reality and in the final analysis it is statistics that compose reality of all kinds. The divine is the most important because it is the only distinction that is uniquely human; it comes straight out of the awful realization of one's divided soul, two or more material contradictions, ineradicable and appositionally creative.

Climactically a reconciliation takes place in philosophy and science. To know oneself is to know more than oneself; it is to know the divine. Here is the reason for the failure of historical religions, which damaged the soul in order to force it to hold delusions about "hard reality" and external gods at the same time

Part Three

SOLARIA BINARIA

"Solaria Binaria" was written in collaboration with Prof. Earl R. Milton

Earl Milton & Alfred de Grazia in Naxos,
working on *Solaria Binaria.*

Since 1924, when the theory of the expanding Universe was first ex-pounded, the phenomena of astronomy have been viewed increasingly as intensely energetic. The notion of an explosive Universe has been abetted by the identification of no-vas, the discovery of the immense energy trapped in the in-ternal structure of the atom, and the detecting of radio noises from vast reaches of space signaling events so extreme as the imploding of whole galaxies. What began as a whisper in scientific circles of the late nineteenth century has become a shout. Yet, for reasons that can only be called ideological, that is, reflecting a constrained cognitive structure in the face of contradictory perceptions, scientific workers on the whole have not heard the "shout".

At the same time as the space and nuclear sciences have had to confront a new set of facts, the near reaches of space have been surveyed and the body of the Earth searched more thoroughly. The results confirm that the wars of the Universe have been disastrously enacted upon battlefields within the Solar System. Without exception, the planetary material that has been closely inspected exhibits the effects of extreme forces unleashed upon it. Mars, Moon, Venus, Mercury - all are heavily scared, Jupiter and Saturn are in the throes of internal warfare. An asteroidal belt that may be called "Apollo" represents a planet that exploded. Nor can we exclude from the common experience this scared Earth.

Consistent with the panorama of catastrophes, and additionally supplying a new dynamic form in cosmogony, there has been developed a body of knowledge and speculation surrounding the phenomena of stellar binary systems. The first binary star orbit was computed in 1822, but not until the past three decades has sufficient information become available to speak about binaries systematically. Since the first discovery, a large proportion of observed stars have come to be suspected as multiple star systems.

Moreover many cosmogonists speculate that the Solar System itself was once a binary system, or at least is now a kind of fossil system, with Jupiter exhibiting star-like traits. It may be pointed out, for instance, that the distance between the principal bodies of the Solar System is comparable with the distances between the separate components in many binary systems.

Hence it becomes logical that a cosmogony of the Solar System should be modelled after the theory that it was, and is, a binary system, a *Solaria Binaria,* accepting and applying for the purpose of the model what is known and thought about the observed stellar binaries elsewhere.

The explosive or catastrophic Universe poses basic problems to chronology. The span of astronomical time has been increasing dramatically even in the face of time-collapsing explosive events that reduce drastically the constraints upon time as a factor in change. Great stellar bodies exhibit rotations and motions that accomplish in hours phenomena that would on a gradual timescale be accorded millions or billions of years. It appears that one has to work with a paradox: even as one studies a Universe that changes over billions of years, one studies local events where changes are measured in microseconds. Consequence, which is the last hope of causality, is often strained in the straddling of time.

When the Solar System comes to be viewed in the light of newly discovered universal transactions, the idea necessarily arises that it has developed under time-collapsing conditions. Time measures - radiometric, geological and biological - that have been painstakingly manufactured to give billions of years of longevity to the system - must submit to a review of their credibility.

The need to generate a new chronometry is enhanced by current reassessments of legends and knowledge that ancient and prehistoric human beings possessed. The authors would not have ventured upon this reconstruction of the recent history of the Solar System were it not for the fossilized voices whose shouts about their catastrophic early world and sky sound louder even today than the shout heard in contemporary science about the exploding Universe. Those anthropologists, archaeologists, and scholars of ancient humanity who believe that these shouts must have been mere whispers confront the same impasse ideologically as those scholars who overlook the larger meanings of explosive cosmogony today. What the ancients said, and did not say, about the world are to be taken into account. Both their concepts of time and their visions of events deserve consideration.

This consideration and the others advanced before direct this monograph towards resolving the cosmogony of the Solar System into a model of a Solaria Binaria, the last stages of whose quick and violent quantavolution have been witnessed by human eyes. The model stands as plaintiff, confronting the model of uniformitarian evolution as adversary. It may be well to stress in the beginning that a prerequisite of scientificity is the ability to suspend judgment on a case being tried. This is especially painful when one is expert on the matter at issue. Even so, a scientist who cannot suspend judgment must be deemed as incompetent as the judge who cannot suspend judgment while hearing a case in a court of law.

Chapter One

THE SOLAR SYSTEM AS ELECTRICAL

The sun as star, radiates energy into the space surrounding it. Stars can be conceived to have originated from electrical cavities in the structure of space. Space, to our mind, is an infinite electrical medium. It is electrical in that it is everywhere occupied by a charge, which, when it moves, assumes the character of electrons, that is, "negative" charge. The movement energizes and carries material into the cavities which become and are the stars.

Such electrical cavities or stars are observed in the millions, and inferred in the billions, in a fairly random distribution about the Sun. Materially, a star is an agglomeration of all that has accompanied the in-flow of electrical charges from surrounding space. The cosmic dust which astronomers see throughout the galaxies is matter yet to be forced into stellar cavities, or matter that has been expelled after a star dies. This dust is detected in greatest amounts in the vicinity of the most highly active stars.

Once in the cavity, the material cannot readily escape; it acquires increasing density because of electro-chemical binding and electrical accumulation. A cavity or star is increasingly charged but during its lifetime it cannot be more charged than the medium around it.

The life history of any new star may normally proceed as its cavity acquires first matter, and then charges continuously until its charge density reaches equilibrium with the surrounding medium, which is to say that the cavity has then been filled. Thereupon the star releases or mixes its material with the medium until it no longer possesses distinction as a body. This "normal" procedure is conditional upon the star's transacting with the space around it in a uniform manner.

The majority of stars seem to transact quietly with their surrounding space, whether they are small red stars, or giant red stars. They end their existences as they lived, quietly, passing their accumulated material into the medium of space, eventually becoming indistinguishable from the medium itself.

However, the fact that the star is in motion within the galactic medium poses an occasional problem. It may journey into regions of the Galaxy which present it with greater or lesser electrical differences than it has been used to. Then quantavolution occurs. The star becomes one of the types to which astronomers pay the most attention - the variable stars, the highly luminous stars, the binary stars, the exploding stars.

It was in one such adventure in space that the original Super Sun lost its steady state, fissioned, and became Solaria Binaria. The system then consisted of a number of bodies, acting first as small "suns" with a primary partner.

In recent times, this Solaria Binaria encountered a galactic region whose characteristics rendered the lesser stellar partner of the system unstable. In a series of quick changes the binary was transformed into today's Solar System.

Bruce (1944) sees the process of stellar evolution as a cyclic build up of an electrically charged atmosphere above the star. As we see it, galactic potentials will determine the nature of the "surface" presented to the outside observer. As the star journeys through galactic space, its surface nature changes in response to differences in galactic potential. A change in the local galactic environment can lead to an instability which results in catastrophic electrical redistribution of the whole stellar atmosphere and sometimes of material found well beneath the star's surface layers. In short, the star becomes a nova.

The Solar System is probably the descendant of a Super Sun, a body containing at least eleven percent more material than the existing Sun, which became electrically unstable and underwent a nova explosion. In his cosmogony Bruce argues that binary stars form by division of an original stellar nucleus. When the star becomes a nova, the returning nova discharge, transacting electrically with the normal outward flow of stellar wind off the star, induces the outbursting star to rotate. A possible reverse jet blast from the explosion might also cause the rotation to occur. Stars then, should have maximum rotation during the nova outburst. Fission of the star into a binary would then logically happen most frequently by rotational fission (Kopal, 1938) immediately after a nova outburst. Close-binary pairs are indeed found among the post-nova stars (Clark et al., 1975, Cowley et al.,1975).

When the Super Sun erupted as a nova it divided into a close binary pair, whose primary became our present Sun; and its companion was a body about ten percent the size of the Sun (see Lyttleton, 1953), henceforth to be called Super Uranus. Enveloping the binary was a cloud of solar material constituting at least one percent of the Sun's material.

Also created in the fission were the seeds which grew into the so-called "inner or terrestrial planets"

Turning our attention to the Sun itself, we observe an opaque layer called the photosphere. This layer is regarded ordinarily as the Sun's surface. The spectrum of the photosphere shows radiation produced when the atoms, ions, and electrons of the photosphere collide, and therefore the spectrum reflects the state of atomic collisions there. The light is emitted during the collisions. It appears that the photosphere is a region of plasma and atoms where the motion of the material is chaotic, randomized. Collisions occur after short journeys, after short mean free paths of electrical accumulation. The electrical field is small.

Above the photosphere lies the transparent solar atmosphere, which is difficult to observe. First comes the chromosphere and then the corona. The spectrum of the chromosphere represents the release of the internal energy of excited atoms and ions. Light is emitted not so much at the moment of collision among atoms, but it is cast off by rapidly accelerating atoms moving to and from collisions, that is, between rather than during collisions. The chromosphere is a region of directed, vertically moving electrons descending into the photosphere, and atoms and ions escaping into the corona and the solar wind. The mean free path is long, not short. The electrical field is large, not small as in the photosphere. Perhaps the key to star behavior is the distinction between the photosphere and chromosphere.

The photosphere, thus, is a region where the transmission of energy is observed. The chromosphere is a region where the transmutation of energy is what is observed.

The photosphere glows brightly with a silver color (Menzel, 1959). Blemishing this visible face of the Sun are dark, slightly cooler regions called sunspots, the average spot lasts less than a day (Abell, 1975, p527). Viewed by telescope, the whole photosphere, except where sunspots obscure it, shows a granular appearance. These granules are bright patches, hot tufts of gas that live for only a few minutes (Juergens, 1979b).

The photosphere and the behavior of the solar atmosphere which lies above it can best be explained using a model based upon electrical processes. Bruce (1944), and later Juergens (1972) and Crew (1974) have shown that photosphere granules have the properties of a large number of parallel electrical arcs.

Where the atmospheric pressure falls to a value equal to one percent of the atmospheric pressure measured at the Earth's surface, collisions between gas atoms can no longer dominate the exchange of energy between the atoms. Instead it is the electrical processes that govern the energy exchanges in the solar gas.

Instabilities in the arc discharges lead to a build- up of charged regions in the solar atmosphere. These eventually produce electrical breakdown; sudden discharges occur, causing bright faculae and the temporary extinction of some photosphere arcs. The result is a sunspot (Bruce, 1944).

The upper atmosphere of the Sun is the apparently intensely hot corona. The gas atoms of the corona have been stripped of several electrons by collisions with inflowing energetic cosmic electrons. The removed electrons are drawn towards the Sun so other ions can flow outwards into the corona allowing the coronal ions to recede into the solar wind.

The corona seems to be constantly ejecting its contents into space as the solar wind. The fraction of the solar output represented by the solar wind is about one-millionth. Haymes states that the whole corona is lost and replaced in about one day.

Some of this material flows past the Earth's orbit as a cloud of energetic protons and helium nuclei, accompanied by electrons. In every second 100 million solar ions arrive above each square centimeter of the Earth's atmosphere.

Sudden explosive eruptions, called flares, occur above the solar surface. Energy in the form of light, atoms, and ions, is accelerated away from the Sun. The energy in a single flare could supply the Earth's population with electrical power for millions of years. A large flare releases in an instant about one-fortieth of the continuous solar output.

Flares start near sunspots, with associated faculae, and develop over hours. They move as if driven by an electrical potential difference between the Sun's surface and the higher atmosphere (Zirin, Obayashi). Once accelerated, the flare gases escape the Sun and modify the solar wind significantly. The cause of flares is baffling to conventional theories, which underplay electrical forces in cosmic processes. The presence of magnetism implies an electric source. As we shall thrive to show, the Sun once had an electrical connection to its companion, within which energy was released that created and sustained life within the binary system. Today's flares represent an undirected remnant of the inter- companion arc of yesteryear.

The solar wind consists of coronal gases which have been boiled away from the hot solar atmospheric discharges. It conducts the Sun's electrical transaction with the Galaxy. It is the Sun's connection to the Galaxy. The electron-deficient atoms (ions), by escaping from the Solar System, increase the negative charge on the Sun. This brings the Sun towards galactic neutral and thus, in time, would end the Sun's life as star.

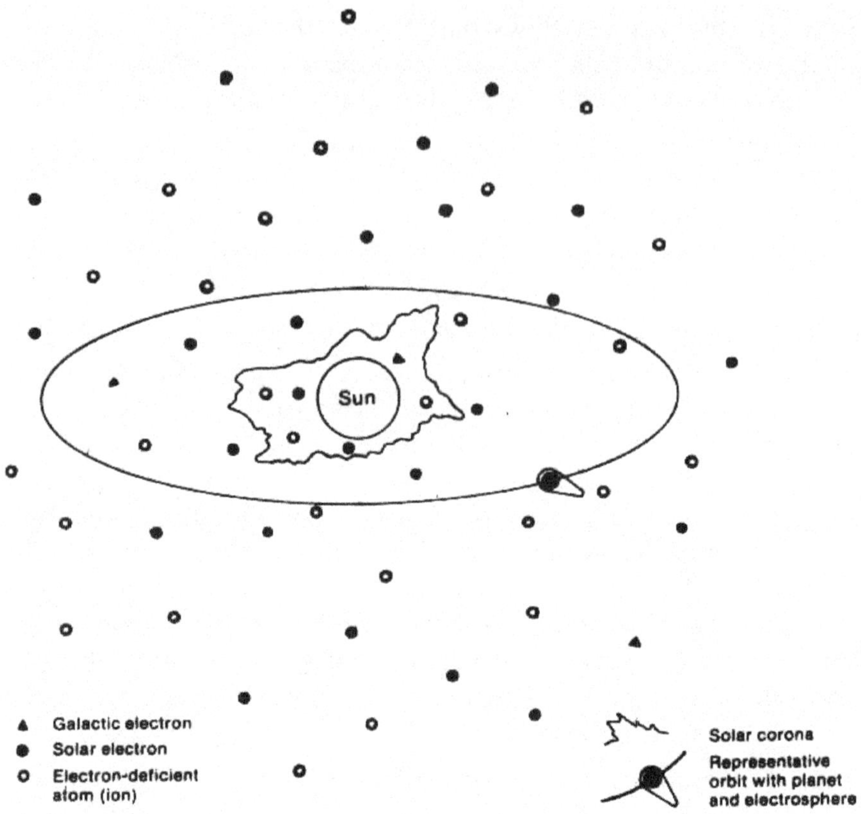

. *The Sun's Connection to the Galaxy*

Outward-flowing solar wind ions carry an electric current between the negatively charged Sun and the more negatively charged galactic space around it. The solar wind flows through a "transactive matrix" (see Technical Note B) of solar electrons , which permeate the interplanetary space but do not flow through it as do the ions. Inward-flowing galactic electrons, travelling at velocities close to *the speed of light,* carry energy from the Galaxy to the solar "surface" where it is released and radiated as light and other electromagnetic waves which constitute the solar luminosity.

It follows that in the past, when the Sun was less negatively charged, more current flowed from the Sun to the Galaxy. Thus the present flow of solar wind is less than the flow in ages past when the Sun was more out of equilibrium than it is now. The Solar wind varies with the ongoing "evolution" and "quantavolution" of the Sun.

Regarding the processes which power the Sun, most astronomers believe that there is an energy source deep in the solar interior obscured from view behind the opaque photosphere. If this belief is correct then the interior of the Sun must be hotter than the photosphere.

Knowledge of the conditions within the Sun is inferred as the consequence of the physical forces assumed to be governing the stability of the Sun (Smith and Jacobs). It is usually inferred that near the center of the Sun the gas is sufficiently hot and dense enough to bring about nuclear fusion on a large scale.

A thermonuclear Sun is an attractive theory since the Sun seems to be composed mainly of hydrogen. By compressing itself into a nuclear-powered core the Sun might radiate energy long enough to accommodate the gradual evolutionary processes believed necessary for the biological and geological developments that have occurred on the Earth.

However, thermonuclear fusion processes must dispose of large numbers of neutrinos, and a vastly insufficient number of neutrinos have been detected on Earth in experiments specifically designed to capture this normally elusive solar sub-particle. (Parker).

Fatal, furthermore, to all theories of an internally powered Sun is the minimal temperature of the photosphere. How can the "surface" of the Sun remain cool when it is blanketed by hotter regions below and above whose temperatures reach millions of degrees (Parker)?

The usual answer is that the Sun's atmosphere is heated by turbulence within the Sun's outermost interior layers below the photosphere (Wright). Somehow this process which, overleaping the photosphere, heats the Sun's atmosphere is supposedly divorced from the flow of radiant energy from the Sun's interior. Since such separation of processes is unknown elsewhere, this explanation is untenable.

Consistent with the electrical phenomena of the Sun's atmosphere, we propose an external source of solar power. The Sun's light and heat output arises from the energy released by a flow of highly energetic electrons arriving from the Galaxy. This electron current is enhanced by the flow of energetic solar wind protons away from the Sun.

Outward-flowing solar wind ions carry an electric current between the negatively charged Sun and the more negatively charged galactic space that surrounds it. Inward-flowing galactic electrons, travelling at velocities close to the speed of light, carry energy from the Galaxy to the solar "surface" where it is released and radiated as light and other electromagnetic waves, which constitute the solar luminosity.

At the edge of the Solar System, escaping protons, accelerated to high energy by the drop in electrical potential between the Sun and the Galaxy, become galactic cosmic rays and flow in all directions towards other stars. The protons expelled by other stars arrive in the Solar System as cosmic rays. About six cosmic rays impinge upon each square metre of the Earth every second, but these few energetic particles carry inwards about one- twentieth of the energy flowing outwards with the solar wind. That electron-deficient cosmic ray atoms continuously flow to Earth enhances the probability that the Earth is electrically charged.

Where the solar wind ends is yet to be determined. It was once believed the wind stopped inside Jupiter's orbit, later near Pluto, but today the wind is deemed to flow well beyond Pluto (Haymes). Somewhere the "galactic wind" meets the solar wind; there a boundary exists where the flow of incoming cosmic ray protons balances the outflowing solar wind protons. This is the edge of the Sun's discharge region, the limit of the Solar System.

To conclude, a star is born when an electric cavity forms in the charged medium of space, and matter rushes along with the charged space to fill the cavity. Then, after the cavity fills, the star dissipates into charged space, spilling out its matter simultaneously. No tombstone marks its demise; no derelicts travel forever through space. Indeed, existence is an attempt to achieve nothingness. Pockets of lesser negativity become existence by seeking to accumulate enough electric charge to emulate universal space, at which time they are capable of disappearing into nothingness.

CHAPTER TWO

THE SUN'S GALACTIC JOURNEY & ABSOLUTE TIME

Conventionally viewed, the formation of a solar-type star and planets from a cloud of gases and cosmic dust takes on the order of several hundreds of millions of years. After accretion, an Earth-like planet supposedly takes another one or two thousand million years (1- 2 giga-years or aeons) to develop a stable lithosphere, which when formed allows the much slower evolution of a viable biosphere from the materials and energy available at the planetary surface (Oparin). To us, these processes seem too slow and rely too much upon random occurrences to be viable.

.However, the processes forming stars and planets and leading to living things may proceed much more rapidly. Our cosmogony employs electrical cavities, charges and forces to accomplish change. These produce changes which are much more powerful and are highly selective.

Electrical force, as measurable by the repulsion between two electrons, compares with the apparent gravitational attraction of the same two electrons in the ratio of 10^{36} to 1.

What "gravity" is supposed to accomplish in aeons, electricity could quickly accomplish before the eyes of the earthly observer. Driven by the powerful motivator, electricity, quantavolution becomes not only possible - but also essential. Furthermore an understanding of electricity's role provides a powerful new and unified explanation of most observable phenomena.

We assert, in line with the past chapter, that stars take their properties less from the material which they contain and more from the electrical difference between the cavity, which creates the star, and the surrounding medium of electrified space.

Translated into more common astronomical language, the luminosity of the star depends upon its galactic environment rather than upon the amount of material which it contains. The more luminous the star, the more it transacts with its companion, and so the companion completes its orbit more rapidly.

All of the stars in the Galaxy are in motion. Since there is no standard of rest all we can detect is the motion of one star relative to another. Two streams of stars are observed moving past the Sun parallel to the Milky Way (the arms of the Galaxy). The two streams move oppositely at a relative speed of 40 km/s, the outer stream moving towards Orion, the inner one to Scutum. These motions apparently reflect differences in the motion of consecutive galactic arm segments in the Galaxy. The stars in our - the Sun's - "arm" we assume move with the Sun at 275 km/s towards the constellation of Lyra near Cygnus, which is a motion away from the stars of Puppis. Looking only at the net motion of stars close to the Sun we detect the drift of the Sun within its arm of the Galaxy. This analysis reveals a motion of 20 km/s towards the constellation of Hercules (away from the constellation of Canis Major) (Mihalas and Routly).

Neither of the Sun's motions is precise but they should suffice for our purpose. The Sun's motion within its arm carries it four astronomical units per year. It takes nearly 22,500 years for the Sun to drift one lightyear from its present position. But, when the galactic revolution motion is considered, the Sun is moving up to fourteen times as fast. In the extreme only 1,107 years are required to displace the Sun one lightyear, so in ten thousand years the Sun moves nine lightyears, and in one million years it travels about 904 light years.

If our hypothesis is correct and the stars derive their properties from the space in which they are embedded, then a look at the stars presently in the Sun's wake will tell us how the Sun appeared in ages past. Unfortunately the path of the Sun over the last million years, within which we believe Solaria Binaria developed and collapsed, is not wholly within measured space. Luminosity assumptions need to be made during the first two thirds of the binary's lifetime.

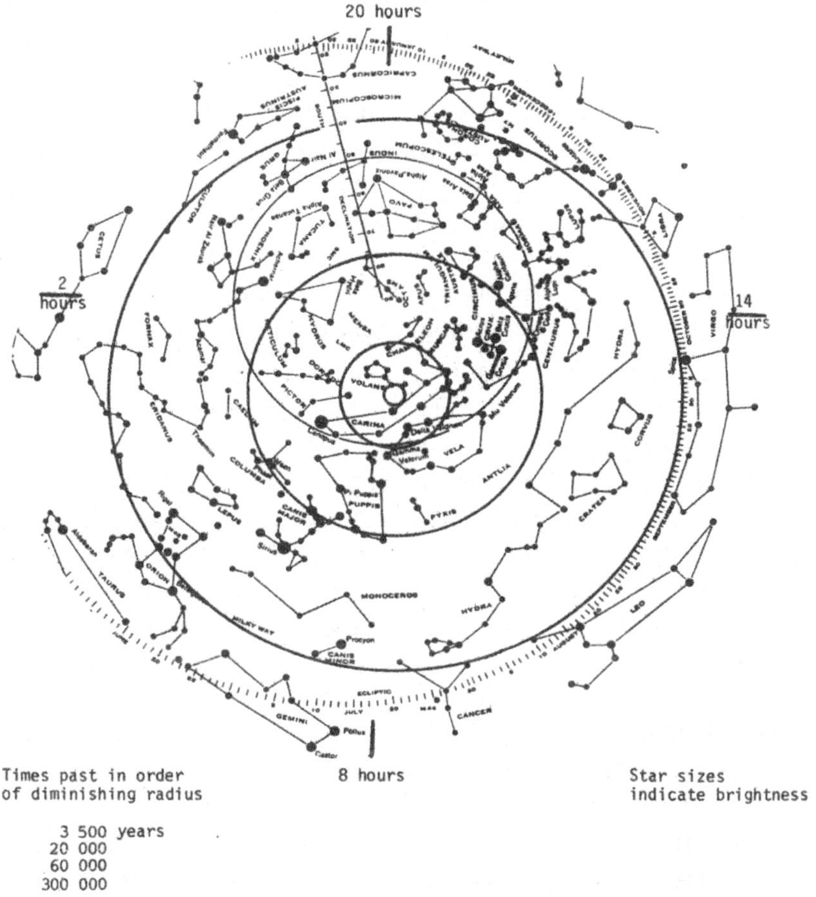

The Sun's path traced backwards through the stars of the Galaxy passes through a cylinder of space whose axis stretches from the center of the Sun through the point on the celestial sphere with co-ordinates 8.4 hours of right ascension and -62° of declination. The edge of this cylinder, chosen to have a radius of 13.25 light-years, is represented for different eras by the series of circles converging onto the solar antapex.

For star systems close to the Sun, adjacent stars are about 10.3 light- years apart. Given such a low star density, a rather large volume must be examined around and along the Sun's wake to ensure that some stars are included. We have constructed, therefore, a cylinder thirteen and one-quarter light years in radius about the Sun's path. Moving for ten thousand years through this cylinder the Sun will "encounter" about 5000 cubic light-years of space. In such a volume there would reside about nine stars or star systems at the average local star density.

Over the sixty-five light year swath through space covered by the Gliese Star Catalogue there are only fifteen star systems. The Sun entered the region included within the Gliese catalogue about 74,000 years ago. It appears that along the Sun's path, the actual star density is only twenty seven percent of that expected. Within that volume, our analytical sample of stars is reasonably complete . Beyond it, many of the stars located along the cylinder do not have published parallaxes and so they cannot be located in time; they cannot be used in the analysis.

The stars occupying the space inhabited by the Sun through the current era (the Period of Solaria) and during the time of the Late Quantavolutions are in this sample. Here, we find the nearest star system, the Alpha Centauri triple. The largest star is very similar to the Sun. Its first companion is 23.5 astronomical units away moving along an elliptical orbit (Menzel et al.). This star is slightly cooler and fainter than the Sun. The second companion is located almost two degrees away in the sky. It is over 550 times more distant than the separation of the closer pair. Frequent eruptions superpose bright emission lines on its otherwise faint class M spectrum. It is a flare star; its flaring might be associated with some intermittent transaction with the pair of distant companions. Unfortunately the a-Centauri triple is the only occupant within the space transited by the Sun during the series of quantavolutions preceding the historical period. It gives us no clue to an understanding of that space besides learning that solar- type stars can exist there.

Extending the Sun's line farther into the past to the limit of the Gliese catalogue we find no stars as luminous as the present Sun until we go back 54,000 years. Then along the path are positioned three stars that exceed the Sun in luminosity. The closest, an F3 subgiant, is five times more luminous; the second, the primary star in a triple system, is only 1.44 times brighter. Its two companions are very faint. The last of the three brighter stars exceeds the Sun's output eight-fold.

At the 75,000 year limit we reach the edge of the reasonably complete star sample. So far there are no conflicts with our theory. If our calculated course is correct, the Sun's past behavior, as mirrored in the listed stars' present behavior, would show significant variation in luminosity over the tens of thousands of years represented here. Noteworthy, there are no highly luminous stars thus far along the Sun's trace.

In our analysis more distant stars cannot be located in time along the Sun's path. Yet we can place, although uncertainly, several bright blue supergiant stars at locations surrounding the antapex in all directions and at distances corresponding to times between one-half and three million years ago. Several of the stars are components in binary star systems. Within or on the periphery of this highly transactive region of space, the original Super Sun may have parturitioned to give birth to Solaria Binaria.

Although proof is hardly forthcoming from this analysis, at least evidence disproving the hypothesis is absent. We are encouraged to retain the idea that the behavior of star systems depends, if only in part, upon the celestial charge level of the space through which they pass. It seems as if this electric charge is contained not only by material residing in the space (stars, atoms, and electrons) but also, in part, as a charge embedded in the space itself, what we shall call a space infra-charge. Literally, the space infra-charge means that a vacuum (empty space) contains normally unavailable electric charges (here electrons) which generate the structure of that space and affect the behavior and properties of all matter occupying the space.

CHAPTER THREE

SUPER URANUS & THE PRIMITIVE PLANETS

About one million years ago our Sun, then a Super Sun, underwent a nova eruption because of a sudden or unendurable change in electrical conditions. Solaria Binaria was instantly born. The Sun fissioned and in a huge blossoming cloud there would have been found a diminished Sun. Within a concentration of gases from the old sun would occur an admixture of chunks of the old Sun's interior material (nucleus), including a body that became the binary partner, which we here call Super Uranus. Between the new Sun and Super Uranus lingered other fragments of the fission and great quantities of the material that were to be absorbed into the planets. This impressive electrical quantavolution occurred in a matter of hours. The separation of the two bodies increased rapidly.

The original Super Sun was a star transacting quietly with the electron-rich space around it. It became unstable when its galactic journey carried it into a less electron-rich region. Here, the enrichment presumably was rapid and of great magnitude. The Sun for a short time became relatively too electron-rich. The resulting nova, which is an explosion of electrons that forces (requires) a material accompaniment, created Solaria Binaria. In an explosive expansion the binary was born, not just from the solar atmosphere but also from the refractory materials normally hidden within its interior.

The nova explosion had propelled what temporarily was excess charge away from the Sun. This of course would be illusory, for the Sun remained a region of relative electron deficiency. Thus, the initial dismemberment of the original Super Sun quickly halted: the expansion of the plenum of material, now surrounding the Sun, ceased both because of the Sun's continued need for electrons and because the charged surrounding medium continued moving in upon the cavity.

The Sun now seeks its lost charge. The easiest way to get that charge is to launch into the plenum electron-deficient atoms (ions). The proximity of super Uranus distorted greatly what otherwise would have been a radial flow of ions (as in the original transaction between the Super Sun and the Galaxy). A strong electrical connection coupled the Sun and Super Uranus; a lesser connection joined the Sun to the plenum.

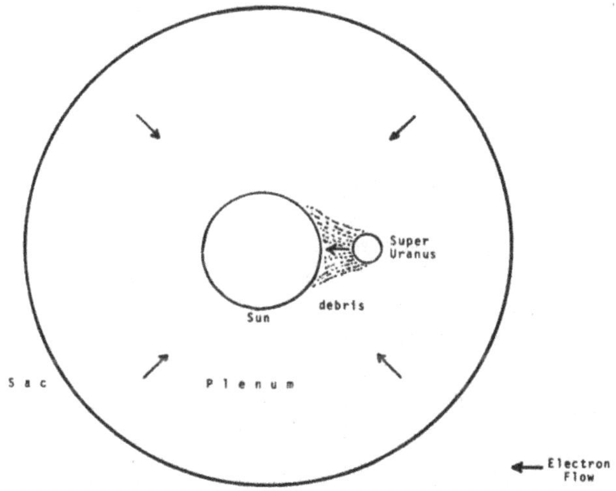

At its birth Solaria Binaria was embedded at the center of a plenum filling a sac of electron deficient matter. Electron flow into the sac from the Galaxy was augmented by electron redistribution within the plenum and among the components of the binary system.

The strongest electrical transaction occurred between the principals; accompanying this electrical flow, and highly influenced by it, was the transfer of material from one of the principals to the other.

Ionized gas atoms would be induced to flow between the principals. This flow of countermoving electrons and electron deficient atoms would constitute a strong electrical current. As a consequence an intense magnetic field would be generated surrounding the current. This magnetic field would pinch the flowing ions producing a relatively narrow electrical flow channel. Collisions between neutral and electrified atoms would transfer the influence of the magnetic field to all of the gas between the principals; the result is a magnetic bottle.

The Sun's rotation began consequent to the nova discharge, creating Super Uranus. Super Uranus thereafter wheeled about the Sun in close orbit. The magnetic field produced by their electrical transaction was instrumental in locking the rotation of the Sun to the motion of Super Uranus about the Sun. Strongly coupled together the pair rotated looking like an ever expanding but otherwise rigid dumb-bell. The gases and the planets as they formed remained trapped along the gaseous electrified axis between the principals.

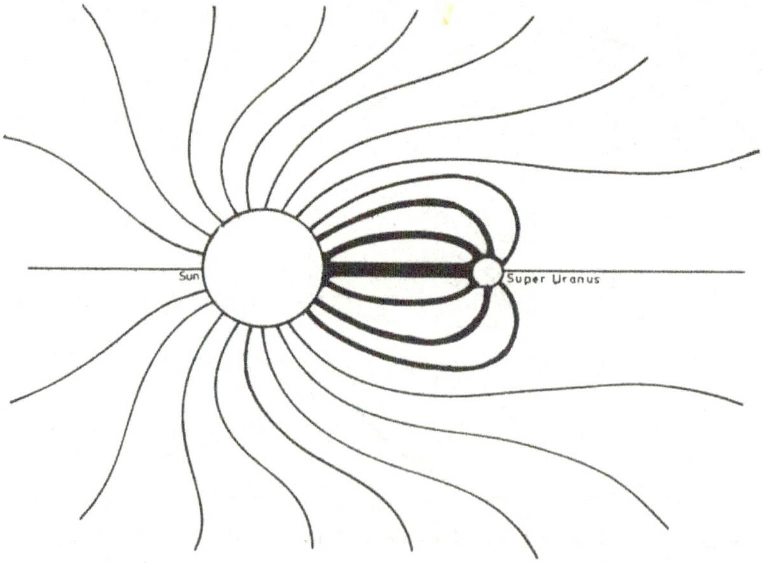

Material Flow vCoupling the Sun, Super Uranus, and the Electrfied Plenum

The creation of the Sun's companion, Super Uranus, greatly distorted the electrical flow between the electron-deficient Sun and the Galaxy.

The Sun's daughter, Super-Uranus, like it's parent, was short of electrons compared to galactic space outside the sac. The electrical flow coupling both the two stars and the stars with the Galaxy caused and directed a significant material exchange between the pair of stars.

The evolution of Solaria Binaria was such that the two principals were slowly driven apart, in part by the momentum of the flow of mass from one to the other and in part from increased repulsion caused by the growing level of electric charge in the whole system by the accumulation of galactic electrons. All the while the angular momentum (spin) within the system was being transferred from the primary to its companion. If matter was transferred mechanically from the heavier Sun to the lighter orbiting Super Uranus, the spin of the binary would decrease, but if the transferred matter is electrically driven, acceleration would be expected to accompany the transfer, thereby potentially increasing the spin of the binary.

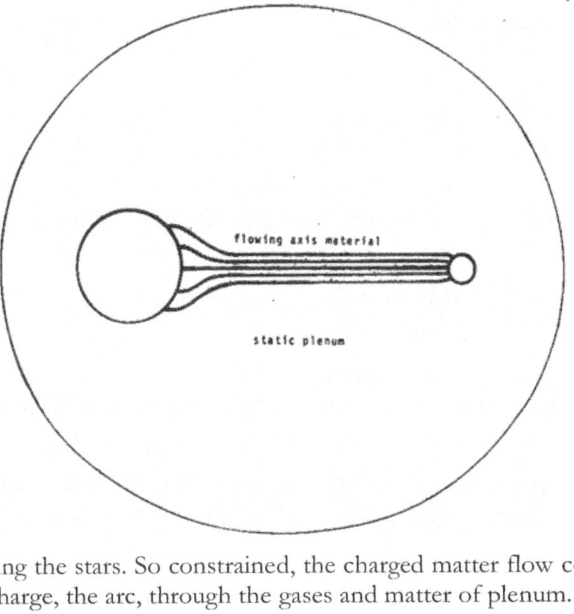

Flow of Material Between the Sun and Super Uranus under the Influence of a Self-generated Magnetic Field

Electrically charged material flowing between the Sun and Super Uranus generated a strong magnetic field about the axis between the two stars. The effect of the magnetic field was to squeeze all material flow into a thin tube joining the stars. So constrained, the charged matter flow constituted a potent electric discharge, the arc, through the gases and matter of plenum.

The pulling apart of the principals was reflected in an increase in the binary's period of revolution. That is, Solaria Binaria wheeled more slowly about its center. In its earlier stages, Solaria Binaria would have looked to a distant observer as a close binary with an unseen companion.

Within the sac, where the two stars were located, the energy flow may always have been similar to what we observe now. However the outer parts of the sac were transacting intensively with the cosmos and thus were radiating so as to appear markedly hotter. The perimeter of Solaria Binaria, then, would have appeared to radiate as an early- type star and not like the Sun does now. Its period of light variation, radiation emitted, and flow of mass would have attracted the attention of astronomers elsewhere to Solaria Binaria.

As the solid-wheel binary evolved, the Sun eventually was separated from Super Uranus by 105 gig meters (about 0.7 astronomical units). Before the next great quantavolution the primitive planets Mars, Earth, Apollo, and Mercury ended up between the two principals in the region between 61 and 96 gigameters from the Sun. At such separation this would bring the planets Mars and Mercury closer to Earth by factors of four and six respectively. Even so these planets would produce visible discs which were only about one twenty-seventh the size of today's Moon. If they could be seen (which we doubt) they would still be observed almost as points of light in the sky.

The planets were originally debris from the Super Sun nova. They travelled out in the trail of Super Uranus, held in the electric and gaseous flow. They settled into their original positions rather than moving on because they were electrically less negative than Super Uranus. They distributed themselves in their magnetic cage along the axis in accord with the principle of maximum mutual repulsion (elsewhere known as "the principle of least interaction action").

From the beginning Solaria Binaria was enveloped in a cloud of solar material (gases and solids). As the binary evolved this sac became extended along the lengthening axis from Sun to Super Uranus. Compressed by the magnetic field generated by the flowing electrified gases, a stable gaseous tube surrounded the planets; indeed these gases pervaded the entire planetary region, enveloping all of the planets in a single sac of gases.

Within this dense gaseous sac, the contents of which the ancients called the aether, and we will call the plenum, the planets could receive biologically necessary temperatures from the axial electrical discharge connecting the Sun with Super Uranus (de Grazia, 1981).

CHAPTER FOUR

THE SAC & THE PLENUM

In the sac was the whole system of Solaria Binaria; the Sun, Super Uranus, the primitive planets, and the plenum (of gases and solids) of solar origin that nurtured the planets.

As the binary widens, the sac becomes conical in shape, narrowing from the size of the Sun at one end to about the size of Super Uranus at the other. Viewed from the outside the ancient plenum would have been opaque to light. Not so with the gas of the Earth's atmosphere today, which is eight kilometres thick if the atmosphere is considered as a column of gas of constant density. This atmospheric layer is of trivial thickness compared to the radius of the Earth, yet its importance to the environment is unquestionable. Even this negligible atmospheric layer removes 18.4 per cent of the incoming sunlight, mostly by diverting it from its original direction of travel.

Some of this scattered light returns to space, but most of it is redirected several times to produce the blue sky so familiar to us. Atmospheric scatter is enhanced near sunset when the incoming light traverses an atmospheric column tens of times longer than near noon. The setting Sun is notably fainter and its color redder because of the increased scatter. If the atmospheric column were as little as 1280 kilometers thick (at the present surface air density) all of the sunlight would be deflected from its incoming direction. Light would still be seen but only after scattering several times; no discernible source could be identified with the light. So it was in the days of Solaria Binaria.

Around 14,000 years ago, Super Uranus was located about as far from the Sun as the orbit of the planet Venus today. This would provide the plenum with a volume of about 10^{20} cubic kilometers. If the plenum contained as much as one per cent of the atoms in the present Sun, the gas density would be several times that found at the base of the Earth's atmosphere today.

Neither star would be seen directly, and only a dim diffused light could reach the planetary surfaces.

In the earlier stages of Solaria Binaria the plenum was impenetrable to an outside observer; all detected radiation came from the surface layers of the cone- shaped sac, an area up to fifty-five times the surface of the Sun. The luminosity of the sac would arise from the transaction between inflowing galactic electrons and the gases on the perimeter of the sac.

The plenum, at formation, was electron-rich relative to the stars and the planetary nuclei centred within it. These latter electron deficient bodies promptly initiated a transaction to obtain more electrons by expelling electron-deficient atoms into the volume of the plenum. The charge differences within the sac were modulated with time. In other words, the plenum was losing electrons from its perimeter to its centre. In response, the size of the sac collapsed under cosmic pressure. In time this charge redistribution might have diminished the volume of the sac by as much as tenfold, compressing the cone of gases into a cylinder or column of smaller diameter.

Running along the axis between the Sun and Super Uranus was an electrical discharge joining the two principals. Moving with this electrical flow was matter from the Sun that was bound for Super Uranus. Some of this matter would be intercepted by and incorporated into the primitive planets.

Induced by the electrical flow, a magnetic field was generated which encircled the axis and radially pinched the gases. The pinch effect is self-limiting in that the more the current, the more the pinch. An infinite current in theory pinches the current carriers into an infinitesimal volume, extinguishing it. Material would be extruded at both ends of the pinched flow by the pressure induced in the pinch.

This circular magnetic field, a magnetic tube, would induce randomly moving ions of the plenum to circulate along the field direction. The circulating motion of the ions eventually would be transferred by collision to the neutral gases. The result would be that in the outer regions flow would be dominated by revolution around the circumference of the tube. Everything here would eventually revolve uniformly.

The innermost regions of the column were dominated by flow along the axis. Considerable transaction occurred at the junction of these two separately moving regions of the column, the central and the peripheral.

At the perimeter of the plenum, optical effects would show to an outside observer an apparent absorption shell associated with the hidden binary within. Like many of the close- binary systems, the stars of Solaria Binaria would not be resolvable in a distant telescope, but the binary nature of the system could be known because observable differences would be produced as the dumbbell revolved.

With passing time, the level of light would increase. In the beginning, the light is scattered and the sky is a dim white. As the plenum thinned electrically, the sky bodies would emerge as diffuse reddish patches. During this process, the sky would brighten and become more blue. Thus, as they emerge, around 14,000 years ago, Super Uranus and the Sun brighten and whiten while the sky becomes darker and bluer.

Seemingly, aboriginal legends describe the heavens as hard, heavy, marble-like and luminous. Earliest humans were seeing a vault, a dome. Probably in retrospect, to the heaven was ascribed the human qualities of a robe or covering, and, by extension, part of an anthropomorphic god. Thus, the Romans saw Coelus, the Chinese T'ien, the Hindus Varuna, and the Greeks Ouranos. Vail (1905/1972) presents ample evidence that day and night were uncertain and that the heavens were continuously translucent. When Hindu myth says that "the World was dark and asleep until the Great Demiurge appeared", we construe the word "dark" as non-bright relative to the sunlit sky that came later. Heaven and Earth were close together, were spouses, according to Greek and other legends. The global climate of the Earth in the plenum was wet; all is born from the insemination of the fecund Earth by the Sky, said some legends. There was so much moisture in the plenum that, although the ocean basins were not yet structured, the first protohumans might confuse the waters of the firmament above with the earth-waters. In some legendary beginnings, a supreme deity had dispatched a diver to bring out Earth from the great primordial waters of chaos (Long, 1963).

The earliest condition was referred to as a chaos, not in the present sense of turbulent clouds, disorder, and disaster, but in the sense of lacking precise indicators of order, such as a cycle that would let time be measured. T'ien is the Chinese Heaven, universally present chaos without form. The gods who later give men time, such as Kronos, are specifically celebrated therefore (Plato).

Sky bodies were invisible. Legends of creation do not begin with a bright sky filled with beings, but speak of a time before this. When the first sky-body observations are reported, they are of falling bodies. The earliest fixed heavenly body in legend is not the Sun, the Moon, the planets, nor the stars, but Super Uranus, as will be described later on.

Nor was the radiant perimeter of the sac visible. It lay far beyond discernment as such, and was in any case practically indistinguishable from its luminescence. The electrical arc would have been visible directly only in its decaying days, being likewise sheathed from sight by the dense atmosphere of the tube. That the arc or axis appeared along with the sky bodies before its radiance expired is to be determined in the next chapter, where its composition and operation are discussed.

As the binary evolved, the plenum came to contain an increased electrical charge; it expanded, leaving less and less gas in the space between the principals. Thus it became gradually more transparent.

CHAPTER FIVE

THE ELECTRICAL AXIS & ITS GASEOUS RADIATION

The binary electrical system was distinguished by an electrical flow between the principals. This was a highly energetic discharge which generated chemical, and probably nuclear, transformations among the gaseous constituents of the plenum. It provided the heat and energy for life to emerge.

The gaseous plenum initially seems to have contained an excess of hydrogen. The combination of electrical arc with hydrogen rich gases favored the quick production of organic molecules and of biological systems (Miller and Urey) within the plenum.

An electrical current flowing along the axis of the binary of the order of 10^{14} amperes (100 tera-amperes) would produce sufficient magnetic field intensity to constrain the plenum gases up to 64 000 kilometers from the axis. Electrical breakdown can occur in dense gases where the electric field intensity is of the order of 25 kilovolts per centimetre (Schröder). The longer the electrical column the greater the potential difference required between the principals in order for breakdown to occur. Once breakdown occurs the voltage drop at a place decreases from kilovolts per centimeter to tenths of a volt per centimeter. Along the discharge column the voltage drop varies considerably. It is greatest near the electrodes and is very small at most points in the body of the discharge.

Thus, the discharge should have produced a small region of hot gas centered along the electrical axis. Surrounding the gases of the discharge was a large opaque mantle of cool gases. Within the cooler gases were the electrically charged planets, which had been repelled from the arc but caught up in the magnetic tube.

It is difficult to estimate the energy that would be released by a discharging arc unless the voltage drop across the arc is known. Since we do not know this value for Solaria Binaria, we must define the arc's parameters in terms of other criteria. One must be preoccupied with the thermal constraints upon the Earth and its developing biosphere.

It is known that transverse heat flow is hindered by the presence of a strong magnetic field (Kapitza). Also the gases insulated, diffused, and rendered uniform the intermittent blasts of the arc by the time the radiation reached the region occupied by the planets. Nevertheless, the closer the Earth to the arc, the more energy it would have received. For some time the Earth gained its energy almost entirely from the arc source. It is reasonable to assume that Earth's temperature must soon have devolved below 325 K.

A lightning-bolt leader moves about 300 kilometers per second. Thus late in the history of Solaria Binaria it would have taken about 350,000 seconds for the leader to work its way along the 105 gigameters between the principals. The return discharge propagates faster, taking only 2190 seconds or 36 1/2 minutes.

Did this arc, so necessary to life, persist even into the time of human awareness? As Super Uranus receded from the Sun, and the planets redistributed themselves farther apart in its wake, it is logical to assume that the intensity of the arc declined and its low became intermittent.

Hence, at around thirteen thousand years before the present an observer on Earth would have seen a great flickering and coiling axis or column of fire.

In an electric arc the gases "burn" in a relatively narrow column. The higher the gas pressure the narrower the discharge column and the more difficult it becomes to sustain a uniform current through the discharge

(Somerville). Strong arc discharges, such as lightning channels, seem to bend into a helical shape. Such bending seems to generate a condition within the arc which can terminate the discharge.

Revolution of the gases around the longitudinal axis of laboratory discharge columns tends to stabilize the discharges. When they do, the rotating gases are said to create a radial "gravitational" field (Somerville).

The rotating gases surrounding and driven by the magnetic tube in Solaria Binaria would act to keep the electric discharge going when it otherwise would have gone out. The arc pulsed regularly responding to some natural rhythm between the forces leading to extinction and the forces promoting resurrection.

The electric arc operating in Solaria Binaria is a cosmic discharge of long duration. Bruce, in the course of seventeen letters about *Cosmic Electrical Discharges* (1958-1964), has documented examples of smaller arcs of shorter duration and of longer arcs lasting to millions of years.

He proposed that an electrical discharge liberating energy comparable to that ascribed to the quasars was capable of transforming elliptical galaxies into spirals. This is the grandest of the cosmic lightning discharges; in its wake the spiral arms of the galaxy form with their "metal-rich" stars. Bruce speculated that in the enormous temperatures generated in these discharges, nucleosynthesis transmutes smaller atoms into larger ones. It is this latter possibility that leads us to postulate that nuclear transformations were accomplished in the arc of Solaria Binaria. If they were, they probably occurred most vigorously at the beginning, when the discharge current was greatest.

Despite the many problems with laboratory experimentation in this area, some supportive work has taken place. Using a pulsed high current arc discharge, Russian workers produced beams of 40 kiloelectron volt deuterons at instabilities in the discharge (Somerville). This achievement is consonant with certain proposed nucleosynthetic processes that occur in low energy flares above star surfaces (Canal). Zirin gives a mechanism for the generation of solar flares resembling processes which might occur within regions of a pinched electrical arc.

Even more closely related to the situation in Solaria Binaria is Joss' speculation that X- ray burst sources result from thermonuclear flashes. X- ray burst sources are episodic; in some, bursts are much more frequent. Many burst sources can be inactive for weeks. X- ray sources, steady and bursters, are associated with binary star systems. If, particularly, the burst sources are due to thermonuclear reactions in close binary star systems, then we can be confident that these reactions occurred in Solaria Binaria and that they were instrumental in shaping its chemical and biological structure.

Inasmuch as these thermonuclear events were part of the earlier history of the electrical axis, no human would have observed this part of his ultimate creation.

At one region of the Earth, the axis might be expected to appear as a kind of rainbow of fire or "neon-tube" glow across the sky ending at Super Uranus. In another region the arc might appear more short at the horizon and stretch to the red star. In the opposing hemisphere, the arc might be visible alone, first, and then might reach for and finally attain the Sun, with the axis blossoming at the Sun, thus "creating" it, or vice versa. The flickering of the arc, when slowed down enough to be noticeable, might resemble red coiled snakes, intertwining and crawling brokenly towards the great red god.

The snake and dragon accompany very early gods and goddesses. "The Serpent of the Jupiter-type myth is always seen to be a creation of the proto-Saturn god" (Tresman and O'Gheoghan), that is, Uranus. The color red is widely used and sacred in archaic, perhaps Paleolithic Uranian times (Wreschner). However, the abundance of such symbols is countervalenced by their generality as referents. Lacking specific applications to phenomena, they are unreliable indication of the electrical axis.

Certain symbols associated directly with Saturn (of the time of Super Saturn) are also suggestive of the arc. These include the courtly long-gowned figure of the god, the tree of life (including the Christmas tree), the sacred mountain , and others (Talbott, D. N., ch. 8) that convey the image of the god atop a cone-shaped or pyramidal design on top of the World.

In Iroquois legend, at the beginning of things, the Chief of Heaven, in a fit of jealousy towards his spouse, uproots the tree whose flowers illuminate the celestial world. The Sun and Moon did not exist at the time. He cast his wife, "Fertile Earth", into the hole and replaced the tree (Eliade, 1967).

Among the Nagdju Dayak of Borneo, the Creator couple, dwelling as birds in the tree of life, fight and damage the tree badly. Some time after the first humans are born of their efforts, the tree is destroyed.

More closely correlative with the axis in the linguistic frame of modern science is the concept of the "Central Fire" that occupied early Greek philosophy. This has particularly descended through the fragments attributed to Philolaos, the Pythagorean (Dreyer). Philolaos was the first of the secretive Pythagoreans to publish a book and his treatment by Plato leaves little doubt that he represented a considerable school of archaic science.

Some thirty-two attributes of the "Central Fire" are to be elicited from Philolaos and Heraclitus, as presented by Rose, all of which can be accommodated to the theory of the electrical axis of Solaria Binaria. The Central Fire was thought to have been a layer of fire above a layer of air. It is the center of the world. It is cone-shaped. It never sets, and has always the same location in the sky. It is "alone", the "highest", "unmoving", "stable", the beginning of everything. The Earth orbits around the Fire, but Earth does not rotate upon itself. Nor has the Earth any obliquity.

The Fire is not counted among the numbered bodies of the celestial sphere. It is not called Saturn or by any other name except that it was termed the "Mother of the Gods". It is called the "hearth of all", the "residence" of Zeus, his "throne", his "tower", his "fortress". It is a "divine ruler and teacher. It is the "altar", the "bond", and the "measure" of nature. The

Sun borrowed light from the Fire; the Sun orbits around it. The Moon, planets, and stars orbit the Fire.

Heraclitus reported it as "an ever-living fire, kindling itself by regular measures and going out by regular measures". He said that "it advances and retires" (Rose, 1979). Earth turns always the same face towards the Fire. A Counter-Earth exists, which is closer to the Earth than to the Fire, and obscures the Fire from view.

The match of Solaria Binaria's axis of "electrical fire" (as electric discharges were called until the nineteenth century) with the attributes of the Central Fire in Greek cosmogony is close. The mention of celestial bodies can be explained as reflecting later observation of some traits.

CHAPTER SIX

THE MAGNETIC TUBE & THE PLANETARY ORBITS

The arc along the axis between the principals created a magnetic tube, which surrounded the discharging gases.

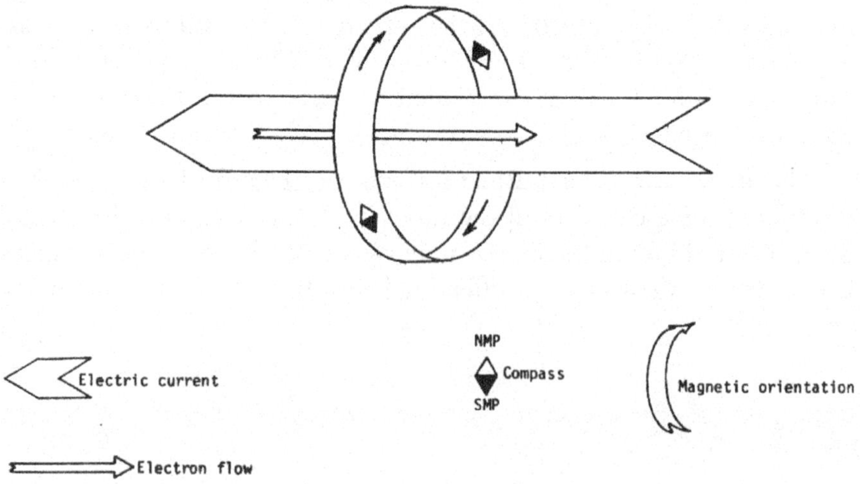

Magnetic Field Associated with an Electrical Flow

An electric current is always encircled by a magnetic field. By convention the direction of the electric current is opposite to the motion of the electrons contained in that current. For a moving electron the magnetic field is directed such that, if the electron flow follows the thumb of the *left hand*, the north magnetic pole of the magnetic field created by the electron flow is orientated around the motion in the direction of the curled left fingers.

The circular magnetic field surrounding the axial electrical discharge was also responsible for the production of much electromagnetic radiation within the plenum. Radiation would be emitted as ions and electrons were forced to spiral around the curved magnetic field lines of the magnetic column. This radiation process is called Bremsstrahlung, or braking radiation. It is emitted whenever charged particles are retarded (decelerated). By radiating, the particles lose energy to the surrounding gas. An enormous glowing gas cloud soon surrounded the arc- an afterglow- like the great ion trails left by large meteors.

As they lose energy, the spiraling particles move with smaller and smaller radii around the magnetic field line. By emitting braking radiation, motion across the magnetic field is greatly reduced. Since motion along the magnetic field is unaffected by the presence of the field, it eventually dominates. Thus the gases surrounding the discharge tended to flow around the magnetic column.

The same radiation as described here is used by biologists to mutate rapidly growing species such as Drosophila, the common fruit fly. In Solaria Binaria, radiation was abundant precisely when needed to explain periods of great biological change.

As the binary extended, the arc weakened, in consequence of which the magnetism around the arc declined. With diminished magnetism, more and more material could exist in a non-magnetized state. Although this material had all been magnetized earlier, the magnetization would begin to decay as soon as the surroundings allowed it. The decay of magnetized Earth rocks has been documented by Nagata. There is probably a positive correlation between the ease of magnetization of a material and the duration of its remanence. So, as the magnetic tube weakened, the now orbiting bodies would lose their magnetism differentially depending upon their composition. Earth's decaying magnetism of today is a remnant of its stay in the magnetic tube.

In the opaque plenum, and because of their different revolutional phases, no planet could be observed from another planet. Even when the plenum was clearing, the planets so positioned would be difficult to discern. They may not have been perceived until the Age of Jovea (around 6000 years before present) when, freed from the tube, their dim radiance could be isolated against a darkened sky.

The magnetic tube, unlike the arc and the sac, was not directly observable, and so could not enter into human awareness, even in the final days of Solaria Binaria. Its presence, nevertheless, allowed later men to see the arc. Its grasp upon the gases amidst the revolving planets was instrumental in converting the energy moving away from the arc into a huge visible column of "flaming, twisting" light -- again a possible source of serpentine imagery in early symbolism.

The magnetic tube may have played a part in generating cosmic sounds. Archaic Greek philosophers, especially the Pythagoreans, employed the phrase "music of the spheres" to designate what has since been regarded as an unreal belief in celestial and planetary sound. That the violent forces within the tube would have emitted acoustical waves is unquestionable. In the manner of whistling atmospherics (Hines), such sound would be trapped by the magnetic field and propagated along the magnetic tube.

In the late times of the tube, when celestial bodies could be distinguished visually, the sound might have inspired the Pythagoreans to the invention of their sacred musical scale, which was also related to their sacred theory of numbers - both sound and numbers constituting theophanies. The sounds might first have been involved in the earliest sacred music. The magnetic tube worked its wonders by an invisible hand.

CHAPTER SEVEN

RADIANT GENESIS

The physical history of Solaria Binaria may be divided into three major periods according to the intensity of quantavolution occurring: a primary period of violent changes and rapid development, extending perhaps to a quarter of a million years; a secondary period of relative balance among the elements within the system, extending almost to the present; and a shorter tertiary period of system breakdown, when Super Uranus, the planets, the sac and plenum, and the electrical arc with its magnetic tube underwent abrupt transformations.

A biosphere was generated during the primary period and produced its main forms. That is, there was first a time of radiant genesis, a proto-zoic stage, followed by a time of the escalation of basic biological types, a palaeo-zoic stage. Then occurs a meso-zoic period of formal and ambient stability, which coincides with the secondary period of relative balance in physical history.

The prevailing theory among scientists conjectures that a sequence of chance chemical combinations occurring over time produces the "self-replicating molecule" deoxyribonucleic acid (DNA).

In Solaria Binaria, the sac is the vat of chemical evolution. Its gases are hydrogen-rich but contain, by inheritance from the body of Super Sun, all simple ingredients found in life forms. The energy sources which catalyze the process are ultraviolet radiation, electric discharges (lightning bolts), and ionizing particles (from cosmic rays or radioactivity).

The composition of the plenum gases varied significantly over time, though for a long time the gas density remained fairly constant. Electrical forces produced electrophoresis among the electrified atoms throughout the system; in electrified gas mixtures the components apportion themselves within the mixture in relation to their ionization potentials.

The effect of the discharge was to reapportion the plenum gas mixture, changing the local percentage of hydrogen relative to the heavier atoms. This would effect greater efficiency in producing organic compounds in certain regions within the plenum (Dayhoff et al).

After the nova the plenum occupied a large volume; it was honeycombed with variously electrified domains producing a state of great electrical disequilibrium. Held together by pervasive cosmic electrical pressure, the gases of the plenum assumed the smallest volume consistent with their charge density. In reaction to the nova, electric flow within the plenum worked to equalize charge densities within the sac, while maintaining an outward radial gradient of increasing charge density in concession to the external demand from the continuing cosmic transaction.

The result was an initial implosion of the sac, as charges were redistributed, superposed upon a much slower expansion of both the sac and the rest of the system as galactic charge accumulated. Consequently, over most of their history, the Earth and the other primitive planets were immersed in a dense plenum of gases which was opaque to radiation; this gas was at least as dense as the present atmosphere at the Earth's surface.

The nutritive soup from which living forms emerged was not wholly the primitive vapors of Earth (conventionally the oceans and atmosphere) but the total surface of the planets and the volume of the sac. Appropriate temperatures were available in most of this volume within thousands of years of the nova of Super Sun.

If we consider only that portion of the plenum which enveloped the planetary region (a cylinder 35 gigameters long by 100 megameters diameter) we have a reactor volume which is sixty million times the combined volume of the Earth's atmosphere and oceans, in which life otherwise is believed to have been generated. The energy source for the plenum was the electric arc. The early arc may have liberated about 10^{23} watts to the plenum compared with 3×10^{13} watts received as ultraviolet radiation by the Earth's atmosphere, or with 3×10^{11} watts received as lightning discharges (see Chalmers for data).

If Solaria's plenum at the edge of the central flow zone is compared with the outer surface of the Earth's atmosphere with regard to energy density, Solaria's plenum will have had an advantage by a factor of 500,000.

At the other extreme, if the energy is spread throughout the entire volume of both reactors, the advantage in energy density still is with Solaria fifty- fold.

Should the initial photolysis not be the rate-controlling step, then the immense volume factor greatly favors a more rapid biosynthesis in the plenum than supposedly occurred in the Earth's atmosphere and oceans aeons ago. Furthermore, a highly electric environment may speed up generation time, and therefore the intergenerational opportunities for mutation. As we see it, the plenum was an ideal reactor in which living systems could be synthesized and sustained.

Evidence that the generative environment was highly magnetic can be inferred from the sensitivity of many living organisms to magnetism. Both animal and plant life respond to strong magnetic fields (above 100 milliteslas), showing modified growth or behavior (Kolin).

To summarize, in regard to the time available for the origin and development of species, the Solaria (SB) model is 2000 times less "effective" than the Evolutionary (E) model. With respect to the volume of the life-generating region, the SB model is six million times more effective. Considering the energy density, SB is five hundred thousand times more effective following the establishment of the binary arc. Actually, before its establishment, the nova phase, lasting for months, would have organized the Solaria Binaria system to the equivalent stage of two billion years (2 aeons) of conventionally ascribed Earth history. Hence the SB model, assessing energy density, would well exceed by a millionfold the E model. Since mutagens work upon mutable forms, and branching of species is an exponential concept, the effectiveness of Solaria Binaria in quantavoluting life is multiplied again by the volume of the life-generating region. So, even on a short time schedule, Solaria Binaria appears to be millions of times more capable of producing the species of today.

Still, even this might not be enough to originate and develop the species. The first stages of life are of such low probability, and the alter stages of higher but still low probability, that a "guiding factor in life development " must yet be sought.

The model of Solaria Binaria might only serve to supersede conventional theory of the evolutionary process, and not to discount it and provide an alternative positive theory, were it not for its electrical features. Life begins by microscopically mimicking its gigantic progenitor, the sac. It has no choice. Every atom, in endeavoring to hold its electrons or gain others, seeks to surround itself with the smallest and densest complete electrical perimeter possible. This is usually an octet of electrons. Whenever necessary, atoms aggregate into molecules where a compromise sharing of electrons will lead to a higher density electrical perimeter. From here the molecules proceed to more complicated systems that ultimately come alive.

The concept of life therefore is an extension of the concept of the "cavity" with which our book began. Life is a way of gaining, hoarding, and begrudgingly doling out electricity. In countless numbers organic molecules determinedly build themselves micro-sacs of chemicals in reaction to electric gradients, capture raw materials, manufacture compounds within the sacs, fire themselves with ever accumulating electric charge, until, incapable of continuing this process without bursting their sacs, they force out unused parts. Usually these are excreta.

In critical cases, they are replications of themselves -- if not exactly so, then in fundamentally similar ways. No cell divides itself in mirror-like fashion, uniformly, in the beginning. But every deviant is a candidate for the first exact mitosis.

The step from excreta to exact reproduction is critical. The sac of organic electrical activity is not "intelligent" except by human prejudices, ex post facto. But the sac can most efficiently -- effectively and reliably -- excrete if it separates its ingredients on the binary principle of "one for you and one for me". Least change, least imbalance, and therefore longer life ensure if the sac polarizes uniformly prior to excretion, setting half of its contents opposite the other half and splitting itself down the middle, closing the gap at the instant of its division.

Excretion becomes reproduction.

Sacs that thus form cells which divide offer more chances of survival and conquest of space by numbers than sacs that either hold their accretions until they burst or bifurcate inequitably from an electrical standpoint, thereupon having to internally reorganize their electrical accommodation upon every mitosis. One notes the terrific speed with which life can develop and reproduce under rules of uniform mitosis. Within a few thousand years the plenum might be filled with such cells. Indeed, perhaps large areas were filled with them.

Life's arrangement of electrical signals is perhaps its chief embedded characteristic. "Electrical potentials occur in all cells studies thus far, although their biological importance is recognized in only a few cases" ("Cell and Cell Division", Ency. Brit.). The surface of cells is negatively charged. The cell membranes are 6 to 10 nanometers thick and are highly resistant electrically (from 1,000 to 10,000 ohm/cm 2). They produce voltage gradients which drive the biological functions (as noted ahead) and produce a cell interior that is more highly negatively charged than the surface layer of the cell. That cells are so electrically arranged is understandable when one considers charged cells in a charged universe. In metaphorical language, the overall picture of the cell, and the image of the primordial cell, then, is one in which a peculiar combination of chemical compounds survives by erecting an electrical screen to admit nutrients and to repel destructive invaders, then organize its internal components to sustain itself and to resist random escape from the community.

Several varieties of cell growth and transformation are observable. The "main" type of self-duplication ensues as a permissible, organized, collective escape, or excretion, providing for the maintenance of a complete defense system. Cell division would operate by an electrical signal system. The members are an electric grid (as in a vacuum tube), and acts as a gatekeeper among the elements in and surrounding the cell and during mitosis.

Cells make macro-molecules, including genetic molecules, which do not exist elsewhere in nature and are not allowed exit through the cell membrane. Inasmuch as macro-molecules are concentrators of electricity, this synthesis permits the cell to sustain longer than otherwise would be possible its quest for additional electrical charge. The cell thus builds a higher concentration of charge than is available elsewhere in the plenum mixture. This process is the essence of metabolism.

Metabolism concentrates electricity in the macro-molecules, thus depleting of its nutrients the medium trapped in the cell. (The analogies of cell as sac and of nutritive medium as plenum are close and possible homologous.) The cell responds by excretion of water, ions and gases (by-products) and ingestion of electron-rich nutrients.

Strain is imposed upon the cell membrane, for it must both contain the increased material and at the same time defend the cell against penetration by electron-deficient atoms and molecules. The membrane signals the cell nucleus concerning an imminent site of charge deficiency and leaking. Then the genetic macromolecules of the cell, which are the only ones capable of dividing themselves more or less equally, and have been so doing since their last episode of cell division, respond to the signal of impending disaster by completing their synthesis, and by lining up on the two sides of a perimeter membrane that is being electrically trenched through the nucleus at the future site of fission. Actually, the division line-up is provoked by an electric polarization of opposed centrioles, each representing a focus of peak negative charge on the edge of the nucleus.

Midway between the two centrioles, the newly forming perimeter constitutes an electron- poor trench. Following the genetic molecules, the other materials of the cell are drawn electrically to flow in equal amounts to either side of the perimeter-to-be, pursuing the two centrioles. By contrast the cellular material that is to constitute the cell wall itself flows into the trench from both sides. Thus, without breeching its old perimeter membrane, the cell has doubled its surface and has divided. Electrical forces move the two new cells apart. Never are two cell membrane in contact even in a densely packed tissue. Some 15-20 nanometers of intercellular space, filled with a sugary fluid, separate them.

From the self-reproducing cell to the hominid of a few thousand years ago requires passing by many landmarks in the organization of life.

Close to the solar nova and birth of Solaria Binaria at the beginning of the Period of Radiant Genesis, one may position groups of critical developments: the provision from solar debris of chemicals and transmutations in the plenum; and pre-biotic organic molecules (amino acids, sugars, nitrogen bases, plus other compounds).

Cell membranes, left- handed symmetry of organic macromolecules, proto-enzymes, porphyrins and nucleotides - these developments would readily follow. The cell probably took in the latter three constituents after proto-proteins had been formed independently in the plenum. Some cells, instead of dying, began to engage in mitosis, whereupon self-duplication, as described here, would soon follow.

Large cells would ingest small cells, or form around them, performing two types of action: digestion, the beginning of animal behavior, with the breakdown of the electrical defences of the smaller cells, and in other cases the formation of cell colonies using the membrane of the host cell as a supermembrane or skin of the smaller internal cell or cells. Large cell colonies would float in the magnetic tube and, later on, settle upon solid bodies.

From the development of the cell, the mode of basic change in life forms ever thereafter can be surmised. Time after time it happens that some portion of the excreta of the organism is retained within the sac of the colony and supplied with the coded electrical signals that connect with the master genetic material so that its descendant in the next generation can draw upon its experience and existence. The developing special organ excretes within the organism and returns signals to make demands, denote satiety and share directiveness in the behavior of the full organism.

The plenum of Solaria Binaria was the creator, cradle, and mutagen of life. The broad sculptures of plants and animals were completed during the first half of its existence. If fossils represent the basic variety of life, the phyla and the orders came into being then. No new general forms have originated in recent times (Brough). Despite great waves of extinction, slightly over one million living species are named today. The fossil record should show millions of ancestral species to provide the present number, but in fact shows only about one hundred thousand species. This contrast has excited comment: why were large changes peculiar to early existence; why were small changes more common in recent times (ibid.)?

Set up in this manner, the questions seem to accept answers from Solaria Binaria theory. The plenum promoted creation initially, as would be expected, promoted it less when the binary was stabilized, and became quite destructive and conservative as it exponentially decayed and collapsed.

The agents of these change may be identified. The first period provided an immense number of prototypes and access to abundant nutrients, so testing their viability (Ayala). The second period provided a stable environment of abundant nutrients but an end to the easy method of forming combinations. Further, the more distinctive and specialized the species, the less likely its electrical transformation would eventuate in new designs of life.

In the final period, environmental disasters extinguished many species, but also promoted very many, already genetically deviant individuals to the status of families, genera and species.

To acknowledge that a great many of these lesser, less creative designs have emerged in the later history of Solaria Binaria requires a theory of genetic realization. The genetic material can carry far more instructions for the construction and behavior of any organism than are required at any given time (Ayala). Under lower (but higher than present) solar system quantavolutionary conditions, suppressed instructions can be triggered. It is conceivable that every living species carries in its genetic code instructions for metamorphosis (monsterism). Cosmic rays, nuclear explosions, radiation fall-out, meteoroids, electromagnetic typhoons, encounters of Earth with large bodies (comet, meteoroid), viral epidemics, and "silent" significant changes in electrical discharges within Solaria Binaria and the Solar System may be the means of suddenly extinguishing some genetic instructions and releasing others, quantavoluting a species into a similar but substantially modified species that is altered anatomically, physiologically and behaviorally.

A final short period follows the period of evolution; it is an epoch of explosive quantavolution that comes down to the present. It witnesses catastrophes of life forms, quantavolution through genetic realization, and the rise of Homo sapiens. On the physical side, it carries the record of the destruction of Solaria Binaria and the advent of the Solar System. Though short, this period contains the full human experience.

CHAPTER EIGHT

DESTRUCTION OF THE SOLAR BINARY - INSTABILITY OF SUPER URANUS

A number of peoples claim that the primeval chaos was present before the creation. It was a plenum - dark (compared with what followed), uniform, dense, and housed a Demiurge who had not yet acted and a world of things and beings that were potentially activatable. The Ngadin Dayak, a people of Borneo, insist that "at the beginning, the cosmic totality was undivided in the mouth of the coiled water-snake", possibly referring both to the togetherness of the chaos and the omnipresence of the electrical axis mundi. In the Hindu Vedas, Dyaus-Pitr (Dyava), "Father Sky", can be identified with the age of first man and an unbroken plenum. He married Prthivi, Earth. The world was dark and asleep, "says" Manu (a Hindu Noah) until the Great Demiurge "appeared to scatter the shades of darkness".

Coelus, or "Heaven", was the most ancient Latin god of the sky. His name means "covering". Ouranos was "Heaven" and the first god of Greek legend; this Heaven was at first a calm and settled person, married to Mother Earth, Gaea. In China, The Chinese legend pictures Heaven as T'ien, at first a marble- like ceiling, unbroken. P'an Ku, a creator god, began pounding on T'ien, breaking large chunks off with his hammer and chisel until the skies showed through. According to the Iroquois of north Eastern America, the Chief of Heaven was persuaded into marriage by "Fertile Earth" (Awenhai) and impregnated her by his breathing.

In many places, a theory of the Cosmic Egg is used in connection with the earliest god, who is Heaven; it explains how God and the World were born. Thus the Hindus asserted that a seed was laid and became the Golden Egg. The Cosmic Egg is often said to have existed from an age before it revealed itself. We construe from this that the earliest humans are present on Earth as the troubles of Solaria Binaria heighten and that they were newly human for a short time before Super Uranus, that is, the Cosmic Egg, appeared.

The Egg can take related forms: a shell, a seed etc. Fuliginous powers that break out of the Earth to assume living forms are a type of creation widely believed; usually an external force inseminates or provokes the Earth, as in the Egyptian image of Nut and Geb, Heaven copulating with Earth.

Hesiod's version of the Greek creation myth has Ouranos or Heaven squeezing down upon Mother Earth, oppressing her until she cries out in agony. We interpret this suffocation of Gaia as an increasingly disturbed atmosphere, with many extinctions and quantavolutions in the biosphere.

Ouranos goes increasingly mad, taking up his children and hurling them beneath the Earth, Gaea in desperation urges her brood to revolt against the Father. We interpret his sons plunging to Earth as a bombardment by heavy meteorites, released into the plenum by the "unsettled" Super Uranus and encountering the Earth.

We assume that primordial observations gave rise to all of these legends. As part of the continuing emergence of life and intelligence from the chaos befalling Solaria Binaria, the Cosmic Egg myth is imagery for the vision of the first great "sun", Super Uranus, as it emerges from the Heavenly gloom. It emerges shortly after human self-awareness, and in a time of troubles for mankind. This Uranian Age was, both in legend and in our astronomical theory, a time of disturbances.

As is typical of evolving binary systems, the principals, here Sun and Super Uranus, move apart with time. During the phase when a strong electrical current flowed between the two stars the components remained relatively close together, while the whole system charged negatively in transaction with the galaxy.

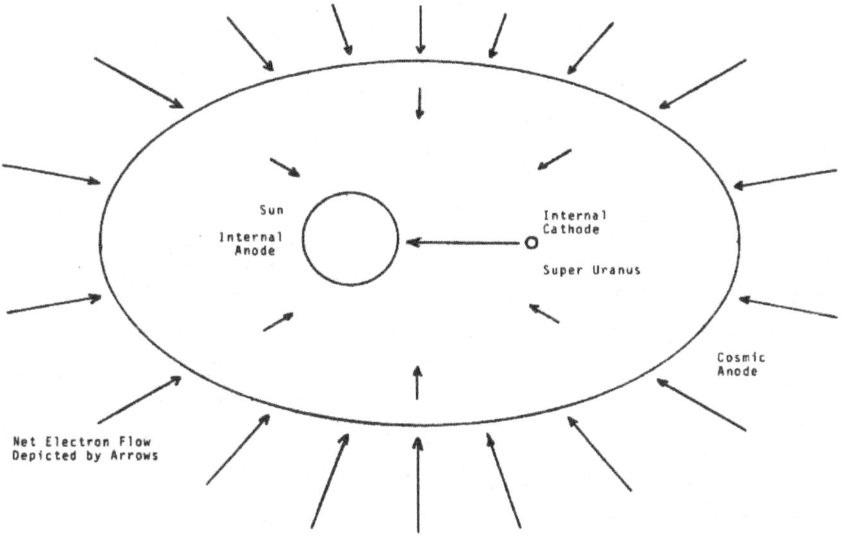

Transaction Between Solaria Binaria and the Cosmos: Dense Plenum Phase

Originally Solaria Binaria transacted with the Cosmos as a unit. Electrons flowed from the Galaxy into the sac of the binary, liberating energy to the gases along the perimeter of the plenum. Charges so delivered to the binary as a whole were subsequently redistributed among the parts of the enveloped binary producing secondary energy releases as they impinged upon some particular component (star or planet) within the dense plenum.

This charging drives the Sun and Super Uranus apart so that the current flowing between them weakens and from time to time falters; the two become more isolated electrically within the ever-diluting gases of the plenum. The importance of the electrical axis of the binary diminishes.

While the two bodies, the Sun and Super Uranus, were transacting vigorously, they were quite luminous, though not sufficiently bright to be perceived as celestial bodies. As the arc between them began to falter, they and the arc remained luminous, but the latter less so than before.

Though the central arc was sputtering, the surrounding gases in the magnetic tube sustained an afterglow and so were not always extinguished between discharges. But, as the arc decayed further, the discharges became less frequent, so that eventually even a long afterglow could not maintain continuous luminosity throughout the magnetic tube. At times the sky briefly darkened. This was the first light and darkness experienced by humans, who may then have deduced the concept of contraries, good-bad, yin-yang, or light-darkness, the basis for religious dualism and human thinking processes.

Relative time may have been invented in the period of Super Uranian instability. If the arc pulsed regularly, the earliest humans would have responded to it, first subliminally and later consciously. Possibly, when the glow of the arc darkened and lightened perceptibly and in rhythm, a notion of periodicity would be imparted to humans, they would have a clock. The experience of the first abrupt darkness would be terrible, the unexpected gloom of even a minute, provoking fears of a shutdown of light. If the pulsing were regular, but interruptions occurred, terror would ensure with the interruptions. Frenzies of fear attending eclipses, historically and recently (Corliss), may be traceable to primordial experiences with a degenerating axis. The birth of Super Uranus, emerging from the plenum sky, would be both terrifying and reassuring.

Whereas before, the two stars transacted internally to produce the arc, while the opaque plenum transacted at its perimeter with the Cosmos, now each star transacted separately with the galaxy through the thinning plenum. This new galactic connection could occur because the plenum density had fallen as it expanded and both bodies were still far from electric equilibrium with their galactic environment.

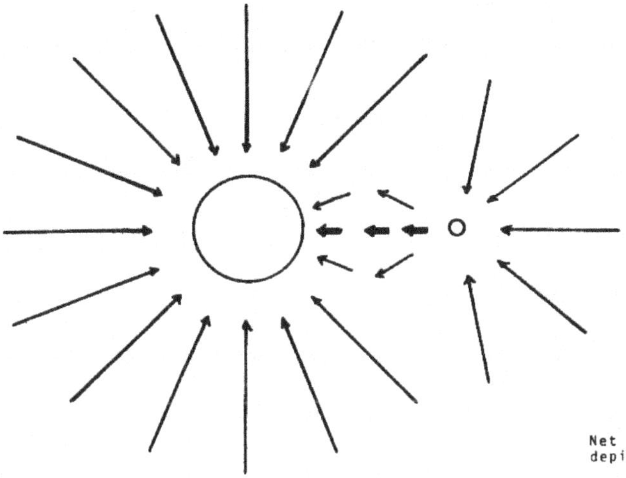

Solaria Binaria as the Plenum Thins and the Stars Separate

Late in the development of Solaria Binaria the gases of the plenum had been thinned to such an extent that the transaction between the Cosmos and the binary ceased to liberate the major part of its energy at the binary's perimeter. Thereafter the cosmic transaction deposited energy individually at the two stars . The inter-star transaction (the electric arc) continued for some time after each of the stars attained a separate connection with the Galaxy.

Thereupon, Solaria Binaria would be observed to be a semi-detached binary star system from the vantage point of another star system.

To the pre-nova turbulence of gases and bombardment, a duration of 1000 years may be assigned before the human creation is connected with the turbulence; subsequently another 2000 years is assigned before the climactic nova of Super Uranus.

The outer layers of Super Uranus and its space charge sheath were the first places to react to instability. At intervals, a shell of material expanded explosively away from Super Uranus. To the outside observer this small star had become a nova of low intensity. Weak outbursts are not uncommon in under-luminous close binary systems.

Super Uranus can have erupted many times. The shock of its recurrent explosions propagated through the plenum, damaging the planets nestled within it and electrically thinning it further. Not all of the ejecta was gaseous. Fragments circulated within the system for a time, encountering explosively other bodies differently charged. Some fragments fell back upon Super Uranus, which was diminishing in brightness and may be also in size because of its outbursts.

The rude disturbance of the hitherto peaceful atmosphere of the Earth was noted fearfully by the rapidly developing human culture that was spreading throughout the World. Men perceived the heavens to be alive and exercising a control over earthly affairs. Heaven both inflamed and frustrated man's desire which seized him in the course of his very creation - to control himself and his environment.

Chapter Nine

QUANTAVOLUTION OF THE BIOSPHERE: HOMO SAPIENS

The difference between the certainly catastrophic age of radiant genesis and the catastrophic recent record of explosive quantavolution clearly rests in the extremely powerful and rich environment of the first period and its vast domain of the plenum. The period of collapse of Solaria Binaria was incomparably poorer in genetic capabilities; to extinguish, yes; to capacitate, also yes; to create, no. The many millions of mutations and environmental changes occasioned by the instability and destruction of the system were paltry by comparison with the possibilities of the first period.

Therefore, when one approaches the subject of the genesis of Homo sapiens, one need not expect grand changes of a bio-physiological type; these do not exist. With protein "chains" as the basis of comparison, humans and chimpanzees "share more than 99% of their genetic material" (Washburn, p203). A comparison of the earliest fossils of hominids with the similar parts of modern humans does not demand an acknowledgment that the two are of distinct species; and judging from remains alone, the hominid may have equal or greater capabilities than the modern human. For example, the brain case of hominids, which may contain 500 cc. is not so small theoretically as to preclude intellectual competition with a modern human brain. Though larger by far on the average, modern mankind does offer braincases that, while intellectually competent, are akin to the hominid's in relative size. This is quite apart from the presently unresolvable issues of the intensity of convolution of the brain and the percent age of brain tissue ordinarily utilized.

The brain originated in a world of lower electrical levels and greater electrical differences. There may be a functional problem today in a world where electrical levels are higher and electrical differences much diminished. The brain was possibly originally more stable, that is, instinctive, perpetuating the less anxious hominid.

An environmental de-electrification would seem to occur as the Earth's interior increased its supply of electrons (relative to its cosmic surroundings) simply by the steady accumulation of charge. In a changed environment, the repetitive correlating signals that constitute a large part of the exchange between the two hemispheres of the brain would encounter increased environmentally induced resistance; so they would bunch up and interfere with one another. That is, fewer transmission lines would be available to the same number of messages.

The messages between the brain hemispheres propagate relatively slowly, by direct current through chemoelectrical diffusion, so to reflect a slightly diminished electrical constant, enough to furthermore "encourage" crowding of signals and a more frequent desynchronization. The effect would be both delay and confusion - delay in microseconds in assessing a neural trigger for an information or command bit, and confusion in overburdening the channels with combined but incompletely co-ordinated messages.

The simplest change would be a general constraint upon instinct. Instinct is a non-learned activity and response, unfettered by self-awareness. Homo sapiens is the least instinctive of all animals, hence the least predictive and most responsive to internalized planning. Very many, perhaps all, human actions and physiological processes can be internally constrained or modified unconsciously (psychosomatism) or consciously. The extraordinary achievements of Homo sapiens, it is argued, are entirely due to the operations of an instinctual incapacity upon an otherwise normal primate constitution.

This instinctual incapacity is closely connected with and may have given rise to the generalized anxiety or fear characteristic of humans, especially "intelligent" humans. Human fear, resting on top of animal fear, was originally fear of oneself, fear of the inability to act and react instinctively under conditions of the mental division of the self into several differently aware parts.

The transformation of hominid to human with respect to instinct delay, which leads to self-awareness, which then promptly adduces symbolism, ideology and recall, is most likely to have been accomplished by contradictory pressures - one to diminish instinctive response and the other to increase response. Together they produce continuing anxiety and a number of mechanisms to cope with it.

Some of the pressure to diminish instinctive response may be attributed to an increase in electrical resistance between the two hemispheres of the brain, distributed throughout the corpus callosum, the large membrane occurring between the two hemispheres. This membrane would increase its resistance to the passage of messages between the right and the left brains, which are in fact electrified and responsive to changes in the external and internal environments.

Signals that must "wait" and may get out of phase would necessitate momentary verification of otherwise instinctual responses, a delayed reaction, and even conflict and aborted decisions. This is enough to set up the unique pattern of human behavior in an otherwise pedestrian mammal.

Thereupon two paramount qualities of the human mind would result; the need to think before acting, and the analogizing of experiences and events, leading to synthetic combinations of all types. In addition, we admit the possibility of a change in the functioning of hormonal glands, such as the adrenal cortex. A continuously higher level of secretion and induced stress - a new constant might have been provoked by the disasters of the time of humanization and/or by a new, stronger and persisting electro- chemical stimulation. The brain would be permanently stressed towards anxiety and action. Taken with the decline in the correlation of the hemispheres, this contradictory stress would further humanize the person with the evermore-poignant auto-instructions to "look before you leap" and that "he who hesitates is lost."

Promptly there would emerge a conception of the self, a continuous fear of loss of self-control developing out of the need to compromise with oneself, an aggression against those who provoke difficult decisions or restrictions of the self-conflict or who "cause one to have to think," and the need to talk to oneself (one's other self), which leads quickly to talking to others to engage them into talking to one's self - which leads in turn to talking to "the most important people in the world": the anthropomorphized gods. The self would project its hopes and fears to the external world, but especially and exactly to those features of the external world from which the most impressive experiences emanate - the heavens.

Thereupon the human mind is structured and in place. The devising of culture was practically instant. Words, operations and thoughts establish social contact on a level unknown to "hominids", and a "social contract" comes into being. Society helps people to talk to themselves; people talk to themselves through other people.

The social process, the instant culture, is not only formed of the present. It accrues memories. It recalls. It is obsessed with its own creation simply because it is so unbelievable and dramatic (traumatic). Since this scenario was enacted only 260 memorial generations ago (de Grazia, 1981), the transmission of some valid and reliable information in decipherable form need not be surprising.

The view here conforms to the theory of genetic realization. It may be maintained that hominid is as old as the end of the period of radiant genesis; further it may be maintained that hominid had a genetic potential for becoming the modern human. A large change is not necessary to differentiate the human from the hominid.

CHAPTER TEN

NOVA OF SUPER URANUS & EJECTION OF THE MOON

When the heavens were broken open, as by P'an Ku, the Chinese creator god, Super Uranus appeared in the north, immense and egg-shaped, probably resembling a giant eye, too, atop the sky. At first the white of albumen, it became yolk-red, and radiated heat. It was probably the primordial light of the beginning lines of the Hebrew Genesis, where the present celestial bodies appeared only on the fourth day of creation.

To the south, less luminous because it was much more distant, was the Sun. There were now three sources of heat, the magnetic tube (powered by the arc), the Sun, and most prominent of all, Super Uranus. Depending upon the earthly observer's location, either Super Uranus or the Sun could be discerned through the thinning gases. In general, northern observers glimpsed Super Uranus, southerners saw the dimmer but larger Sun.

During its period of instability Super Uranus erupted regularly, outbursting charged matter into space. This outburst would not normally escape from the domain of the star which generated it; most of it would in time be reabsorbed into the star, returning the "released" charge. Charged debris falling back could help induce conditions for another electric compression and outburst. And so the star erupted cyclically.

In the Age of Urania each body might be said to possess an autonomous sac and plenum, immersed in the now diluted plenum of the whole system.

The plenum of Solaria had by this time become so tenuous that the individual bodies had established around themselves electrospheres - regions of charges, gases and, from time to time, solid matter. These spheres, or regional environments, were to their central bodies what the plenum had been to the system, in that they defined the limit of the body's influence upon nearby matter and charges.

The resorption of erupted pieces occurs so long as they do not exceed a certain critical size. Judging by today's Solar System, this would seem to be about 22 kilometers in diameter or a volume of about 6 000 cubic kilometers. Pieces that could not be resorbed could become satellites of their parent body - as had Super Uranus and the primitive planets - but in Solaria they could be transferred from the realm of one body to another whenever the two electrical plena involved were contiguous. In these special circumstances escape was possible; the smaller sacs could leak gases and pieces into adjoining sacs. This is how Super Uranus bombarded the planets. Its outbursts electrically charged the Earth's sac and filled it with debris of diverse sizes.

In its climactic explosion, Super Uranus ejected a large chunk of its material down the magnetic tube towards the Sun. This element, to be termed Uranus Minor, was preceded and accompanied by gaseous blasts and water. A vast, brightly glowing space-charge sheath surrounded Uranus Minor as it hurtled towards the Earth.

At the time of the eruption the Earth is revolving around the arc, moving counter-clockwise (viewed from Super Uranus). The globe is oriented with Africa (the old north) facing the explosion; the magnetic poles lie on the rotational equator, Greenland leading and Antarctica following. At this time the Earth had a continental crust everywhere. The continents that survive today were bunched around Africa.

Uranus Minor, moving from Super Uranus towards the Sun, encounters the outside edge of the Earth (the Pacific side) in passing. An intense transaction occurs between the two. Electrical polarization distorts the shape of both bodies and their sacs, and the Earth's magnetic axis wrenches out of line, which causes the world to shudder. The sudden movement loosens part of the lithosphere; torrents of water (or ice) flow (or slide) across the surface. Fiery blasts strike the area which is now the West Central Pacific Ocean, opening massive craters, some deep enough to release mantle material previously thirty kilometers below the surface.

At perigee the transaction between Earth and Uranus Minor reaches a maximum. The crust on the side closest to the intruder cracks and fragments. Explosively, as much as half of the Earth's continental material rises into the sky, leaving exposed much of the upper mantle. This extraction initiates the reshaping of the lithosphere to produce the structure we study today.

Whereas before this event the entire Earth was topped by a thick granite layer, basalt was now exposed. In its brief encounter with the Earth, which we estimate to have lasted many hours, Uranus Minor peeled a deep swath of crust (and some upper mantle material) from the Central Pacific and to lesser depth from the great seamount area west of the Americas. Much of the shell of granitic crust was lost when the Earth encountered Uranus Minor. The Earth was in orbit around the electric arc at the time when Uranus Minor passed close by it.

The Earth wobbled eccentrically as mechanical, electric and magnetic forces acted upon it. Surrounding the wounded surface was a rampart of devastated granites. Within it were thousands of seamounts, unable quite to explode into the sky, and now frozen like pulled taffy. At its center was an abyssal plain where the surface of the Earth's mantle appeared scoured of its covering; it was then, and is now, the deepest basin of the Earth's surface. The blow-off was so great that it pulled the great central magnet of the Earth 436 kilometers towards it, making the shortest circle line of the Earth's magnetic field of today pass through the Society Islands. The Earth's center of gravity was pitched five kilometers towards the great Pacific depression (Baker, 1954, p5).

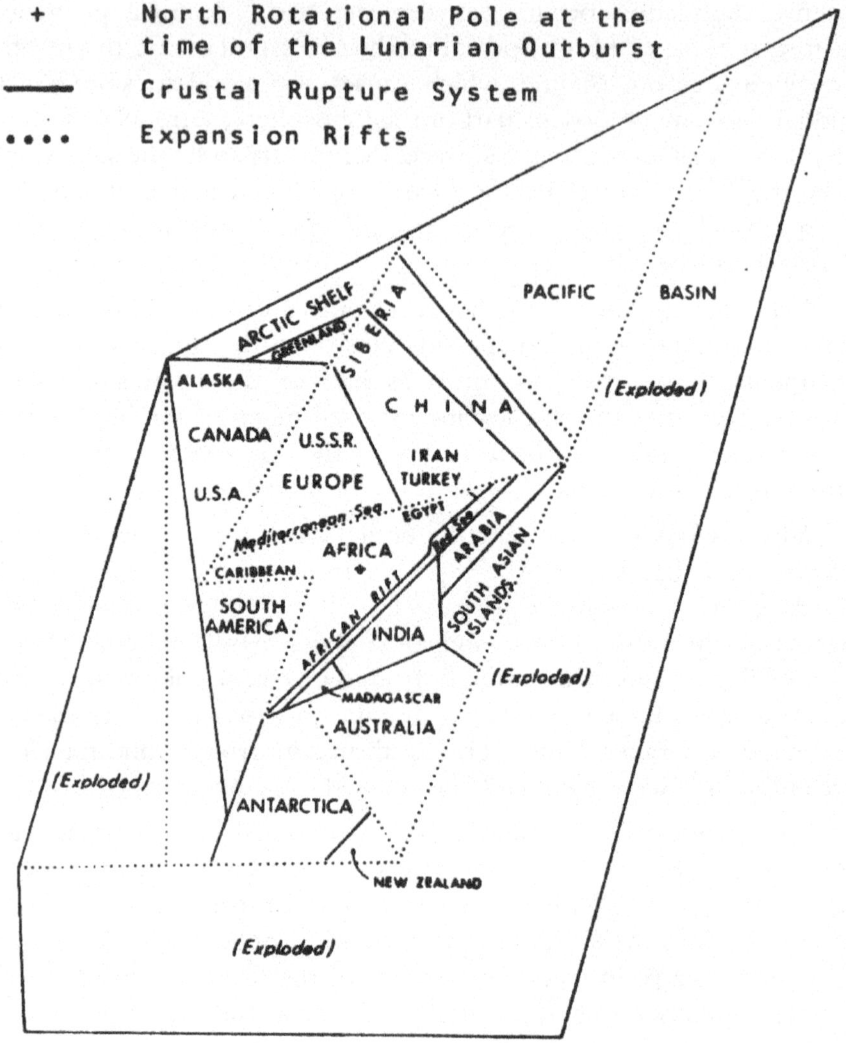

The Surviving Land from the Age of Urania

Prior to the eruption of Super Uranus, which hurled the large fragment Uranus Minor down the magnetic tube past the Earth , our planet was covered by a complete shell of granitic crust. Much of the crustal layer was lost when the earth encountered Uranus Minor. That crust which remains was once clustered around the ancient North rotational pole, which then always faced towards Super Uranus. It was ruptured and rifted by the close passage of Uranus Minor.

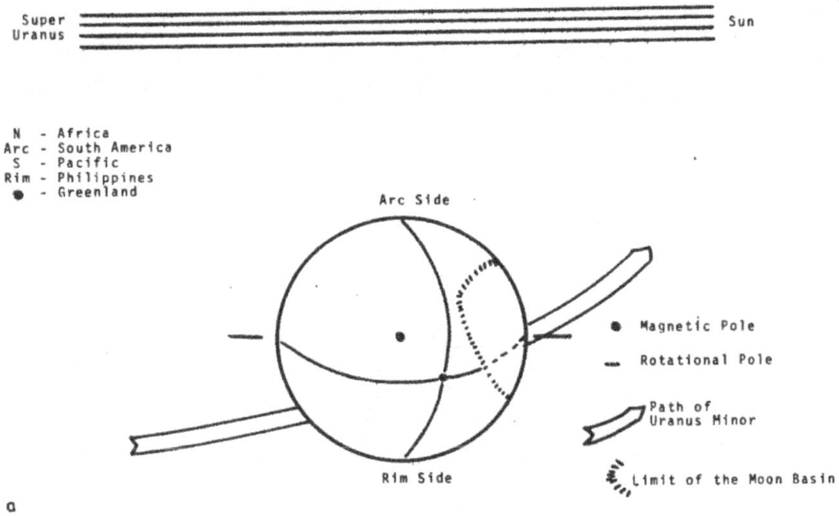

The Encounter of Uranus Minor with the Earth (1)

The Earth was in orbit around the electric arc at the time when Uranus Minor passed close by it. Then the Earth's surface was irrotational with respect to the arc so that one point on the crust (magnetic south) was always leading on the orbit (see above) and the point on the opposite side (magnetic north) was always trailing (see over page). Though the crustal arrangement of that time placed the lands differently, the lands on the present globe, which did not take up their present positions until after the catastrophe described here, would be oriented as indicated on figure 11 and 12 relative to the arc and the Sun. Uranus Minor met the Earth on the trailing side and departed on the leading side , tearing away crust and creating the Moon Basin where shown.

In this one brief event, the entire original Southern Hemisphere of the Earth's crust was electrically ejected into the sky. The unbonding of the crustal granite and mantle from the subsurface magma involved a large transfer of energy. If the inter-body transaction is translated into thermal terms, the heat would have been perhaps impossible for the Earth to support without vaporizing the biosphere and the globe itself.

However, the transaction consisted of a trade of material for electric charge. Uranus Minor, much more heavily charged, deposited charge upon the Earth. The new electrical energy was incorporated by the molecules of the Earth. Their internal (atomic) bonds were stretched. A very large amount of energy was required by the chemical bonds and supplied athermally by a huge column or front of lightning bolts blasting a swath into the sky during the pass-by.

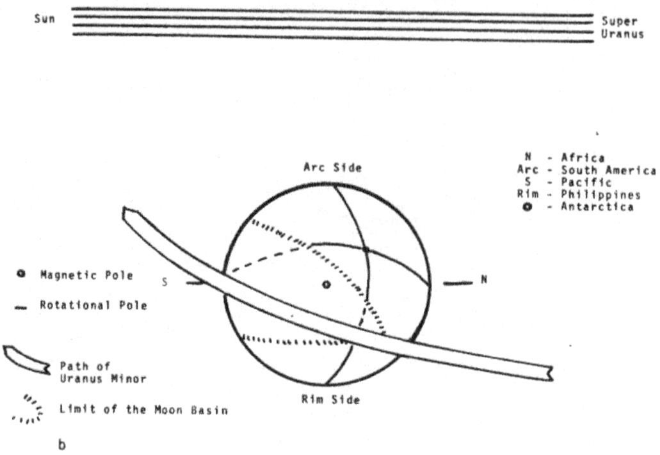

The Encounter of Uraus Minor with the Earth (2)

That heating which occurred was concentrated at the interface of crust and mantle and at the bottom of the moon basin. In adjusting its figure following the ejection, the remaining land mass fractured and the Earth expanded by about twenty per cent (de Grazia, 1981; see also Meservey, p611). This represented a radial expansion of nine per cent and a corresponding atomic expansion throughout much of the Earth.

The remainder of the continental mass that had covered the Earth fractured into the complex ocean-ridge and land-rift system viewed today. The separated blocks were electrically repelled and squeezed apart. They rafted speedily towards the Moon basin. Lava welled up from below the fissures and widened them. Thousands of new volcanoes were instantly activated.

The constellation of fractures exhibited in the world map (See p.232) probably occurred within a day's time (de Grazia, 1981; Manson)

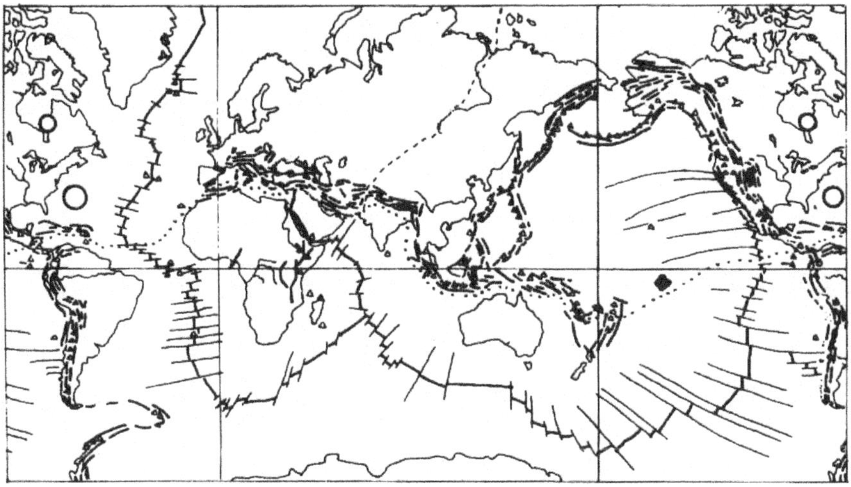

The fractured Surface of the Earth

Following the removal of the Moon from the bulk of the Earth by the action of the passing Uranus Minor, the surviving broken continental crust of the Earth shattered and rapidly scattered, taking up new positions around the crater surrounding the Moon Basin in what is today the Pacific Ocean deep. The troughs, ridges and faults of the current crust were sculptured as the Moon was torn from the Earth and as the Earth recoiled and recovered from that devastating encounter with Uranus Minor. The volcanic and mountainous rim around the Pacific Ocean was created in this same catastrophe. The most prominent astroblemes were also products of the Age of Urania, but not all of them were blasts accompanying the ejection of the Moon.

Possibly, certain minor fractures branched out or lengthened in the following months or years. Some fractures were not fully consummated, such as the African-Near East rift and the trans-Asian rift. Others have been covered in part by subsequent torques of the crust, as in the case of the San Andreas fault, which was buried in the "westward" movement of North America, or the Red Sea-Adriatic-Rhine Valley rift, which was partially overridden by the Alps. It will be noted, too, how the Atlantic Ocean crack probably shot out from an Arctic base, travelled swiftly but against resistance, and then branched off to circumnavigate the South Pacific area, sending four continents on their separate journeys: South America, Africa, Australia and Antarctica.

A deluge of water fell from Uranus Minor as it passed. These waters more than replaced the water carried away with the lost crust, fell into the hot fissures and onto the volcanoes blasted into existence at the passage. There the waters exploded, expediting further cracking of the continents.

A cubic kilometer of Earth's atmosphere at present contains ten thousand tons of water. If the Uranian deluges were precipitated by electrical activity of the Earth's electrosphere, only 540 tons of water per cubic kilometer would be required in order to achieve the oceanic levels that we estimate occurred in the Uranian Lunar periods. In the course of a millennium an annual rainfall of 1.2 meters depth would have descended upon the Earth's surface, a rainfall that a substantial section of the world's people enjoy today. This would be adequate to fill the basins up to the continental slopes, about half of the present ocean volume.

Thus the geography of the modern world was established: separated continents, ocean basins, global fractures and ridges, mountainous ramparts around the Pacific Basin and a global- circling welt (the old rotational equator), which forms the other great mountain chain of this planet.

The Mohorovicic discontinuity, which is found beneath the crust throughout the world, marks the level at which the ancient crust was served from the underlying solid. The continents continue to float, moving today ever so slowly, but only eleven and one-half thousand years ago that motion was initiated in hours and rapidly completed. Three thousand years later the continents were almost at rest and located close to where they are now found. They came to a halt because an electrical equilibrium had been established among them, at both their prows and their sterns (see Harrison, 1966). The ocean basins by then held water roughly to the base of the continental shelves.

Within the basin, the heavy heat from continuous mantle extrusions evaporated the waters, forming dense clouds that filled and rose above the abysses into the atmosphere above the continental blocks. The early continents, until eroded, were large buttes surrounded by the new paved basins located five thousand meters below the surviving land masses.

There are indications that the drop of five kilometers into the abyss from the continental shelves was known to the ancients. Ouranos, the Greek Super Uranus, cast his rebellious sons into Tartarus, "a gloomy place in the Underworld, which lies as far distant from the Earth as the Earth does from the sky; it would take a falling anvil nine days to reach its bottom"

(Hesiod). One notes the gloom (the dense clouds below the habitable plateaux), the position (below the human world), the precipitousness (the metaphor of an unimpeded falling object). Ancient sailors spoke of falling off the edge of the world (a fear also present in the modern child). The concept of hell, which John Locke said was so persistent that it must have represented some human experience, may have arisen from the era of the great chasms; hell is straight down, is burning, is fulminating, is sulphurous. There was, too, a world of the antipodes that could not be reached, possibly across the abyss (Dreyer, pages: 7, 37, 213, 220).

The disposition of the atmosphere is a crucial problem for our model. The atmosphere would have been sucked up under a gravitational model, and unquestionably much atmosphere, and also water, was lost in the eruption into space. However, the electrosphere was already operative, as indicated earlier, and was ionizing as well as electrically repulsing, and hence returning, the gases that sought to leave the Earth.

The reduction in atmospheric pressure was short-term but unquestionably fatal in a great many instances. Even at the antipode of the catastrophe, the air would have rushed towards the scene of the disaster. Some help would have been derived from counter-winds electrically repelled and driven to the antipode. The electrosphere, containing a mixture similar to, but richer than that of the weakened plenum, would have originated downdraughts at the antipode.

Assuming that the pre-Lunarian atmosphere was three times the present density at sea level and taking as the short-term extreme the habitat of people in the High Andes today, the atmospheric pressure might have been reduced to one-sixth for a short time (see Gray, pp63ff,hite, p763). This would not eradicate life.

The departing Uranus Minor is deflected slightly from its path by the Earth. It then crosses the binary axis so as to approach the Sun on the forward side (orbiting directly). Its passage by the Sun causes an electrical transaction which increases Uranus Minor's negative electric charge and ejects it into an orbit beyond the surviving binary pair. It now becomes the planet we know as Uranus, or possibly the planet Neptune.

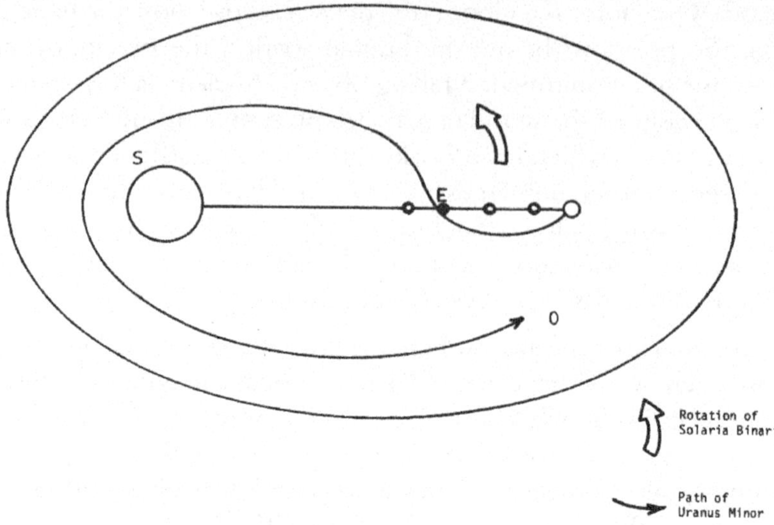

Fragmentation of Super Uranus

In schematic form the relocation of Uranus Minor is shown after its explosive ejection from the solar companion, which to that moment had been Super Uranus but then had become the smaller Super Saturn. The released fragment, Uranus Minor, first travelled sunwards along the magnetic tube, where it passed close by the Earth (E), tearing crust away and forming the Moon in the process. Thereafter, the still electronrich Uranus Minor moved into the vicinity of the Sun (S) before escaping into the outer regions of the system beyond the orbit of the new companion (0), where it is likely located today as one or the other of the two most distant major planets.

The granite and mantle material removed in the passage of Uranus Minor past the Earth is strewn along an arc between the retreating intruder and the gashed Earth, there mingling with ejects from Super Uranus and travelling with Uranus Minor. A portion of this debris escapes with Uranus Minor, but most of it, amounting to about one-fiftieth of the Earth's volume, is left behind in the space near Earth. For a time some of it fell back upon the Earth as stone and dust.

Fission of the Earth-Moon Pair

This simple diagram illustrates the minuteness of the Moon compared to the Earth's bulk, despite its having removed half of the Earth's crust when it departed.

The rest, partly molten, was assembled by electrical pressure into a rapidly cooling globule, minute, despite its having removed half of the Earth's crust when it departed.

A new sky god/goddess, the Moon, is born, child of Mother Earth, Aphrodite-Urania. She is worshipped after her father retires from Earth's view.

Eventually the Moon orbited the electrical axis, repelled by its excessive charge to a greater distance from the axis than the Earth. From its removal as a piece of the Earth to the present, the Moon can never have been free of the Earth - if it had escaped it would now be an inner planet of the Sun, on an independent orbit and far from the Earth. This leads us to conclude that so long as the Earth remained in the magnetic tube, the Moon remained close by.

It orbited then, as it does now, with the Earth. Though considerably closer to Earth than today, seen only in a daylight sky, the Moon was not a significant object in the sky. Likely it was larger than the disc of Super Saturn but it was incomparably fainter.

In the period of accretion, debris and lightning would be striking the Moon from the plane of the swath in large part. Far-flung cultures portray the goddess of the Moon as a spinner, the first spinner (de Grazia, 1981). In ancient spinning, as in its modern survivals, threads are held and fed from the one hand to the spindle held by the other hand (Suhr). The spindle grows fat and round. Such myths may represent the accretion of the Moon and its assumption of a globular shape (see Baker, 1954, p18). The process would have endured for generations and would have been most impressive at first, especially while the Moon was candescent. Afterwards, it may have been visible only on occasion during electrical discharges.

Physical and astronomical evidence is abundant on the manner and recency of the Earth's parturition and the birth of the Moon. Here we cite only salient examples from several scientific disciplines.

The Moon's overall density approximates the density of Earth's mantle material; so it is natural that many scientists have suggested some connection between the Moon's origin and the Earth's missing crust. The similarity between the chemical composition of the surface rocks of the Moon and the Earth enhances the believability of this hypothesis, especially with regard to the amount of the noble metals - gold, platinum, nickel, etc. (O'Keefe, 1973).

That the Moon rocks are more impoverished in the volatile elements (zinc, cadmium, lead etc.) than Earth rocks indicates that the Moon material has been subjected to more heating than has Earth material. The Moon rocks were formed under reducing conditions: ferrous iron is common on the Moon, while it is rare on the Earth, where oxidized ferric iron is found (Arnold). If the Moon agglomerates electrically in a depleted plenum all of these differences are explained. The lunar material is heated as it is wrenched off the Earth. In space it is dispersed into an oxygen-poor dilute gas, where it accretes electrically again, liberating heat. The forming Moon is unable to conduct this heat away efficiently (as had the Earth, which accreted in a much denser and more electrified plenum at an earlier time).

The Moon's internal structure testifies to the rapidity of its formation (Wood). Despite a high surface heat flow, the Moon's interior is relatively cool today (below 1300 K); the Moon seemingly accreted as a conglomerate, like stew chunks in sauce (de Grazia, 1981). Its sixty-five kilometers of anorthosite crust reveals that it was melted or, better, metamorphosed electrically, at the time of its agglomeration.

Its over three trillion craters of one meter or greater in diameter (Short, p48) show that it underwent extremely heavy electrical blasting and debris bombardment after its emplacement. As a result we are not surprised that some rocks returned to Earth from the Moon show strong and fairly stable remanent magnetization (Strangway et al.) despite the weakness of the lunar global magnetic field. If electrical events magnetized the Moon rocks their in-situ magnetizations should be quite disorganized, sporadic, and of varied strengths.

Several in-situ observations testify that the Moon's formation was very recent. Lunar samples do not match modern theoretical expectations about primordial planet composition (Wood, pp71- 5). The oxygen isotope ratio in lunar samples is identical to that in samples of terrestrial oxygen (Epstein and Taylor). The amount of Helium-4 (a product of radioactive decay) found in the Moon's rocks is exceptionally low (Heymann et. al.), indicative of the Moon's youth (Cook, 1972, p18).

Cracked crystalline surface rocks show evidence of shock metamorphism and rapid cooling (Douglas et al., 1970). The bombardment has been extensive and repeated, while some debris is of recent origin (Quaide et al.). Using the conventional time scale, recency means about one four-hundredth of the Moon's age; using our time scale it means very recently (de Grazia, 1981; Baker, 1954; Velikovsky, 1969). The thin lunar atmosphere is accumulating now; gases trapped when the Moon agglomerated are still escaping from orifices in its surface (Cook, 1972). Periodic eruptions are reported, notably in the craters Alphonsus and Plato, and within Schroter's Valley (Menzel et al., p229; Wilkins and Moore, p235, p263). Moonquakes, frequent though weak, may be taken as evidence that equilibrium has not yet been attained within the Moon and within the Earth-Moon system (Latham).

CHAPTER ELEVEN

THE GOLDEN AGE & NOVA OF SUPER SATURN

When the skies opened after the lunar disaster, the new great god, Super Saturn, was visible dimly, through the clouds, to Earth's inhabitants. They had already been "religious" for millennia and might readily once more identify the sky objects with human forms and actions and project their hopes and fears upon the heavenly objects newly visible. Magic, spiritualism and animism, which have been regarded sometimes as substitutes for, and predecessors of, celestial religion, were derivative accompaniments of the human preoccupation with celestial behavior; they were forms of homeopathic social medicine for the "great disease".

Thousands of years were required for the reconstruction of the Earth's surface and the recovery of a biosphere; the possibilities of monumental and record-keeping cultures were low for many generations. Saturn would have ruled therefore either from 11,500 (or from about 8,000) down to 5,700 years ago.

The concluding date of 5,700 BP is related to proto-historical times, the time of the Great Deluge of Noah in the Bible and the First Dynasty of Egypt. The date is likely to be fixed with exactness someday. A Sumerian prism, for instance, names ten kings who ruled before the Flood, and says, "then the Flood swept over the Earth. After the Flood swept over, kingship aga in descended from Heaven." Wiseman, the editor of the prism, adds that "there is actually a line drawn across the text to separate the postdiluvian events from those occurring before the Flood".

At this time, as the Saturnian era moved towards a close, the disc of Saturn appeared three times larger than today's Sun. Its orbit about the Sun took about sixty-four present days. Earth was still wheeling in orbit between the Sun and Saturn, such that from Earth Saturn looked about four-fifths as large as the Sun. In a time close to three present days Earth completed its circuit about the arc.

Granted an elliptical orbit similar to that for binary stars of 64-day period,

Saturn and the Sun would apparently expand and diminish in size (cf. Talbott, D. N., p181). With this movement, time-keeping would be suggestible and simple. The Earth would also enjoy two seasons, cooler and warmer, each of more than thirty days duration.

Super Saturn sat atop the sky, a dull red disc three times the size of the present Sun. Because of the Earth's offset from the arc, the sub-solar position on Saturn's face was askew 21 per cent from the center of its disc. The arc impinged upon about five per cent of Saturn's face. Below Saturn it widened like a tree until it passed Earth's horizon, where it was something like 15 degrees wide.

The insistent worldwide legendary connections between the Pleiades, the Deluge, and Saturn as god and planet (Cardona, 1978b) point to the likelihood of the Deluge as having occurred at the time when Super Saturn had masked this star group, and of the Pleiades having emerged for the first observable time in the sky at the zenith following the clearing of this place by the actual bodies and the debris of the nova. The Pleiades were widely taken to be the remnants of the Deluge nova by the myths.

Synchrotron radiation emitted by the planets Jupiter, Saturn and Uranus has been detected and cosmic ray sources have now been associated with these planets. Saturn, like Jupiter, emits much more energy that it receives from the Sun (Milton, 1978; Hunt and Burgess).

The heat excess given off by the gaseous planets is an indication of their electrical nature. The emission of X- rays by a stellar source is often taken by astronomers as an indication of a very recent thermonuclear nova. Following our theory, the X- rays emitted by stars and these gaseous planets of the Solar System come from electrical transactions.

At the time of its nova Super Saturn broke into at least three major fragments; these pieces, constituting the present Jupiter, Saturn and Neptune, all receded from the Sun following the fission. Jupiter, the largest surviving piece, took up a position near the Earth's present orbit; Saturn Minor, the intermediate-sized part, blew into space beyond Mars' present orbit; Neptune receded beyond Jupiter's present orbit after depositing much water into the Earth's electrosphere. In the nova, charge was expelled from Super Saturn into the plenum and dispersed in surrounding space; it was as if the mass of the system had been reduced two- and-one-half-fold; so, the planetary motions slowed considerably.

At the moment of the nova the electric arc was interrupted long enough to free Mars, Earth, possibly "Apollo," and Mercury from their million-year captivity along the axis of the binary partners. Thereupon they orbited the Sun independently for the first time, moving along a plane close to that of the old binary. Their new, roughly co-planar orbits, were similar to, but much more closely spaced than the orbits of these same planets today.

Brahma as Super Saturn absorbs, regurgitates and reabsorbs as the ages pass (Mullen, p15). So does Kronos, identified here with Super Saturn. As with Super Uranus, Kronos' instability made him less than an ideal father. In Greek myth, the great god Kronos (Saturn) swallowed at least five of the children born to him out of his sister-wife Rhea. She then hid her youngest, Zeus, and fed a wrapped stone to Kronos. When Zeus matured, he led a revolt that ended in the banishment of Kronos, after he had vomited up all his children alive. The interpretation here is that Super Saturn was absorbing fragments that remained from the Super Uranus debacle or ones erupted later by Super Saturn itself. Then during its instability, it exploded back a number of them, before as a nova it fissioned into four major parts, corresponding in the myth to Jupiter-Zeus, his brothers Hades and Poseidon, and Saturn Minor, the distant planet of today.

That Saturn possessed satellites was known to the ancients (Tresman and O'Gheoghan, p36). That Saturn was regarded as a Second Sun, a

Night Sun, is shown by Cardona (1977, p33) and others. That it was bright is claimed by numerous ancient texts and authors, surveyed by Jastrow, Mullen, Greenberg and Sizemore, Velikovsky (1973, 1978a), Cardona (1977), Tresman and O'Gheoghan, and Talbott (1980). That it became exceedingly brilliant just before the Deluge of Noah is implied in Hebrew legends (Tresman and O'Gheoghan, quoting Ginzberg). One Hebrew source has the star "as bright as one hundred suns". That widespread cultures held Saturn, the god, responsible for the Deluge is made clear by several authors already cited. That Saturn the planet was deemed responsible for the Flood is equally plain (see especially Velikovsky, 1978a, 1979). Velikovsky appears to have been the first to claim that Saturn became a nova, an idea that he found buried in Jewish rabbinical commentaries on the Deluge. That Jupiter was a prime element in the nova and subsequent events is evidenced in many of the same places; Ea, the Akkadian Saturn, reproaches Enlil, or Jupiter, for having caused the sky-waters to fall (Mason, p77). Trisiras, a son of Prajapati and a saintly Saturn figure, was a three headed god with heads resembling the Sun, the Moon and the fire, which we interpret respectively as Saturn itself, the celestial crescent and the electric arc. Indra (a Hindu Jupiter) slew Trisiras with a thunderbolt, whereupon Trisiras' three heads shone with brilliant energy until they were cut off, and then flocks of birds flew out of them and his fever left his body.

The powers that acted in the Heavens were manifested to humans amidst increasing disaster. In terror, self-abasement and pleading, man created a Uranus-Heaven religion and hoped for cosmic tranquility. Unanimously the legends of the world acknowledge the human to be an imperfect creation; some have him created more than once, as for instance, the Olmecs of Mexico; they sensed faults in their powerlessness against cosmic forces. They projected then retrojected their faults to the behavior of the gods. These notions of imperfection were indelibly imprinted. Perhaps never until the past two hundred years did humans believe that they and the natural environment might be benignly controlled through human intelligence.

What may have most bothered the early humans was their inability to manage their internal psychic systems. They were interminably made anxious and self- reflective by their lack of self-control. From their ungovernable alter egos arise the huge variety of traits and behaviors of the gods; they are artifacts of human conduct analogized to objective features of the natural environment. In every aspect of nature could be found some physiognomic and behavioral parallel with the self and with the primary human group with which the self identified. This intrinsic, practically congenital, confusion of the inner and outer worlds of mankind was a two way transaction that led humans to emulate the most extreme and complex manifestations of nature, with results upon human nature and culture that were in modern perspective often richly "constructive", but frequently "self-destructive" as well.

The first religions were in the broadest sense "monotheistic." Heaven was worshipped as the active power. As we set forth earlier, from the very beginning, humans have tended towards a supreme god. The Chinese, with T'ien, may have been the most persistent in abstracting a monotheistic idea from the Heavens and using it through a succession of specifically powerful heavenly forces. The I Ching gives this sequence: the First Principle is Heaven (T'ien) eternally present, chaos without form; the Second Principle and First Sun, giver of time, called "The Arouser"; the Third Principle, and Second Sun, an orderer, "The Limiter". "The Arouser" appears to have been both Super Uranus and Saturn, a merging of memories over time also to be found in other cultures.

The Chinese Craftsman God and His Paredra

Fu Hsi and Nu Kua measure the "squareness of the Earth" and the "roundness of Heaven" with their implements. The god is depicted enthroned atop a serpent-like column arm-in- arm with his mate. It is tempting to suggest that this picture illustrates the situation during the Age of Saturn, with the god- star perched stop the "fiery" electric arc, which rose above the world and faded in the distance into the golden sky.

The ancient Persians and others asserted that God created Saturn (whence Saturday) on his sixth and last day of labor, before resting on the seventh day (whence Sabbath = rest) (Cardona, 1978). The implication is that all things - the separation of Heaven and Earth, the other celestial objects thus revealed, the biosphere, the advent of the Moon, and mankind - were all accomplished before Saturn appeared.

Various scholar identify the Biblical Elohim as Saturn. More likely, the story adopts the designation of god employed during the Saturnian Age (as for example, it was assertedly retold by Moses in Genesis). But "Elohim" at the beginning of Genesis is behaving like the great inactive demiurge brooding over the Pangean chaos, who then becomes activated as Super Uranus in the troubled phase, and creates the world, as mankind, born on the sixth day, received and perceived the Cosmos. Whereupon a more detailed account begins, relating the Hebrew experience with Saturn as distinct from the more general, aboriginal human experience.

When the gods changed, humans bowed to the changes. This repeated behavior over thousands of years is a significant motif in religious history. The lamentations over the death of Saturn were worldwide. Because Saturn "died" in what was a historical period, although little of its civilization remains, the hysterical and obsessive mourning shows mankind affected by, not affecting, a real tragedy. Thousands of years after the death of the second sun and the end of his age, the Roman government was acting to suppress infant sacrifice to Saturn.

The parallels between Saturn and Christ as a Saturnian figure are numerous: the passion of Christ is historically and psychologically a reenactment of the character, the unjust death, and the resurrection of the god who had died some four thousand years earlier as Osiris-Saturn. Frequent efforts philosophically to cover over the deep trench of tradition connecting the two gods have failed to divert the mainstream. Such efforts have built a distinctive existential character for Christ. The Age of Saturn in cultural terms was probably what is usually designated as upper Paleolithic and Neolithic. It would be the age of Atlantis and other civilizations lost to view in the disasters that followed.

Saturn, who was generally accredited with bringing agriculture and other useful arts to mankind, was the first Lord of the Mill, a sky wheel grinding out material and spilling it upon the Earth, gold, salt, sand, and stones. Before it sank in a cosmic maelstrom, it ground salt into the sea (de Santillana and von Dechend). The Saturnian Deluge was caused by a salt-water tree cut down by a tapir, according to the Cuna Indians (ibid). In Hindu myth, the gods were churning the heavenly waters and ground salt into the seas (Tresman and O'Gheoghan).

The Churning of the Sea

Vishnu sits atop mount Mandara accompanied by his wife Lakshmi. The mountain of the world is being spun as the great snake Vasuki is pulled to and fro by the devas (grasping the snake's neck, to the left) and the asuras (holding its head, to the right). Together they churn the sea of milk, producing the liquid of immortality. Unfortunately their churning became so violent that it threatened the Earth, whereupon Vishnu, as the avatar of the turtle (seen here below the mountain), came to save the world by assuming the role as its pivot. Still the world was threatened by the heat of the churnings until Indra sent the Deluge from Heaven to quench the fire..- courtesy *Larousse Encyclopedia of Mythology*

Salt domes are among the most common of the mineral intrusions that are scattered over the Earth's surface. Small ones are about one cubic kilometer in extent, but some are several hundred times this extent. The largest known salt dome is in Calliou Island Bay, Marchland Louisiana; its size is unknown but it is estimated to be 43 kilometers long, 20 kilometers wide and 6 kilometers deep; if so, it contains about 10^{13} tons of salt. Globally 950 large salt domes are being used for mining. In some regions these domes are associated with petroleum and natural gas deposits. Other places have sulfur deposits associated with the salt intrusions. Since most salt domes have been found scattered widely from the three major salt-dome fields known today, it is not unreasonable to think that only a small fraction of the buried salt has been discovered. These immense deposits of salt in the ground suggest a non-marine source of all salt. The salt in the seas can be explained in terms of salt falls from space.

Whence comes "The Golden Age" of Saturn? Saturn is golden. So is the light on Earth. The interruptions of absorbing the children were spaced out, perhaps half a dozen marking disasters over the 2300 years that followed the lunar period. Meanwhile life on Earth may have been easy in most places. There was no ice age. Travel by boat was easy, for the breezes were mild. Antarctic may have been mapped in this age, since an ancient map showing its outline beneath the present snow has been found and since no later age would have been able to produce it because the coastline was invisible (Hapgood, 1966). The northernmost and southernmost regions were quite habitable, even tropical. The continental shelves and slopes had become livable. There was a plentitude of moisture and all-year warmth.

The plenum and electrosphere were still insulating. Further, Earth was holding its own surface atmosphere despite the thinning of the plenum under Saturn. The plenum became increasingly more transparent.

The arc was less brilliant and more intermittent, yet the fire was there, binding Earth to its great god. Kronos is addressed in an Orphic Hymn as "you who hold the indestructible bond", while his Babylonian alter ego held "the bond of heaven and earth" (Tresman and O'Gheoghan). The "central fire" is also represented, likely, in the innumerable pillars, phalluses, bonds, pinnacles, pyramids, and one-legged gods (Talbott, D. N.). Even the jagged sickle, with which Kronos castrated Uranus, may have symbolized the electrical axis being severed from Super Uranus. Manu, a Hindu creator god of the Flood, stood on one leg for thousands of years, contemplating his world design. All of these images are closely connected with Saturn, often by name as well as by symbol. Yahweh is imagined in Jewish legend as radiant atop Mount Zion, and Kronos-Saturn ruled Mount Olympus before Zeus toppled him.

Saturn is reputed to have lit up brilliantly and preserved its light for seven days before the Deluge hit the Earth. This may be interpreted as a set of pre-fission flare-ups climaxing with the nova that destroyed Saturn. If the Earth were still only 14 Gm from Saturn, the debris expanding away at 200 km/s would encounter Earth in a little more than 19 hours. If the 40- days/nights period were of present duration, the problem of depositing so much water on the Earth would be practically impossible. Millions of heavy cyclones would be needed, even one per 30 square kilometers all over the Earth. If they were concentrated at the poles by the electrosphere high above the Earth and funnelled down there, the damage might be less. But then the tremendous erosion would be visible today. We think that a longer span of time may have been required, and "a day" was a translated memory of a longer regular interval unknown to us, a magical cipher, or an historical error.

The problem of deluging the Earth is nearly as difficult to cope with as the recent eruption of the Moon from the Earth. In both, almost unimaginable physical phenomena must be conjectured. The biospheric aspect, which often comes first to mind, can be rationalized on the equally incredible capacity of living populations to renew themselves. Even postulating

Manu and his tiny crew or Noah and his family as sole survivors, a thousand years of exponential growth could fill the land to overflowing.

The astrophysical aspect is more intimidating (see Kofahr). To launch the waters and other debris from Saturn is readily conceivable, given the nova. To guide it and land it requires the invention of low-probability solutions. Even if the dynamics thus far presented can be accepted with respect to Earth, how does one explain the absence of water on Mercury, Moon, and Mars, all of which would have been in or near the rush of water? If they were inundated, where has the water gone? There is almost no sign of water on them, or its having been on them in oceans. Where, too, is the salt?

Two types of probability occur. The two planets may have burst out of the magnetic tube ahead of the Flood churning down towards the Sun, whereas the Earth was entrapped. Or else, the Flood can have descended the tube by a passage occupied at the moment by the Earth alone. In this case, the Moon occupied a special position besides, for it was a considerable distance from the Earth towards the perimeter of the tube. In any case, the Earth would still be luckier than Mercury, or Mars,

for both of these planets have ruined surfaces and no biospheres.

Three different Jewish legendary statements refer to a diminution of the Moon in size (Tresman and O'Gheoghan). This would occur presumably after the Deluge, when the Moon followed the Earth out of the old magnetic tube and was repelled by Earth into a larger, but still captive, orbit. Two legends imply that the stars multiplied then, an expected improvement of visibility in the star-system sac. The age of Jupiter, now upon the world, introduced mankind to the light of the stars in the darkness of the night.

Cetus or Seth, the Devil Dog

The Glospie Stone of prehistoric Scotland. The arrow indicates the Cetus beast, the "Devil-Dog" Seth. (Source: Spalding Club). For a complete analysis, see Beaumont (1949)

CHAPTER TWELVE

THE JUPITER ORDER

The planets following the interruption of the magnetic tube, were freed. Instead of wheeling with Jupiter, now the binary component, they orbited the Sun independently, their motion close to the plane of the old binary - now the plane of the reconstituted Solar System. In their free orbits the planets avoided one another and Jupiter because of their electric charges, which produced repulsive forces when they came into proximity. Regularly they passed through, or close to, the axis between Jupiter and the Sun. Then Jovian thunderbolts were experienced.

These could be the now occasional visible discharges of the dying binary, catalyzed by the presence of a charged planet in the path of the discharge; more likely they were locally generated discharges between the planet and its electrosphere, induced by the planet's voyage through the electrified region within the invisible arc-discharge between Jupiter and the Sun. Either way the planet was zapped by Jupiter as it came into opposition with the Sun. From the Earth, for the first time humans might see the other planets swinging on their journeys around the Sun.

Planet Jupiter, now viewed as Ruler of the Heavens, struggled to restore and maintain the arc-connection to the Sun - for a time the arc flared with occasional visible spurts, but mostly the electric connection was dark. It is in this era, possibly, that the existence of a Counter-Earth was proposed, a dark body which obscured the celestial fire.

The Earth emerged from the magnetic tube following the Saturnian Deluge (about 5700 BP) with its rotational axis forcibly relocated. While in the tube, it was constrained to maintain a magnetic axis along the tube's perimeter. Freed from the magnetic tube at the time of the Deluge, the rotation of the magnetized Earth was brought into alignment with the weak magnetic field generated by the relative motion of the electrically charged Earth-Sun pair. From the Earth the charged Sun is seen to flow in a loop around the Earth in one year, representing an electron current flowing counter-clockwise along the ecliptic (as seen looking down on the North pole). Such an electron current creates a magnetic field, which enters the Earth at its North magnetic pole. This field parallels the Earth's magnetic field.

A small tilt and a relatively diffuse plenum made the variations in such sunlight as was released very noticeable on the Earth in the altered system. Seasonal differences in the earlier era were minimal compared to variations in climate now existent on the Earth and during the year. The Earth is now a drier, harsher habitat. The cosmic thunderbolts of Jupiter function in clear skies. The skies are clear because the plenum has greatly diminished in density; the heavens are transparent and the stars are seen.

We see in the Universe countless instances of stellar and interstellar binary currents produced by the discharge of accumulated electrical charges. Also, at any given time there are several million electrical discharges in the photospheric region of the Sun, each about one or two thousand kilometers long and lasting ten minutes (Crew), "Mega-lightning" of hitherto unappreciated voltage has been observed by satellites in the Earth's upper atmosphere, but has not been detected yet by ground observation (Turman).

When legend reports the electrical activity of Jupiter the god, it tells of the electronics of the planet Jupiter. Mountains are levelled or melted, sky monsters felled, citadels destroyed, the Earth scorched, and armies sent fleeing; all the work of the king of gods. Every lightning stroke to Earth becomes a theophany, as in lightning-obsessed Etruria, which gave the name Jupiter (Jove-pater) to the Romans. The sacred manifestations consecrate the cosmic bolts that were memorialized and discussed for thousands of years.

The solar year under Jupiter may have had a succession of different lengths. First occurred the Saturnian year, to which we have assigned a 64-day duration. Then it increased to 156 days when Jupiter receded. The Mayans possessed a 260-day sacred calendar that was central to their religious and cultural life, even while using a more modern and exact calendar (Coe). We attribute this sacred calendar to the Jupiter-Earth synods of this era, to the time before 4 400 years ago.

Jupiter is the most phallic of the great gods. The association of electrical stimulation, phallicism, and thunderbolting is strongly linked to the religious rites in vogue at the time of Jove (Ziegler). Phallic worship is common among Jupiter-type deities (Tresman and O'Gheoghan). The Amun temples in Egypt are liberally decorated with images of the ithyphallic god Min. Shiva (the Hindu equivalent of Jupiter) emasculated himself when the realized that his creative ability had left him. The analogy of this legend with the end of the visible electric arc is plain.

At the Saturnian Deluge we suspect the Earth was around 96 gigameters from the Sun. It moved outwards constantly after that. In the course of these changes, Earth's axial tilt was altered again and again (see Dachille, 1963; Warlow), causing glacial retreats and advances in the extreme latitudes at each period.

A new ice age may now have begun, centered at the rotational poles. The new ice will remain, advancing and withdrawing on occasions between and during encounters with celestial bodies, up to the present time. Today we cannot yet deduce whether the ice caps are increasing or decreasing (compare Kukla and Matthews with Gribbin, 1976; A. Brown).

Today the clouds above the surface of Jupiter are very cold (150 K) yet the planet is very active electrically (Sutton, Gurnett et al.). Jupiter's "magnetosphere" is enormous: if it were optically visible, its size, viewed from the vantage-point of Earth's orbit, would be comparable to the disc of the full moon. The ion and electron currents detected within this magnetosphere represent radiation levels which would be fatal to humans (Panagakos and Waller, 1974, pp15ff). The radio noises generated within this region are received at the Earth, as are "cosmic rays" (mainly protons) of Jovian origin. Jupiter is, so far, the most demonstrably electrical of the planets. Jupiter is like a miniature Solar System, with its planet-sized Galilean satellites, its asteroid-sized satellite family and its entourage of comets. Everywhere the electrical imprint is there, and not always just by implication.

The three inner Galilean satellites, Io (resembling Earth's Moon in size), Europa (about nine-tenths of the Moon's size), and Ganymede (eight percent larger than Mercury) orbit in 1: 2: 4 resonance. When any two meet on one side of Jupiter, the third is located oppositely behind Jupiter or at quadrature to the pair (see Peale et al.). The resonance is seen less clearly in the motion of the fourth satellite, Callisto (slightly smaller than Mercury).

The surface of each of these bodies is distinctive (see Smith, B. A. et al.). Io, seemingly, is close to being molten. It lacks craters, but shows over a dozen caldera-like scars, which were likened to active volcanoes when an eruption was observed during the fly-by of the Voyager 1 spacecraft (Morabito et al.). Photographs of Europa show it to have lobate scarps resembling those on Mercury (Smith, B. A. et al.). Perhaps Jupiter zapped it, causing it to shrink upon loss of charge. Callisto, the outermost of the four, is one of the most cratered objects in the Solar System (ibid), likely the result of thunderbolts striking it. Ganymede, the second closest of the four, shows a banded surface, pocked with ancient craters, then overlaid with younger bright-rayed craters, which stand out prominently (ibid.): distinct electrical scars resembling the rayed craters of Earth's Moon (also seen on Mercury), which Juergens (1974) ascribes to cathode behavior when interplanetary discharges occur.

An electrical flux-tube through which a current of millions of amperes flows between Io and Jupiter (Stone and Lane) has been linked to Io's eruptions (Gold, 1979). The Voyager 1 spacecraft was aimed at this flux-tube (Krimigis et al.) as it encountered Io. It missed the tube by seven megameters in what was labeled a navigational error; but more properly, in our opinion, the cause of the miss was an electrical perturbation (here a repulsion of the spacecraft by the tube).

The persistent connection, by the flux-tube, between Jupiter and its satellite Io, is one of the last sites of cosmic thunderbolting between these two bodies has been known for several years since the advent of the radio telescope, when strong radio bursts which correlate with Io's position about Jupiter were detected (Dulk). That a passing spacecraft was located advantageously to photograph the flash of one of these discharges was happy happenstance. The glow was interpreted by some experts as evidence of volcanism. Apparently, to think that we have witnessed directly the fire of the gods, a cosmic discharge, would seem to be too frightening (Juergens 1980).

Jupiter discharges only to Io today, but its repertoire and gamut may have been more extensive not too many centuries ago. Additional description of Jupiter's electrical nature, especially as it affects the asteroids and comets, has been afforded by Milton (1982). The present behavior of the dark remnant of Super Uranus is, in sum, fully in keeping with the pure electrical theory of the Solar System and the historical reconstruction of Solaria Binaria.

CHAPTER THIRTEEN

VENUS & MARS

It is no longer fashionable to believe that Venus and Mars are beautiful bodies, with some congenital blemishes. Like Moon and Mercury, like Earth itself, they shriek of more recent disasters. The major question now is " How recent is 'recent'?"

Myth and legend (in its deliberate attempts at science) afford voluminous material about both planets, their transactions with each other, and their encounters with Moon and Earth. Research of the past generation has evidenced that the planet Venus dominated the human cosmogonical mind in the years between 3500 and 2000 BP and that the planet Mars entered upon the competition to catastrophize the human mind in the latter 800 years of this period. Venus had hundreds of names and identities, many of them secret, sacred, and obfuscated. For example, the Hebrew word "shakris" means the Evening Star, Morning Star, to sacrifice, to kill something, to make sacred.

Logically, one initially seeks information about the first appearance of these celestial bodies; when were they born? In the case of Venus, legends of the Near East, Greece, Rome, the Teutons, the Hindus, and the Meso-Americans seem to speak of a special time of birth of a deity with a homologous syndrome of traits (Velikovsky, 1950).

Although the name "Venus" may not originate directly from "venire" (" to come"), as Cicero would have it (Lowery, Grant), it may well emerge even more significantly from Venus, which Bloch translates as "blooming nature", hence, as we see it, something new-born and rapidly expanding. The lotus and lily, two life forms suggestive of blooming cometary images, are employed in widely separated cultures as Venusian symbols.

Clearly conformable to an astronomical operation is the birth of Greek goddess Athene, who sprang fully armed with a shout from the brow of Zeus (Hesiod). The Hindu Devi is remarkably similar in the commotion that she causes when born (Isenberg, p90). Various studies analysed by de Grazia (1981, 1982a) set the time of her birth near 3,450 BP, in accordance with older studies by Velikovsky (1952, pp1- 53, 98- 101). The time coincides with the Exodus of the Hebrews from Egypt under Moses, an event so fraught with catastrophe that it remains the substratum of the Judaic, Christian and Islamic religions.

If Venus was erupted from Jupiter, it conceivably burst from the disturbed area of the Great Red Spot. Although not demonstrable, this is hypothetically feasible. It was Jupiter's greatest discharge, its last attempt to rid itself of ions and gain electrons. It succeeded; it retired; and its offspring was unleashed into the inner Solar System, where all massive fragments had gone hitherto.

Unlike Uranus Minor, Neptune, and possibly Pluto, Venus was of low electrical density and fell victim to encounters with Earth, then Mars, and remained within the inner circle of planets. Both the Earth and Mars took electrical charge from Venus, not without extensive physical "damage" to themselves; they both moved away from the Sun after their encounters with Venus (Ransom and Hoffee, Table 1).

Venus was not rich enough in electrons to be coveted by the Sun. By the legendary and historical evidence, it took hundreds of years to achieve a "safe" orbit from where it would not venture close enough to Earth to endanger it and cause electrical damage. Until then, it may have threatened the Earth about every 52 years. During a seven hundred year period both the Jews and the Meso- Americans observed a great "Jubilee year" on those occasions; the inference from the holiday is that the proto-planet, still behaving as a comet and undergoing continuous electrical transaction, was due to arrive in the vicinity of Earth but prayerfully would design not to destroy the world.

It is significant, as one of numerous details in the Venus mosaic, that the American Pawnee Indians until a century ago celebrated a Venus festival on each occasion of the reappearance of Venus, sacrificing a virgin to the star. But many another celebration of Venus could be cited.

Granted that history and legend impart an aura of youth to the planet Venus, one can search for and find confirmation of youthfulness in the present state and observable behavior of the planet. The 925 K surface temperature measured by landed space probes has not been explained satisfactorily (see for example. Firsoff; Velikovsky, 1978b; Forshufvud; Greenberg, L. M., 1979; Morrison) by any theory other than recent and continuing electrical transaction. The observation that the lowest thirteen kilometers of the atmosphere are glowing (Panagakos and Waller, 1979, p3) and that lightning occurs in the Venusian atmosphere (Taylor et al.) encourage us in our view that electrical currents flow between Venus and surrounding space. The two localities which were photographed show evidence of recent surface devastation. Seemingly the surface of Venus is similar to those of its neighbors even though the latter lack atmospheres (Ksanformaliti et al.; Florensky et al.).

One of the major surprises greeting the explorers of Venus and the theorists who welcomed their data was the demonstration of the slow retrograde rotation of the planet. Several non-electrical explanations have been offered to explain how Venus might have reversed its original forward rotation (Singer, 1970; Ingersoll and Dobrovolskis; Kundt). In a two-planet encounter involving electrical polarization (which induces aspherical shapes onto both bodies) strong "tidal forces" act and can alter spins, or flip planets over, as Warlow shows.

The five bodies of the inner Solar System exhibit a spectrum of spins and orientation, running from fast-direct (Mars and Earth) to slow-retrograde (Venus), with Mercury and the Moon in between. Given a number of encounters among these five bodies over the past several millennia, many combinations of rotational alteration must be expected: it is probable that none of the spins are virginal.

The atmosphere of Venus presents another type of problem. Its carbon dioxide composition is like that of Mars, unstable over a long time due to photolysis by ultraviolet radiation. That most of it has not reacted with the exposed surface rocks is termed "surprising", possibly Venus' atmospheric gases have recently been modified.

The chemistry of the deep and dense Venus clouds has been inferred only by indirect evidence despite the passage of several spacecraft through them. Above the clouds, which seemingly insulate the planet, temperatures resemble those found at comparable altitudes above the Earth. Insolation and heat radiation from the clouds do not betray the hellish heat that was discovered below. Even at the base of the cloud layer (which is twenty kilometers thick), temperature, pressures and winds remain Earth-like (Burgess). It is the descent through the remaining forty-nine kilometers that confounds expectations and confuses the instruments of the descending space probes (making some of them inoperative and the data from others uninterpretable): here Venus does not resemble any environment yet penetrated by instruments.

We find a murky and stagnant inferno under crushing pressures many times greater than those on Earth. There, lightning occurs frequently (Taylor et al.), indicating an electrical instability in excess of that found on Earth, and an electrical glow, noted above, permeates to the bottom of the murk.

Even the edge of Venus' sphere of influence has produced the unexpected. Venus is only trivially magnetic; yet its interaction with the solar wind produces 80 percent of the effect generated by the three-thousand-times more strongly magnetized Earth (Russell, C. T. et al., 1979). Since Mercury, whose magnetism is similarly miniscule, also interacts strongly with the solar wind (Ness et al., p480) we are drawn to conclude that the interface between planets and solar wind is electric, rather than magnetic.

The bow wave in front of the planet and the long tail in the wake of the planet represent the junction of two electrospheres: the solar wind flows on one side of the junction and the planet driven ions and electrons flow on the other side. Electrons traversing this junction behave differently at Venus (and Mars), for there they are slowed, rather than accelerated as happens with Earth, Mercury and Jupiter (Simpson et al: Wolfe et al). Electron-deficient solar wind atoms seemingly penetrate and are absorbed by Venus' upper atmosphere (and Mars' surface). The magnitude of the effect indicates that Venus is farther from equilibrium with its surroundings than are the other planets. This finding may be the "Rosetta Stone," telling us why near Venus' surface the heat is infernal. Any unequivocal evidence of disequilibrium tells us that Venus is indeed young (Van Flandern).

Earth's history of the period around 3500 BP, so far as it is known, provides proof of extraterrestrial damage. The skies were disturbed and celestial motions were reportedly irregular. Astronomical alignments of before this age are out of line with references of the period following. No temple can be adduced whose orientations have remained the same through this time.

For instance, the second rupestral temple at Wadi es Seboua(s) in Upper Egypt was originally orientated before 1500 BC from its rear, through its portal, via a faraway mountain saddle to the winter solstice bearing 30° 9' 0" South of East. It was destroyed by fire. Much later it was excavated and rebuilt. Again it was orientated to the winter solstice, this time, however, to 35° 49' 12" South of East (Roussel). Possibly the axis of the Earth tilted.

There were radical disturbances and some lasting changes in atmospheric electricity, radioactivity, temperature, winds, climate, and solar radiance. The Book of Exodus can be read as a meteorological journal - one encounters electrostatic phenomena, gales, dense clouds, a unique darkness, falls of manna (compound manufacture in the atmosphere), dense rains of stones, fire (often apparently electric), a mass of dead quail (driven down by electrical storm and hurricane winds and said to be poisonous to eat), and radioactivity (see de Grazia, 1983a).

A sharp rise in C- 4 levels occurs about now (and at the time of the Mars incursions 700 years later) (see de Grazia, 1981). It is doubtful that C-14 is a valid and reliable clock, since its formation and absorption rates are so easily altered by changed environmental conditions, and so evidence of this nature, though favorable to the hypothesis, must be discounted.

All human settlements suffered destruction or damage from natural causes. In one study, we read, "In the middle of the second millennium BC the ancient cities of Southern Turkomenia declined and were abandoned by the inhabitants. The South Turkomenian civilization perished at about the same time as the ProtoIndian ... and the reasons are still unknown." (Kondratov) Schaeffer's survey of some 40 most important archaeological sites in the Near East arrived at the same conclusion for the same time.

All contemporary accounts or chronologically assignable legends dealing with the period mention a general natural disaster. The prime case is the Bible. The Pallas Athene instance is also referred to. The

Ipuwer papyrus has strong support now as an eyewitness account of the catastrophe ending the Middle Bronze Age in Egypt (Velikovsky, 1952; Greenberg, 1973; Sieff, 1976, Bimson).

Every culture-complex changed markedly. Several major civilizations declined sharply or fell - Egyptian, Indian, Kassite, Turkomenian, and others of the Near East. One might add the Minoan of Crete and the Chinese.

It follows that every institution, behavioral pattern and natural setting that exists today, if its history is complete, will reveal an inheritance of effects from the (Venus) quantavolution.

For the later time, about twenty-six centuries ago, the Mars case offers a similar set of propositions, although the evidence argues for a level of destruction appreciably lower than that obtained from the earlier Venus-Earth encounter.

Around the solar year 2776 BP, human activities related to celestial disturbances were generated respecting Mars, as well as Venus, and are notable in the Near East and Mediterranean world (Velikovsky, 1950). Enough subsequent benchmarks were provided by legends and practices for Velikovsky to surmise that a large heavenly body, apparently Mars, was threatening collision with the Earth at fifteen-year intervals. The Mars encroachments may have been initiated by Venus, which, pursuing an ever-shortening orbit, perhaps encountered and displaced Mars from its earlier orbit between the Earth and the Sun (Rose, 1972).

The Romans were Mars-worshippers par excellence, and the legends, rites and early reports that tie Mars to the history of Rome are not to be disregarded; "archaeological and epigraphic discoveries" continue to demonstrate that "the legendary guise of the traditional material actually masks a real foundation of authentic events" (Bloch, 1085).

By contest with Venus, Mars seems to be an old god. Much less is made of his origins and birth in the Mediterranean world. When he bursts upon the world scene in the eighth century BC he is already well known. Least personable of the Olympian deities, Ares (Mars) is portrayed as a ruthless warrior. Hercules seems to be one of his more interesting identities. New militaristic nations, particularly the Romans, the Assyrians, and the ancestral Aztecs, forged empires under his inspiration. The Roman dedication to Mars is well known. He was believed to be father of Romulus, their founder. In the old calendar they named the first month after him.

The Romans irreconcilably claimed both Aeneas, Prince of Troy, and Romulus as their founder. Aeneas was, and is, placed in the twelfth or thirteenth century BC with the Trojan Wars, by older scholarship. Recently the Wars have been brought into later times, along with Homer, who sang of them (de Grazia, 1984a). This is but one step in a reconstruction of chronology that eliminates the several centuries of a so-called Greek Dark Age and pulls the disastrous collapse of the Mycenaean civilization down to the eighth century as well (Isaacson). Roman legend has Romulus and Remus (abandoned and miraculously suckled by a wolf in their infancy) founding a town called Rome, which Romulus rules until he is lofted into the air, possibly by a cyclone, to join his father, Mars.

It seems to us reasonable that around this time, Aeneas and his band of experienced and cultivated Trojans might have impressed themselves upon, or been welcomed by, the local Latins of the new town, and perhaps even helped by their neighbors to the North, the Etruscans, who themselves were of Anatolian origins.

It is an age of destruction and movement. The powerful nearby Etruscan state was staggered by natural disasters and a decline. According to Pliny, their city of Volsinium, where stands Lake Bolsena today, was destroyed by a thunderbolt. Both Mount Vesuvius and Mount Etna underwent Plinian eruptions around the same time.

Many peoples were on the march or fleeing - possibly the Etruscan elite had not preceded Aeneas by long. They were from the general area of Ilium (Troy). Southern Italy and Sicily were being heavily settled by Greeks, in trouble themselves and profiting from natural disasters that were besetting the earlier inhabitants. And at that time the Dorians (Heraclids, sons of Hercules- Mars) were moving in upon the hapless Myceneans.

All of these loose connections and temporal references need a thorough analysis; indeed, we require nothing short of a total review of the most ancient history, legends, and the records of excavations. Even then the human record, like the fossil record, is scanty. One may hope for, but not expect, that the thousand-to-one chance will occur, and that a tablet or papyrus describing clearly the behavior of the planet Mars in these times will be found. The earliest Etruscan, then Roman, calendar was of ten solar months. That the later twelve months alternated at thirty and thirty-one days does not fit the present lunation, which is better suited to the 29- and 30-day alteration used in all surviving lunar calendars. The indication is that Mars disrupted the month in its transaction with the Earth and/or Moon.

The Aztec Mars was Huitzilopochtli, born of Mother Earth from a ball of humming-bird feathers that fell from the sky. His color was blue, his totem the eagle, and his weapon the blue snake. The interminable human sacrifices of the Aztecs were in his name or to the Sun in his name; the Sun, "the Eagle who rises", was believed to require the hearts of an unending stream of prisoners-of-war if he was to remain constant in behavior. The High Priest of Huitzilopochtli was called Feathered Serpent, priest of Our Lord; but "feathered serpent" is rendered Quetzalcoatl, who at a much earlier time was the ruling deity of Meso-America and was identified unfailingly as the planet Venus by many scholars.

The Assyrians of the eighth and seventh centuries, "like the Romans, those other stepchildren of Mars, and more than the Hittites, its victims, had a lively reverence for the planet Mars: "Nergal, the almighty among the gods, fear, terror, awe- inspiring splendor", wrote Essarhaddon, son of Sennacherib." Evidence of physical destruction by fire and earthquake is abundant everywhere in these years, and it was "beneath the exploits of Mars ... that Assyria marches to world power", Sieff (1981) declares. Patten et al. argue that the Assyrians timed their major offensives to coincide with cosmic approaches of Mars to profit from the physical disorder and consternation of their enemies.

In general, the same hypotheses that we stipulated earlier for the Venus encounters may be translated for and applied to the Mars encounters seven centuries earlier. That is, we can adduce some evidence from around the world of deep disturbances in the six spheres and of their interconnections in the holosphere.

Juergens (1974, 1974/75) has analyzed reports of various disparities between surface features of Moon and Mars, and explained them as effects of anode and cathode behavior that would be excited in close encounters. He has suggested, for instance, that some of the hill ranges of the Moon are so morphologically anomalistic as to represent a job of "electric welding" done by Mars. To extract from these hills and from the canyons of Mars comparable material is presently impracticable, even though we may conceive of chemical tests to apply to the material thus obtained.

Juergens has gone further in the investigation of electrical encounters between Moon and Mars. On the assumption that the ray-surrounded crater Tycho (the most prominent feature on the Moon under high-angle lighting, that is, near Full Moon) could have been blasted out of the lunar highland rock by an electrical explosion liberating almost 10^{17} megajoules of energy and requiring a transfer of 10^{11} coulombs of charge between the Moon and Mars, Juergens sought a suitable anode site on Mars' surface which might receive this discharge. He found a likely receptor in the mountainous feature called South Spot (now, Arsia Mons). He writes"this spot is an enormous pit 140 kilometers across at the crest of an impressive 17 kilometer rise from the floor of the Amazonis basin to the west."

He observes that this volcano-like structure has no known counterparts on the Earth. Tycho also represents a lunar high point: it is some 1.2 kilometers above the hypothetical lunar sphere.

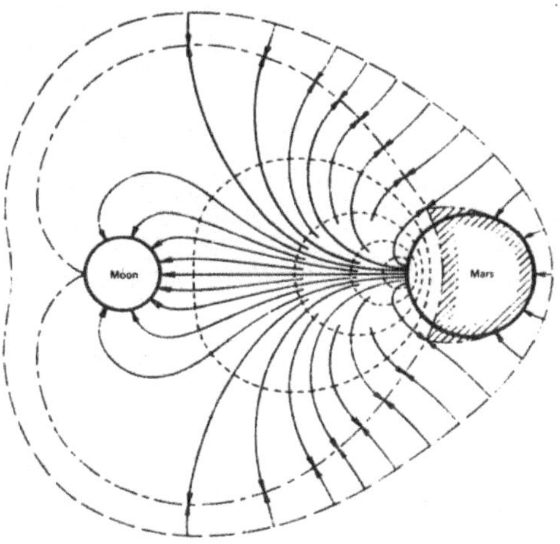

The Electric Field Between Mars & the Moon

During the close passage of Mars and the Moon, *circa 27* centuries ago, devastation was produced on the surface of both bodies, the result of interplanetary discharges across the gap between the two bodies. Here we see the electric field lines (arrows) and the lines of equal electric potential (variously dashed) prior to the breakdown that produced electric discharges from the Moon (the cathode) to Mars (the anode). The outer equipotential depicts the sheath boundary between the electrosphere of the charged bodies and the interplanetary plasma.

- courtesy of the estate of Ralph Juergens

As with Moon and Mars, so with Venus and Mars. Perhaps, in the end, we shall come to regard a famous scene of the Iliad of Homer as an eyewitness account, garbled, to be sure: that is where the goddess Athene, archetype of planetary Venus, engages in awful sky combat with the god Ares, who is recognized as the planet Mars.

Athene drove her chariot towards Ares, bane of mortals, and drove her spear deeply into his belly. Thereupon arose a huge black cloud, and he bellowed like 10 000 warriors and fled into the high heavens.

Some 2700 years later, the United States of America sends Mariner 9 into space and photographs the most prominent feature of Mars, the Coprates Canyon complex.

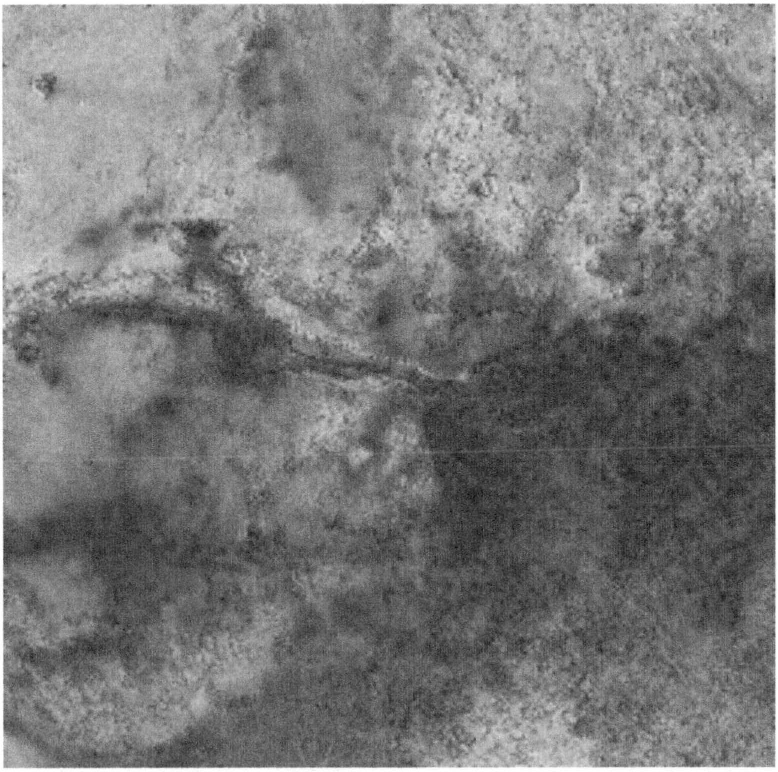

It is a 12 megameter line of seemingly extruded mountains, enormous beyond anything on Earth, leading into a canyon 3,600 kilometers in length, 500 km. wide, and 6,100 meters deep, ending in another string of great volcanic outbursts. Some 8.5 million cubic kilometers of rock have disappeared. The wound stretches across half the circumference of the planet.

Allan Kelly has offered a scenario. The total eruption is a product of the same event, when some very large comet or other massive intruder from space passed too close to Mars. This intruder literally sucked the lava from the interior of Mars to form the huge volcanoes. As it came closer it caused a tremendous bulge, miles high, that burst open along the top and spewed out lava and great chunks of Martian crust; much of this material followed the intruder into space.

There is no evidence of water erosion of the steep stairs of the canyon, no sedimentation, delta fans or stream beds cutting the lengthy lips of the wound. The intruder, be it Athene, Ishtar, or Venus, both repelled and implanted charge onto the defenseless Mars. Leaders of lightning from Venus incised the masses that had assembled in a belt of Mars along half its girth. In the ripping blast that followed, Mars flung them upon the intruder, and, lighter but more heavily charged, withdrew. So the aforesaid "garbling" of history lay partly in not perceiving that Mars laid a mass upon Venus and that the greatest spear thrusts of lightning were exchanged beneath the dense cloud that poured from Mars' wound.

Ancient astronomers and writers appear to have had no difficulty in considering (or perhaps they were really reflecting upon) historic encounters governing the planets; yet we can observe in their own times the strengthening of three psychological defense mechanisms that made historical reconstruction involving quantavolution difficult: denial and suppression of memory, religious and literary sublimation, and abstract philosophy.

CHAPTER FOURTEEN

TIME, ELECTRICITY & QUANTAVOLUTION

Several major observations promoted consideration of the Solar System as a binary development. A growing realization that our star system is embedded in a Universe largely composed of multiple star systems led us to match known characteristics and behaviors of these systems in their varied stages of development with our own system as it might have been, is and might become. Evidence mounts, too, that planet Jupiter has stellar traits, as have, less obviously, the other major planets.

Furthermore, ancient observers and philosophers who were neither primitive nor naive, and who were also reporting the ideas of other experts of hundreds and thousands of years before them, affirm that the bodies of the Solar System and the stars changed their behavior and their motions. These men were cognizant of, and disciplined by, religious systems that were sky-obsessed, and which moved continually between celestial behavior and mundane behavior, in supreme efforts to let happen on Earth what happened in Heaven, and vice versa.

As we have reviewed their ideas and reports, coupled with evidence emerging from legends and archaeological excavations, we found reason to think that they might be living in a world that was strikingly different from our own and that was recognizably a late phase of a stellar binary system.

The prospect before us, then, was to understand an ancient science and tradition which had large heavenly (god-like) bodies hovering over the Earth in what today we would classify as a synchronous orbit. If the Earth were locked between the partners in a binary, much of what the ancients spoke of as their own experiences, or related as the experiences of their ancestors, makes sense.

As soon as we began to draw upon ancient opinions concerning cosmic events, we had to take a position respecting chronology. In this we were encouraged by the binary concept itself to call time into question. Binary systems offer evidence of great forces operating over short times to produce large effects. They illuminate short-cuts in the creation of the raw materials for planetary, atmospheric and biological development. As they transact, the large bodies separate and fragment within the system, creating and destroying worlds while retaining their parts. It may be fair to say that only a binary model can supply those scientists - admittedly a small minority -- who are inclined to shorten natural history with an adequate theoretical instrument.

The binary model suggests and may even require, a short-time scheme for natural history. A short-time Solar System requires high energy and precise interventions at levels of nature ranging from the Galaxy to the atomic nucleus. It is specifically this sort of intervention that is evidenced time and time again in natural history. All "absolute chronometries" become variable in a quantavolutionary world. The rampant inflation of this century, which has expanded the time scale for the lifetime of the Universe from 40 million to 80 billion years, may end in a catastrophic implosion.

As if their technical difficulties were not sufficient to disable long-time chronology (de Grazia, 1981), ancient human voices seem to testify against it. These earliest humans unmistakably assert, among other things, that Heaven and Earth separated, that suns appeared, that gods fought in the skies and invaded the Earth, that the world was repeatedly built and destroyed. They are neurotically obsessed with all celestial bodies and motions, and engage in all known extremes of behavior in imitation and appeasement of the behavior observed in the skies.

It is not possible to claim that this is primate activity, nor hominid, nor that it is primitive, nor finally that it is a collective psychosis of early civilizations. Modern social psychology and psychiatry can document, and even replicate, such human behavior today. The earliest cultures, those that are "guilty" of this behavior, invented social organization, agriculture, manufacturing, science, and the arts. To think that they could do all this without a firm "reality principle", as Freud has termed it, must be an error.

The skeptic of our interpretation may be reduced to postulating that "illud tempus" must have been exciting and stressful, but could not be so very exciting and stressful as they would have us believe. Reasoning similarly, one could assert the contrary: the real foundations of the ancient excitement and obsessions must have been even worse than we are given to believe, because the ancients were used to disasters and hence were less traumatized by them; "war is hell", but less hellish to old soldiers than to recruits. It seems to us that both a priori views -- that the ancients were excitable or that they were blasé -- may obstruct the necessary work of delineating, bit by bit, the experiences of the ancients from the conglomerated assemblage of fragmentary records, legends and geological and archaeological facies, and then exposing them to analysis in the light of the sciences today.

A persistent theme of the ancient voices is quantavolution, that is, that the world and all that is in it owe most of their changes to forceful torsions and saltations. That quantavolution plays a role in the theory of natural and social science has never been denied. But the role has been grotesquely reduced by ignoring it and stressing evolution, by consigning manifestations of it whenever possible to times beyond mind, by framing scientific principles in prejudicial terms, by associating quantavolution with disreputable or outmoded religions and scientific beliefs and by unconscious editing of the evidence.

To our view quantavolution affords an instrument for scientific inquiry as useful as and perhaps superior to that allowed us by evolution. We find that the morphology of the Earth and the patterns and compositions of the skies bespeak quantavolutions. In biology, we see in the decline of evolutionary power over time, in the absence of transitional types in evolutionary branching, in the waves of extinction of species, and in the failure of evolution to provide an internal guiding dynamic, sufficient reason to promote the concept of quantavolution.

The guiding dynamic for quantavolution, whether in biology, geology or astronomy, may be electricity, a "strong force" that has been generally accorded a weak place in most sciences. For several reasons, we believe that electricity is the necessary and sufficient impulsion of cosmology. We noticed that a strong force is needed to accomplish change, whether in biology or astrophysics. Basically, electricity is to "gravitation" (if such exists independently of electricity) as 10^{36} is to 1. In every natural and biological process -- creation, accumulation, structure, function, storage, dissipation -- electrical theory is at home. The smallest observable or inferable operation of a molecule, and the largest explosion of a nebula, can be referred to the unified language and lawful behaviors of electricity.

The behavior of stellar bodies, including the Sun, can be described in electrical terms. The composition of "space" is a plenum of charges and ions, field and currents, winds and relatively stationary matter, of orbiting bodies shifting orbits as they transact, at times attractively but usually repulsively.

The fact that electricity is present in all matter, and an aspect of the existence and activity of all matter, presents us with the opportunity to study all matter and motion in an electrical perspective. Electrical attraction and repulsion seem to operate simply and flexibly in cosmology as well as in microbiology, and can be accommodated to the concept of inertia, the two together constituting a powerful instrument for the analysis of nature.

Finally we would point out one more helpful attribute of electrical theory. Invoking electricity enables us to avoid the mechanical blasting, usually required of gravitational and explosive mechanics, that brings inordinate destruction and thermal excess to situations where we seek quantavolutionary change with a maximum of selectivity and minimal mechanical bursting.

Despite their ubiquity, electrical phenomena have been isolated from the rhetoric of causality. When treated, they have been allowed as only secondary or even tertiary effects; instead mechanical and gravitational processes of enormous magnitude are postulated as the forces playing the primary (causal) role.

Sometimes magnetism (usually not observed directly) is seen to play an intermediary, or secondary, role in the deduced causal train which leads to the observed effect. But our outlook has changed. Once practically dismissed as inoperative in celestial matters, electricity, together with electrical effects, has increasingly been recognized to play a role in cosmic actions

Part Four

THE LATELY TORTURED EARTH

CHAPTER ONE

QUANTAVOLUTIONS

Clarence King was the first Director of the United States Geological Survey. He was liberally educated at Yale University and spent years in field work thereafter. According to the historian Bancroft, he "had acquired a reputation and a position second to no scientist in America." When he returned to lecture at the Sheffield Scientific School of Yale in 1877, he argued against the prevailing opinions in geology and evolution, insisting on the basis of his experiences and visions as a surveyor that the Earth had been lately devastated. The belief in catastrophism, he said, in surprising pre-Jungian language, was a true grasp of what had happened to the World. "Catastrophism is therefore the survival of a terrible impression burned in upon the very substance of human memory."

Because catastrophism is a word that excites emotion and connotes only destruction, the present work and the series to which it belongs prefers the more general idea implied in the word quantavolution. The concept allows a more peaceful invasion of the realms of gradualism, uniformitarianism, evolution, and anthropology.

I do not mean this book to be violent and bloodcurdling. We have far too much of such stimulus today on television, in movies and in other books and magazines. I even go so far as to say that the Earth system has been settling down - this without conclusive evidence. But facts must be faced. The Earth has been severely traumatized in the memory of mankind. In words that I have used before, any place on earth can be viewed as a Quantavolutionary Column:

> Any tube of one kilometer diameter circumscribed anywhere on the surface of the Earth, which reaches as high as the end of the magnetosphere hundreds of miles upwards, and as low as the upper mantle some thirty kilometers down, will have endured within the past 14,000 years radical changes in its absolute and relative orientations, its atmosphere, its rocks and its biosphere, including any long-lived human cultures.

Several principles characterize the theories of quantavolution: Every major feature of the Earth's surface is an effect of quantavolution; hence every feature figured in evolutionary theory is translated more realistically into quantavolutionary theory.

The dominant shape of the most determining events in natural history is a logarithmic or exponential curve where, from a preexisting state, sharp change occurs, followed by a steep exponential decline in the effect. After a time the curve of the effect flattens out, and an illusion may arise that the processes under scrutiny have always been as they are now.

The several descriptive spheres of natural activity: atmosphere, lithosphere, hydrosphere and biosphere, transact regularly, but most emphatically and completely under catastrophic impulsion.

Partly because of the greater force of inanimate being and partly because its own basic nature is identical with the inanimate, the biosphere is as subject to quantavolutionary experience and interpretation as the physical spheres.

The theory of quantavolution depends upon the evidence that catastrophes really happened, for it is upon such abrupt, large scale natural events that the quick leaping changes of quantavolution in the holosphere depend. By the same token, a quantavolutionary theory must show either that large spans of assigned time in natural history are fictitious, or, if they occurred, little of the natural world changed during their passage.

I use the word "exo-terrestrial" in preference to "extra-terrestrial." It is more exact etymologically and less romantic and sensational.

Still, Walter Sullivan, dean of science reporters, could declare in the *New York Times* in December 1981 that serious challenges to the conventional tempo and mode of evolution were arising; they came out of proof concerning links between catastrophe and extinction/genesis of species, out of the capacities of genetic engineering for modeling new life forms, and from the growing tendency to interpret the rarity of so-called missing links or transitional types as the non-existence of said types, introducing therefore the alternative presumption that macroevolution (quantavolution) introduced distinctly new forms suddenly. What Lyell wrote a hundred and fifty years ago, "that no causes whatever have changed the earth except those that still do so under the eyes of man," can be easily updated: today man's eyes are wider; they can see more and can see into themselves.

All quantavolutions imply heavy holospheric events. For periods before human race had quantavoluted (the subject of my work, *Homo Schizo I*), anthropological spheres of existence would, of course, be excluded.

It will be appreciated that, under evolutionary theory, holospherics tend to be less stressed. When large effects are reduced by time to minute causes, the side-effects are proportionately and even exponentially reduced. The more intense and sudden the event, the more spheres will be transacting. The larger the scale of an event, too, the more spheres will enter the action.

Suppose the Earth's rotational speed were to be slowed. This is a mighty event and takes a mighty force; Earth's rotational energy is calculated at 10^{36} ergs. Yet it has been observed (by Danton) to happen recently, if only for a millisecond. No account of effect has yet been rendered; perhaps the effects were immeasurably small, or perhaps the reaction of scientists were too slow. If large solar flares caused the retardation, as seems to have been the case, worse flares or other causes might produce a larger rotational lapse, perhaps a second of time would be lost; perhaps then a minute; why not an hour? Hypotheticals are cheap. The effects of lengthening the slowdown would be heavy. Every sphere of Earth, every force, would be activated in using up the energy surrendered by Earth in the deceleration. One would have holospherics on a grand scale. Ordinary language, the most archaic religious language, and scientific language could each provide the description required.

Now the quantavolutionist reverses the logic as well. We say, "the more affected the holosphere, the greater the force to be sought." The effects are proportional to the original force. When the effects exceed (or are theoretically calculated as having exceeded) a certain intensity, we must even go beyond the Earth into cosmic forces, drastically simplifying. Only in the supra-terrestrial arena, the planetary and galactic systems, are to be found forces large enough to do the Earth what appears to have been done. Only cosmically can truly great holospheric transactions be generated.

One can realize, then, the importance of the "when" and "how long." To say "speedy reactions"is to invite ultimately the cosmos in to explain our terrestrial phenomena. To say "slow reactions" is to keep the Earth within its cocoon in space, travelling evenly and safely. If the Alps tower above Europe, some force must have pushed then up. If the Alps are to arise suddenly, then something besides earthly forces are behind the event. We move into the cosmic realm. If the Alps are to arise over a great many millennia, then the force might be generated in energy measures conceivable from some mysterious, but still earthly, internal force.

CHAPTER TWO

THE GASEOUS COMPLEX

The atmosphere of earth is so delicate that most sudden and violent transactions in space or on Earth transform its constituents and their behavior. Considering what is to come in this book by way of demonstrating terrestrial catastrophes, one may wonder how it happens that life has survived five thousand, much less five billion years. The very fragility of the aura around us bespeaks the recency of the atmosphere as we know it.

For example, in-coming cosmic particles collide with atoms of the atmosphere, giving off neutrons that interact with nitrogen to make carbon 14. Then C14 couples with oxygen to form carbon dioxide, and is often ingested by plants and passed along to animals through the plants. When any plant or animal (living from plants) dies, it ceases to acquire C14 and the C14, which is radiogenic, decays at a constant rate into nitrogen. In the short term, the process is fairly regular. The ratio, in a specimen, of C14 to C12, a non-decaying type of carbon, can be used to date its decease. But lightning, smoke, dust, explosions, vapors and cosmic particle flux can alter the density of C14 in the atmosphere, hence in organic material.

Soviet investigators found C14 deviations in connection with galactic supernovas of the years 1054 and 1700. Judging by the C14/ C12 ratio in annual tree rings in or about the year 1908, when the exoterrestrial Tunguska body exploded with heavy local effect in Siberia, 1% less of the C14 was available in that year by comparison with the year before and after.

In another case, during a period called the Maunder Minimum, 1645 to 1715, when the Sun exhibited no sunspots and the Earth was gripped by a "Little Ice Age," the C14 found in tree rings of the period averaged 20% more than before and after. Grave events disturbed the atmosphere on other occasions. Between 3200 and 3700 B. C. and in the eighth and fifteenth centuries B. C., the quantity of C14 in the air fluctuated heavily.

The atmosphere itself is a model of disorder. It is continuously moving and reorganizing. Everyday its pressure goes up and down. About 99% of its mass blankets the globe at under 19 miles of altitude. This consists of the gases, molecular nitrogen(78%), molecular oxygen (21%) argon (1%) and carbon dioxide (. 03%). Water vapors rarely reach 1% of the total: normally, half of the globe is covered by clouds, which form, reform, and discharge their vapors almost entirely within six miles of the surface.

But the atoms are not in equilibium. In the heterosphere, atomic oxygen, helium and hydrogen are the abundant elements. Some of the helium and hydrogen is on its way into farther space, but is replaced, it is believed, to produce an equilibrium. However, Melvin Cook, a quantavolutionary geophysicist, has asked, "Where is the Earth's Radiogenic Helium?". Cosmic-ray sources are alleged to generate helium at $3x10^9$ g/ year. The same amount is estimated to be generated from the uranium and thorium in rocks of the lithosphere. With an Earth age of $5x10^9$ years, about 10^{20} grams of helium should have passed into the atmosphere by now. The atmosphere contains $3.5x10^{15}$ grams of helium-4; if a steady state, it must have passed out through the exosphere the equivalent of the aforesaid 10^{20} grams.

However helium-4 does not concentrate in the upper atmosphere significantly and "at the escape temperature of 1500°K at the base of exosphere, the rate of escape of helium-4 would be only about 600 g/year, or only about 10^{-7} as great as the replenishment rate from the lithosphere." Only by raising temperatures at the base of the exosphere by thousands of degrees could the helium be allowed to escape in sufficient quantities to permit equilibrium. This can be conceived as possible only by means of a number of immense solar storms that would wreak havoc on Earth or, worse, by large-body encounters wrecking the atmosphere. Cook suggests that the helium-4 is still increasing; the atmosphere is not in equilibrium; and if retrocalculated, a recent beginning or reconstruction of the atmosphere must be confronted.

Geophysicists and meteorologists nevertheless retain the concept of the atmosphere as a whole being in equilibrium. This is probably not so, even in the short run of a thousand years. The idea is difficult as commonsense, considering that all the way from sea level into outer space the atmospheric column is in continuous flux. It is agitated and fed from the bowels of the Earth with heat, vapor, etc. and bombarded topside by elemental particles of all kinds.

Motion is continuous, too, up and down the column and then horizontally with winds produced by thermal changes, such as the seasons produce, and rotational effect that, for instance, disturbs the atmosphere via surface irregularities such as mountains and basins.

Indeed, equilibrium of the atmosphere is probably more of a hope than a fact. What makes the hope into a "fact" is, not surprisingly, the uniformitarian conviction that today's actors and roles are unchanged from eons ago. Given hundred of million of years when animals and plants have been surviving, then the mix of vapors, nitrogen, oxygen, carbon dioxide, argon, ozone, and radiation must have been what they are today. And that spells equilibrium.

The belief becomes so strong that meteorologists, possessed of the "fact" of atmospheric equilibrium, can even take their turns at guarding the portals of uniformitarianism, assuring other scientists that meteorology, too, proves the long-enduring stability of present-day conditions. At the same time, ironically, meteorologists are leaders in the campaign to save the world from the atmospheric ravages produced by a few years of industrialism, atom bombs, and aerosol discharges.

A quantavolutionist may share heartily the meteorologists' fear of the poisoning of our present atmosphere. The quantavolutionist would at the same time point out the extreme improbability of the atmosphere's having been preserved intact, free from radical changes and poisons over long periods of time. Unless, of course, there were, before the present atmospheric system came about, some ancestral system that in its nature involved a true long-term equilibrium.

Two decades ago, dendrochronologists, historians, meteorologists, radiocarbon dating specialists, and astronomers combined in a most unusual enterprise. They delivered a blow to the theory of the constant Sun. John A. Eddy of the National Center for Atmospheric Research conveyed the message: "We've shattered the Principle of Uniformitarianism for the Sun." He presented evidence mentioned earlier, showing that for 70 years between 1645 and 1715 A. D. sunspots were almost entirely absent. It proved to be a period of bitter prolonged winters, when Londoners walked across an iced-over Thames River, when the Northern Lights hardly displayed themselves, and when the 11-year sun-spot cycle was absent. Lapses of the same kind were uncovered in other historical periods.

Other conditions may be expected to vary with sunspots - solar flares, ozone density, radiation diminution, precipitation, magnetic fields, atmospheric turbulence, famines and perhaps even human energy and inventiveness. No doubt the last will be among the most difficult to prove. No simple search of the annals of culture will reveal a closely related trend.

Hurricanes, volcanism, interrupted rotation, ozone destruction, ice ages, geomagnetic field reversals, biological extinctions and even explosions of cometary and meteorological material on Earth can be rationalized up to a point as effects of solar misbehavior. Such a theory is possible, but it would be like hiring a thief to catch a thief. For the Sun would then become sole factor in quantavolutions, in the effort to exclude other bodies from trespassing upon Earth. As we shall see, there is too much evidence of other operative factors to assign the whole job of quantavolutions to the Sun, even though, as a matter of fact, the Sun is the original sire of quantavolution in the solar system, according to the model of Solaria

Binaria, mentioned above, which begins history with a nova of the Sun.

According to the quantavolutionary theory here presented, solar behavior has exhibited only effects of a moderate kind since its gradual emergence as a distinct bright image some thousands of years ago. Before then, the Sun was hidden or a bright prominence in the cloudy firmament. Its indirect influence was of course always paramount.

The sunspots may be a trailing-off effect of the exhaustion of the electrical current and magnetic tube. That is, they may be fairly regular attempts of electricity to jump the gap between the Sun and its binary. In such a case, the sunspots should become less intense and more sporadic with the passage of time, like the plasmoids and bolts of Jupiter.

If the Alaskan muck contains the swept-in plant and animal life of large areas and the species it contains are modern, then one should suspect that sooner or later, as Hibben has opined, humans, even clothed and deep-frozen, should turn up by accident or deliberate excavation. Already, several pre-"Ice Age" settlements have been uncovered within the arctic circle by Americans and Russians. Rodents and mammoths froze quickly while eating warm-weather plants. How abrupt was the climatic change that killed them is unreported, if known. The polar regions were recently near-tropical in climate and ecology.

The bafflement of archaeologists over climate is understandable. They follow the evolutionists. But the attic of climatic evolutionism is stuffed with junk. When a modish dress does not suit the facts, an old-fashioned one is tried on.

Geologist Vita-Finzi practically abandons his search for climatic benchmarks in his authoritative work on the holocene. Lacking the engine of a general theory and a time-table to run it on, freightcars may be switched around at will. In one place he is driven to remark: "On the assumption that every yodel in the Alps had its echo on the coast, pebble bands are equated with glacial episodes, truant beds are eroded away, and the uplift of mountains is delayed to justify the absence of glacial features." He prays that the radiochronometrists will rescue the situation. But I have already concluded in my analysis of tests of time that a rescue must come from elsewhere.

Perhaps a quantavolutionary scheme may do better. It is not written in some law that enough time must be allowed to let humans get away, bag and baggage, from the changing air. Every catastrophe which they underwent would demand a climatic response as one of its effects. Hence there may have been a score of global shifts in climate within a 14,000 year holocene period.

This Pleistocene-Holocene boundary climax is euphemistically carried in the logbook of the sciences as "the end of the Ice Ages". I treat it as the Lunarian climax in *Chaos and Creation* and *Solaria Binaria* because of its apparent connection with the advent of the Moon. Hundreds of titles from many fields are dedicated to it. In oceanography, Emiliani extracts from Gulf of Mexico bottom cores the information that a fresh water avalanche descended upon the basin some 11,500 years ago and he wonders whether this was from a cataclysm such as sank the legendary continent of Atlantis. Tree pollen changed abruptly in the Great Lakes region about 10,000 years ago, according to J. G. Ogden III. "The only mechanism sufficient to produce a change of the kind described here would therefore appear to be a rapid and dramatic change in temperature and/or precipitation."

Oceanographers Broeoker, Ewing, and Heezing gather oceanbed "Evidence for an Abrupt Change in Climate Close to 11,000 Years Ago." Vita- Finzi reports that a group of geosols, or weathering profiles, ended their development about 12,000 years ago; the date is proposed as the holocene beginning for the U. S. A. From Israel, paleo-zoologist Joseph Heller writes of the faunal remains of a Kebaran Site on Mount Carmel:

> What then was the cause of the post-Natufian size crash? (9000-10,000 B. C.) The fact that the crash occurred in certain carnivores and rodents simultaneously suggests that it was not causally related to phases in the evolution of human cultures. Rather this simultaneous dwarfing favors climatic interpretation. Drastic climatic changes occurred in various parts of the world towards the end of the Pleistocene about 12,000 years ago. In tropical Africa, India, South America and Australia, conditions that were extremely arid before 12,500 B. P. suddenly gave way to increase in humidity.

It is generally accepted by pre-historians of Europe that the end of the Pleistocene Ice Ages brought disaster to human races and cultures. The finding is surprising, considering that the warmer the climate, the more abundant the biosphere should be. But if catastrophes were involved, the reduction and retardation would be understandable, indeed demanded.

Ruins of cultures are found in many a harsh climate of the world, in deserts, on high plateaus, amidst perma-frost, and in steaming jungles.

A Woods Hole Oceanographic Institution team reported in the *Scientific American* of March 1982 a set of discoveries which threatens the prevailing theory that oceanic waters are regionally stable, that regional bottoms reflect this aquatic stability, and that world climates can be determined by fossil and chemical balances of the bottom content. Eddies of the great oceanic currents such as the Gulf Stream occasionally break off from these gigantic oceanic flows and set up columnar rings of water that can reach 300 kilometers in diameter, even in this relatively placid age, and endure for 18 months or more. The ring-waters differ significantly in salinity, oxygen content, and temperature from their surroundings. Biological assemblages follow suit. Sedimentation rates are also a function of current velocity. Under such conditions, given several thousand, let alone several hundred million, years false climates can be expected to be inferred practically everywhere. Misleading strata will be exceedingly numerous. Once more, we must warn against the many theoretical structures of climate, hydrology, chronology and paleontology that interlock in varying degrees of poorness of fit.

CHAPTER THREE

HURRICANES & CYCLONES

Winds find a minor place in textbooks on earth features. They erode rock by polishing and pitting it, by making grooves, by shaping and faceting. They make various alcoves and niches in rock walls. They also form sand dunes in deserts, and blow the sand and silt of stream beds hither and yon. A sandsheet in Libya, over a meter thick, rests on bedrock over many thousands of square kilometers and is supposed to have been laid down by winds of the desert.

There are others like it around the world. Such aeolian activity is allotted millions of years to help shape the landscape; the number of millions, one or a hundred, is calculated from estimated past climatic conditions working against various constraints, such as whether landforms exist nearby to provide the material of erosion.

Tornados, cyclones and hurricanes now and then wreak havoc upon soil and settlements. Part of the climatic complex of this age, these storms are localized - the "tornado belt" of the south-central United States, the Japan and China Seas, and so on. Of course, bearing in mind the "many changes of climate over the ages," most places on earth would have suffered such storms in turn. When they occur, part of the biosphere is blown away with some of the natural landscaping. Paleo-anthropology and archaeology debate the relative contributions of the Orient and the Eur-African world to the earliest American cultures, for example, without proper attention to the possibilities afforded travelers by changing winds that come with changing climates, now pushing things one way and then again another way. So that even when the possibilities of cataclysmic changes in early human times are ignored, changing climates would carry culture both East and West.

We cannot figure how often such high energy local events have occurred, until the world is better surveyed with this idea in mind. But one can "think big". With a thousand tornados a year (300 in the U. S. A.) tearing up two thousand square kilometers of sediments and breaking down surface features, an area equal to the total land surface of the world (240 million square kilometers) would be superficially pulverized in about 120,000 years. If a conventional age of 3.6 million years is accorded the Earth's crust, the whole of it would have been scoured, not once, but 30,000 times by cyclonic action. In the short term, not all land would be affected equally, but in the long-term, given changing climates and drifting continents, an assumption of randomized strikes could be tolerated. Where then are the scars of 30,000 tornados in every geological column? Or even in any single one anywhere?

From this we might conclude that we have a great deal of field research to do in geological history so as to obtain a realistic estimate of the number of events. This is also the situation, we may as well say, in respect of meteoroid falls, volcanism, and other high-energy events to be discussed. The quantavolutionary approach to history comes naked as a neonate, without systematic hypotheses, data, or applicable mathematics.

If few such effects are discoverable, it may be because catastrophes acting on a large scale have obliterated almost all localized indications of damage. For instance, if great earthquakes have shattered rock strata, lesser violence to the rock would be hardly visible. The schist dropping deep below the city of Athens is infinitely fractured. Is this tempestite, thermotite, seismotite, hydrotite, turbotite, or what? If the wind god, Aeolus, blew at once all around the world, many sediments would be displaced, losing their local cyclone scars in the process and letting no new strikes penetrate deep into the new strata.

Ancient legends speak of a large role for winds. The sacred book of

Buddhism, the Visuddhi-Maggia, says that when worlds collide the winds "turn the ground upside down. Large areas crack and are thrown upwards. The winds pulverize the ground and it disappears into space, never to return. Thus ends a cycle of the ages." It is the extreme catastrophic typhoon.

The ancient Meso-Americans said that the former world was brought to an end by the great wind god, Huracan. Probably the origin of the word "hurricane" is here. Huracan is also manifestation of the great god Quetzalcoatl, who is also identified with the god and planet Venus. Huracan, the Heart of Heaven, fathered a large number of people, who he then destroyed in the darkness of a storm amidst black rain that fell day and night. So records the Quiche book of *Popul Vuh.*

Then animal gods mangled the bodies. "Air" is rarely missing in the legendary and early scientific classifications such as "earth, air, fire and water." The idea of world destruction by wind is, of course, quite disregarded by modern scholars. One hears the term "marine transgressions" but not "wind transgressions." It is surprising how few pages have been devoted to the winds by catastrophists, too. Again, perhaps the effects of hurricanes and typhoons are quickly concealed by other forces operating. Or the effects may be interpreted as tidal wave deposits.

The splintered bones of some fossil assemblages would indicate aerial rather than water transport. The Cumberland Cavern catastrophic life dump shows no evidence of water transport. Probably as many collections of animals and vegetation have been gathered and flung in heaps by winds as by water. In seeking the origins of some coal deposits, catastrophic winds are a prime suspect, along with rock and water thrusts.

The power of winds to push, pull and lift is great. The Hiroshima nuclear fission-bomb explosion is assigned an energy of 7.9×10^{18} ergs. The measured energy release of a one megaton fusion bomb explosion is in the range of 10^{22} ergs. This is about the same energy as exploded in the Berringer meteoroid crater in Arizona. "In one day a large hurricane releases as much energy as a 13,000 megaton nuclear bomb. Some hurricanes take a week to reach such intensity, others mature in a day or so. And during the time another may be at full blast a thousand miles away." Some hurricanes last three weeks and travel 1,000 miles. (One can bear in mind the immediate transport of resilient living species around the world by such means.)

An ordinary Kansas tornado will approximate $4x10^{18}$ ergs of kinetic energy. Its power in kilowatts is 10^{18} , "which is in excess of the capacity of all the generating stations in the United States." (ca 1959). The wind velocity at the center of its funnel theoretically may achieve 2000 miles per hour. By the Fujita scale, an F-5 wind, indexed at combined forward and rotating speeds of 261- 318 mph causes "incredible damage."

Electrical activity is so vigorous that Peltier's words of 1840 can be used as a model for an electrical cyclone theory. "Everything proves that the tornado is nothing else than a conductor formed of the clouds which serves as a passage for a continual discharge of electricity from above." Observers have been inside of this "enormous vacuum tube, somewhat similar to a geissler, neon or fluorescent light tube, conducting very low density electric current whenever there is a sufficient accumulation of electricity in the clouds to make the jump to Earth." Typhon, the cosmic spectral dragon felled by a thunderbolt from Jupiter, was anciently described by Apollodorus as "rushing at heaven" with hissing and screams, spouting a great jet fire from his mouth. This same Typhon is probably the origin of the word "typhoon."

Cyclones and water spouts (water-bearing cyclones) often appear in groups. An outbreak of 148 tornados was registered in the United States and Canada on April 3,1974. Sometimes associated with a tornado are a number of downbursts of high velocity winds that blow down whatever they strike, whether groves or houses or aircraft. Ted Fujita of the University of Chicago compares the downbursts with giant garden hoses aimed downwards upon circles kilometers in diameters; often they end their work in two minutes.

What might cause a vast number of cyclonic events to appear? A meteoroid bombardment, an interruption of the Earth's motion, a tilt of the Earth's geographic axis, magnetic axis, or sidereal axis: these would do, and also a large meteoroid impact, and a large body passing nearby, the latter, however, being tied almost inevitably to other changes in Earth's motions. Too, a deluge of waters might form into many ribbons, mushrooms, or funnels in descending. The winds and other effects of a heavy meteoroid impact would be simulated if a large number of nuclear missiles were trained upon a single spot and exploded at the same moment.

Winds act faster than water and have the same exponential effect upon the bodies which they may encounter as their speed increases. Wind pressure, that is, increases as the square of wind velocity, up to the velocity of sound at least. A 500 km/ hr wind exerts 25 (not 5) times the pressure of a 100 km/ hr wind; gravel then begins to behave like fusillades of bullets. Kelly and Dachille calculated that the winds created by a large meteoroid impact will move laterally and vertically with the speed of sound. Their effect has to be measured, too, in terms of the amount of debris that they transport. A single such blast, moving horizontally, can strip its area of passage bare down to bed rock, or below, especially if it is loaded with detritus, and may continue its major effects for a thousand kilometers. Only a mountain can stand against it and it, too, will be defaced; an instant ablation corresponding to millions of years of ordinary aeolian erosion will occur. Rivers would be wiped out and set up elsewhere. Valleys would be filled with debris. Great vegetable and animal dumps would be established in many places.

Differential atmospheric pressures define the existence of a wind; two clouds of gas, essentially isolated but lacking an effective "bag" to contain their isolation, interact. Electric potentials are established. Electrical forces thereupon flow throughout the transacting systems laterally and vertically. It is perhaps axiomatic that where there is wind there is an electric current and discharges. And where there is an electric current there is bound to be a magnetic field. And, lacking a better container, an electric current is contained by its magnetic field.

The largest deposits accorded to winds are not those of the Lybian peneplain, nor those of Egypt, but the huge areas of the Earth covered by loess. The term itself was invented for glacier deposits of the Rhine and Danube valleys and elsewhere in Europe. It found itself connected with the "drift", the glacial pebbled clay of North America, where vast stretches of the buff and porous earth, compacted but frangible to the fingers, were found distributed. Here transportation by ice sheets and rivers forming from their melts was imagined. Then, west of Peking, an area larger than France exposed its loess to geological inquiry.

Loess can occur at high elevations as well as on great plains. It breaks down into excellent thick soil in China and its cliffs degrade into natural terraces. Old roads cut through it, sometimes passing through the Chinese countryside thirty meters below the houses and farms on the loess above. In Indiana, the highest lands and ridges in particular have the thickest yellow clay (called drift or loess) and it is free of sand and grave. The loess is not stratified, nor does it contain marine fossils, and land fossils of shells and mammals are only occasionally found in it.

Ignatius Donnelly, in *Ragnarok* (1882), was already ascribing till, drift and loess to fall-out from a great comet, going so far as to deny the very existence of past ice ages, to which most scientists then and still today ascribe these materials. Fifty years after Donnelly, Penniston was advocating the thesis of an exoterrestrial origin for loess. Citing Shapley (later a violent critic of Velikovsky) and Belot for having proposed a solar nova as the cause of the ice ages, he reasoned upon this as a possible source of the material, which, experiencing high temperatures for a period of time, had its silicates metamorphosed in part to quartz, thus arriving at the loess.

Let us rest content here if we have but established several points: The force of wind rises with the square of its velocity, with correspondingly large effects upon the landscape. Hurricanes must be associated with every abrupt and intensive geological event. Cyclones convey major electrical and fire phenomena. In large-scale catastrophic events, a great many typhoons could originate to accommodate changed atmospheric and lithospheric motions or multiple meteoroidal intrusions. Finally, if the sediments of the world do not reflect adequately cyclonic effects, the reason may rest in their continuous erasure by more forceful events which themselves require identification. Furthermore, assigned geological times may be too long; maybe not enough events have happened to flesh out the skeletal ages.

CHAPTER FOUR

MAGNETISM & AXIAL TILTS

Could the earth have even turned over completely without interrupting (interrupting very little) its spin or its magnetic field? The geographic poles would be reversed, and along with them the magnetic field. The Earth could not perform such a movement without an external assist, whether from an upsetting explosion of gases from the Sun or from the attraction or repulsion of a large passing body.

A moment's reflection will rid us of any notion that the action would be harmless. The atmosphere, hydrosphere, and lithosphere would be agitated and produce effects that by any measures would have to be called quantavolutionary. For instance, it appears most likely that the widespread sudden destruction throughout the northern regions of the mammoths and other large mammals occurred in conjunction with a tilt of the Earth's axis in the presence of the exoterrestrial entity causing the tilt. We can say this because a sudden deep vacuum freeze, asphyxiation, thrusting of masses of gravel and bones, and permanent cold ever thereafter, such that the animals are sometimes found still fleshed-out and diagnosed in certain cases as heart-failures or with blood-clotted lungs, must indicate a holospheric event comprising an atmospheric and aquatic withdrawal, the descent of an extreme coldness, and upon the passing of the body, returning tides of water and wind to accomplish quick burial under muck, ice and tundra.

Lest the idea be considered quite fanciful, it should be recalled that several ancient sources refer seriously to a reversal of directions. Herodotus and Plato cite Egyptian sources of occasions when the Sun changed directions and arose in the West instead of the East. A ceiling in the tomb of Senmut of Egypt also pictures a reversed sky tableau such as would occur were the Earth turned upside down. In fifteen spectacular pages, Velikovsky searches out and orders rationally other indications in legends and writing of a reversal of directions that could only come with the Earth turning upside down. The contexts scarcely permit the alternative, a cessation and reversal of the Earth's rotation.

Presently, the evidence for sidereal tilts is considerable, for geographic tilts also some, for upside-down tilts little, for stop-and-reverse rotation very little. There is no way in which astronomical assurances can be lent to geologists on this account. Conversely, there is enough doubt on all scores to let geologists be open to the possibility of several catastrophically effective manoeuvers of "Spaceship Earth".

The possible occurrence of reversals in proto-historical times may suggest additional reversals in pre-human ages. However, Milton and I have presented in *Solaria Binaria* a theory according to which the Earth was in grip of a huge external magnetic field of the solar binary system until perhaps eight thousand years ago; during almost all of geological time, it could not reverse its field. In fact, it is argued that this same magnetic field and its reciprocal electrical current are the present geomagnetic field and current within the Earth, which have been steadily undergoing decay since the grip of the external magnetic field was released. This theory permits us here to explain the principal geological problems connected with terrestrial magnetism.

We would have to assert that the numerous alleged reversals of the Earth's magnetic field in geological history simply did not occur. Obviously there is no evidence to be obtained one way or another by atmospheric testing of the field; any number of reversals (or none at all) might have occurred without leaving discernible evidence.

The geophysicist, however, can search for evidence of the magnetic field in rocks. Igneous rocks have often been imprinted with magnetism when in a molten state; hence they hold myriads of tiny compasses, pointed towards the magnetic pole. If for one set of rocks the compasses point north and for another adjoining set they point south, it is conceivable that the magnetic field had reversed itself on an occasion between the melting and hardening of the first set of rocks and the melting and hardening of the second set.

Magnetic mapping of rocks is almost entirely of this century but has burgeoned swiftly and, some say, chaotically. Persuaded that they can tell the ages of rocks by radiometry, explorers have used time as a reliable indicator of the change in the magnetic field of the Earth. Since the rocks of the world have exhibited a bewildering variety of magnetic directions, many "dated" strata of differing magnetic direction have been assigned to the different magnetic periods, usually forced into a preconceived mold of "normal" and "reversed" magnetic field.

Depending upon the angle of declination, not only have such fields been noted, but they have been asserted to pertain to shifting magnetic poles. Some students have supported the idea that hundreds of field reversals have taken place in the several billions of years allotted to the Earth's history. One catalogue reports 433 paleomagnetic poles for 3 to 4 billion years of Pre-Cambrian time, an average of one new pole per 7 to 9 million years. Since the Cretaceous, says Heirtzler, 171 reversals of the magnetic field have been identified. Others have perceived certain intervals of time to elapse between reversals, 700,000 years, fifteen million years, and so on; several studies claim that the farther back in time one goes, the longer the period between reversals.

Also, field reversal is an indicator that worse things may be happening. An incoming giant meteoroid may dislocate the magnetic field in the course of destroying life and blasting rock. Whatever it lays down or heats to melting point will be stamped with a deviant magnetic imprint as it cools, provided the field has not sprung back into its original figure.

Meteors have pronounced magnetic effects. Passing cometary trains exhibit strong electrical disturbances and can cause the same in transacting bodies as in the space plasma. A large meteoroid, whether impacting or passing close by, will disorder the Earth's electromagnetic field. Also, were the Earth to change its orbital position, it would behave like a comet, with a flaring electric tail representing electrical transactions with the unaccustomed medium of passage.

To conclude these pages on magnetic and geographical tilts, we can state our position: the geographical figure of the rotating Earth can tilt or reverse north and south, with moderate applied exoterrestrial force and with large holospheric damage. It has done so. The magnetic figure of the Earth will tilt or reverse in general accord with a change of geographical figure, but can also tilt or reverse independently depending upon a large electrical exchange between the Earth and a massive agglomeration in space. It has done so repeatedly. The damage is much less. Both types of change - of geographical and magnetic axes - could not have occurred, by the theory of Solaria Binaria, until the binary system was collapsing, which has been placed in time by the present author and again by Milton and myself at less than 14,000 years ago.

There remains a more devastating change, whereby the Earth not only tilts but also emplaces its poles upon a new geographical location. The physical force needed to accomplish such a change is many times greater than that required for the tilt alone, because the rotation of the Earth is both interrupted and altered in orientation. It is known that the Sun changes differentially the rotational speed of its several sections and some sharp movements may occur in connection with solar storms.

But hard geophysical evidence, as presented by Hapgood, Velikovsky and Cook, for instance, supports belief in a recent ice-age finale that shifted the north geographical pole from a position presently denominated by Baffin Island, 20° south of its present location. There is a measurable spring-back occurring all the way from Scandinavia to the Hudson Bay area, a rising area that may be due to a new rotating figure of the Earth, involving a new equator, and possibly to collapse and sudden removal of a burden of ice that had been weighing down the region. (Inasmuch as the great global cleavage passes through the center of this region, one has to introduce the probability of a forcing apart and expansion of the area between the two rising elements of continental rock.)

Surely, if the Moon were to have erupted from the Pacific Basin, the Earth's shape would have been altered, the crust would have been half removed, and the conditions Gold sets for a shift of geographical poles would be satisfied. A great force moving southwestwards would have tilted the globe, removed the crust, cleaved the globe, set the continental fragments into motion, slowed the speed of rotation, and established a new figure of spin, with a new equator and new geographical poles.

The full range of possibilities in tilts has not been completed yet. Two additional ideas remain to be presented. The first concerns crustal slippage. The Earth's shell or crust, contributing about 1% to the Earth's radius, lends about one thirtieth to the moment of inertia of the whole Earth. Apparently, then, if the shell can slip without an identical movement of the mantle and core, the energy required to change celestial and geographical orientations on the shell would be less than that required for a total reversal or retardation of Earth motions.

There are signs that this stratified slippage has occurred in the overwhelming evidence of crustal destruction around the globe as, for example, in the outpourings of lava found everywhere. Even so, the energy required for total shell slippage (following the attraction of a passing body) is formidably high, and where it would be applied is crucial, so that this idea appears, initially at least, to be as totally destructive as any other means of moving the Earth about.

Archaeology affords support to the proposition that the Earth has changed position relative to the Sun and the planets in recent antiquity. In connection with the human drive to build settlements according to the prevailing cosmological observations and beliefs, the compass orientation of the constructions presents highly important issues in regard to changes in the Earth and the sky. That the earliest humans felt compelled to address their dwellings and public places to astronomical occurrences is generally granted. No one has yet found an ancient settlement capable of taking some shape that is not sky-oriented.

Certain claims of "fixed" structures warrant study. The most famous is the Great Pyramid in Egypt. The age of the Great Pyramid of Ghiza is in question. It has been ascribed to around 3200 B. C. and to other times. But no one suggests that it was built after 687 B. C. or for that matter after 1450 B. C. that is, after the end of the Middle Bronze Age. The West face of the Great Pyramid, which Stecchini believes was drawn first and is the basic face, is oriented 2'30" west of true north.

A more important question concerns whether the almost perfect north-south orientation means that no tilt or change of poles has occurred since the Great Pyramid was constructed. The Pyramid may have been oriented to a pre-existing true north, marked by another star. The axis of the earth shifted celestially. But an abundance of stars can be used to mark true north;

Polaris is the most recent star and naturally the Pyramid points to it.

The huge stones placed in circles and lines at Stonehenge, England, can be proven to be only generally oriented to observe solar solstices of the present age. Otherwise they display actual rearrangements of stones, done with immense labor, which can best be accounted for by an axial tilt, that is, by catastrophe.

Here, as at other magnetic settings, the earth scientist needs to take into account human motives, asking oneself: is it likely that the stupendous collective labor required to build these great structures, admittedly astronomical, would have been mobilized if the Earth (and hence the skies) were not exhibiting strange and terrifying changes of motion? Was the human urge to control the sources of his terror implicated?

Attempts have been made at dating Stonehenge by C14 on organic objects found in association with it. MacKie is of the opinion that the dates of Stonehenge and other megalithic astronomical sighting locations would not permit one to claim reorientations of the Sun after 1500 B.C.

Still, Stonehenge, like the Pyramid, is a catastrophized artifact in the first place, and bears also the marks of catastrophic changes in its settings. The C14 dates are not abundant and consistent, nor generally reliable within the span of centuries.

The Mesoamerican sites magnify the uncertainty. There are many of them. All are thought to have been set up after 1500 B. C. Macgowan, (1945), and now we quote Anthony Aveni extensively,

> ... seems to have been the first person to suggest that the plans of a large number of Mesoamerican cities exhibited an east of north axiality. Among those sites which evidenced some orderly arrangement, he observed that the orientations fell into three groups: true north, about 7° east of north, and about 17° east of north; he noted that few sites were oriented west of north. In the 17° group were Teotihuacan, Cholula, Tenayuca, Mexican period buildings at Chichen Itza, Tula, and the pyramid adjacent to the Zocalo in Mexico City. A number of sites of the Peten District seemed to belong to the 7° group. Macgowan suggested that a historical pattern might emerge in the sense that early structures such as Cuilcuilco possessed a nearly true north axiality while the 17° east of north orientation showed up in the later buildings.

Aveni found by transit that fifty of the fifty-six sites surveyed align east of north; the 17° orientations seems to be prevalent in the valley of Mexico.

Yet Carlson, working on centers carbon-dated between 1000-1400 B. C. says that "Olmec culture is well-characterized by ceremonial centers, which are generally 7° to 12° west of north...". This would suggest that tilts of different ages are represented in the two regions, or that the Olmecs, who invented the magnetic compass, may have oriented their buildings to a magnetic north. Almost all of them deviate from true north orientation. Or a shift in the ground beneath occurred.

Teothihuacan was probably the religious center of ancient Mesoamerica, like Rome of medieval Europe. The fundamental Teotihuacan grid as excavated is oriented 15.5° east of north. Of the some 30 symbols that the Aveni group have assembled from elsewhere in Mexico, the orientations of 19 are given. Of the 19, nine are oriented within 2° of the Teothihuacan grid. Of the nine, all except one (carved on an outcrop) are on a floor. Of the remaining ten with known orientations, all range between 35°42 and 80°24'; all are incised upon outcrops except one that is on a broken flat stone whose "axis points toward Teothihuacan"(TEP3), and another (TUI) that is "pecked on horizontal floor of lava field." Most of these are considered as pointing towards the summer solstice sunrise, which is rather insulting to the intelligence of the humblest pecker. Are we to believe that they could not find the point of farthest advance of the Sun? And why should outcrops be carved east and floors to the north?

It would appear that either (a) the carvers were inexact amateurs with biases towards the east, or (b) the larger part of Mexico shifted its axis by 15.5° to the west of north in response to seismism and/or a tilt of the Earth's axis in reference to the solar and sidereal system, or a geographical transfer of poles implying a changed axis of rotation, or (c) the axis of Teothihuacan shifted at an early time eastwards from true north and its new position was assigned sacred and ritual meaning, a Holy North to be imitated, just as the 260-day Sacred Year was tenaciously preserved, without a celestial referrent, until modern times, alongside the 360 and 365-day calendars. In the case of (b) and (c), the extreme eastern orientations of the peckings might have been memorial, without special orientation, to Teotihuacan's gods upon the occasion of faulting, fracture and exposure of new rocks. The geology and the relative dating of the peckings are important in considering these alternatives. Especially, the hypothesis can be entertained of a deliberate attempt to follow a fault line (especially if an electrical current were running) in the outcroppings. (If the Etruscan priests took possession of and catologued all aspects of a spot struck by lightning, similar obsessions may be expected among the equally obsessional Mesoamericans.)

CHAPTER FIVE

ELECTRICITY

Electricity is everywhere. It presents itself in the smallest particle and, some of us believe, commands the behavior of every remote galaxy of stars. It is part and parcel of every natural transaction. Perhaps it is the hunger of protons for electrons that initiates all natural behavior, whatever the scale or intensity.

Earth scientists have been reluctant to admit electricity to their domain. When geologists stood upon mountain tops and St. Elmo's fire flowed from their beards and hammers, they could not well deny this "god's fire". But one searches in vain for a treatise on St. Elmo's fire, one of the oldest and most fascinating phenomena continuously reported.

In fact, there exists no treatise on the full range of electrical behaviors related to geology. This universal presence of electricity in geological events has not excited systematic attention, no more than it has in astronomical events up to recently. If one seeks non-rational explanations of an ideological or psychiatric sort for such avoidance, it may be in the quixotic or miraculous appearances of electrical phenomena. Bordering upon the religious and the occult, these set up psychological resistences among "hard" scientists.

Lately, the ionization of the atmosphere has come to be studied. Even the ground beneath our feet has come to be conceived of as a conveyor of waves of numerous types, ranging from the gross seismic tremors that topple whole cities to the delicate motions of the wire in the hands of dowsers in search of underground water. Ions are electrified particles; they affect the growth, fibers and nervous system of plants, animals, and humans in ways mostly unknown. Many students think that an abundance of negative ions in the atmosphere produces a sense of well-being, but that "excessive" positive ions provoke depression, irritability, and illness.

The Earth's surface contains a charge; it too is unknown in extent and effects. The charge is called negative originally because it is of the kind that comes from rubbed resin, and conventionally because it comes from the ground. On a clear day an electric potential of about 100 volts per meter of height occurs. The charge of Earth tends to persist in the absence of exoterrestrial intrusions, employing the lower atmosphere as an insulator. The charge, in our opinion, will have varied greatly over the human past. Then its variation, as well as its constancy, must have had significant effects upon human behavior and ecology.

The Earth may be presumed once to have been in the grip of a constant heavy charge, for reasons that will unfold below and were also treated in *Solaria Binaria.* It began to lose this charge, both gradually and in series of catastrophic discharges. Today, solar flares excite large surges in the flow of charge from upper atmosphere to ground. Too, thunderstorms may be principally a method of balancing the atmosphere-lithosphere equation by releasing ground electricity.

There persist certain phenomena that may reflect this decline of charge. All over the world there are pathways that were worked out mysteriously (part instinctively and part deliberately) by ancient men and that are followed today. Michell has sought out the English paths especially. He shows that they are often not the shortest way between two points. Rather they have seemingly pursued geodetic "power lines" which thereupon developed as religious routine, ritually followed. As with many customs, people follow behavior that originally had a perceived and sound meaning.

In an atmosphere where electrical and dust turbulence were prevalent, as in times of Exodus and other periods that I have identified elsewhere, and the Earth was discharging at an effectively higher level than it is today, the incitement of electric displays without motors, pumps, and wires was easy: large potential differences continuously presented themselves for exploitation. Electrical effects became essential to political and religious roles and were subjects of jealous contention within and between governments. A full social analysis is presented in my treatise on Moses; what may be stressed here is that the existence and activity of such devices evidences that the Earth was then in a state of heightened electrical activity relative to modern times.

With the settling of the skies, the intensity of electric phenomena diminished. The divine spark manifested itself less and less; the arks were carried more and more up to the mountain temples (e. g. both the Temple of Solomon and the Temple of Jeroboam). The angels, demons, and mountain gods manifested themselves in electrical demonstrations on high with the aid of crosses, trees, and poles. These, too, could not be maintained. Empedocles, when discussing the four elements, fire, earth, air and water, says that fire has ceased to "travel", and no lower forms of fire remain. Plutarch wrote at the end of the pagan age an essay on why the highly placed Delphic oracle had lost its influence; he gave the vaguest of references indicating a failure of electric current, but the question itself is significant. By late classical times, the knowledge of arks and of the exploitation of "god's fire" was largely defunct.

The Tower of Babel was probably erected at a time when electrical perturbations were attributed, if my analysis in *Chaos and Creation* is correct, to movements of the planet Mercury. The arrogance of the builders in attempting to reach the sky was punished, recites the Bible, but in a peculiar way. They who spoke the same language when they began their work were caused to "babel" in many tongues. The Earth shook long beforehand; the tower partly sank into the ground, so say Jewish legends; but also much of the tower was destroyed by fire from the sky. The work had to be abandoned and afterwards the nations spoke different languages.

I offer a scenario for consideration. The Tower of Babel was being built in terror and hope of appeasing sky-bodies, possibly Jupiter-Marduk or Mercury. Conscripts or slaves of many countries made up a work force of 50,000 men. They put together a rough lingua franca from the language of the area to communicate on the job. The approach of a large heavenly body occasioned the build-up of charge and then a flowing discharge through the structure, creating a confusion in administrative orders and a linguistic amnesia especially in the lingua franca. No longer could people understand each other. And then the whole edifice was struck by cosmic bolts, partly fractured, and exploded. To this must be added a sobering consideration: according to recent Bible scholars like Israel Finkelstein, Chief-archaeologist at University of Tel-Aviv, the Tower of Babel, then a functioning astronomical observatory, loomed from its perfectly finished 91.5m above the captive Hebrew scholars in Babylon who were beginning to write down the Books of Moses - and survived their departure intact. It's destruction in *Genesis* may have been an instance of wishful thinking projected onto a hated symbol of oppression, which probably triggered memories of a much earlier event, or events, which occurred in Mesopotamia and elsewhere. It is possible also that Babylonian records of precisely such events were available to the Hebrew scholars working on the Pentateuch.

"Slow lightning" is the geologically and biologically effective discharge of terrestrial electricity. A "slow lightning flood" may be conceivable, too. The curious vitrified forts of Scotland may be a case in point. They remind us of the Ziggurat of Babylon. Their stone and mortar are fused solidly with the clifftops to which they adhere.

It is probable that many thousands of burnt eminences exist around the world whose tops have seen the fusion of rocks, perhaps even Troy IIg, the "Burnt City" so-called. The famous site, whether or not it was the real Troy, is on an eminence. While not high, the city would have had many small reservoirs of water, whereas the ground outside might already have been dried out. In Troy IIg a sulphurous color suffuses all outdoor spaces and passageways of the town. A deposit of lead and copper melted and flowed around the town. (It is possible that this melt had been scavanged after Schliemann reported it in the 1880's and the discoloration was all that was discoverable when the Blegen expedition re-excavated the site in the 1930's).

No human hand could have or would have set such a fire. The heat was fierce. The ash was far too abundant for a deliberate fire from local materials, and carried a red color. In places it was like calcinated rock, a meter or more in depth. No one would have wanted to destroy precious metals (not so mention even more precious metal left in abundance in the scorched houses and the Treasure of Priam, found on a Wall). Noteworthy is the absence of human and animal skeletal material in the ruins. Either they turned to dust from the heat, or the electrical build- up was sensed, as it is by animals before earthquakes for example, and they fled from the hill onto the plain where the sensations were absent.

Perhaps a heavily charged cosmic body was approaching or was near the Earth with an opposite charge or inducing one to collect on Earth; this would cause numerous discharges. Every eminence, one might imagine, would offer an exit for lightning, especially if it held the slightest metallic component, and were not surrounded by damp lowlands. Buildings are not needed.

The "slow lightning" may shape not only eminences but also subterranean cavities. Von Fange writes that "The same phenomenon has been observed in the mounds and barrows of the British Isles. Some have at one time been filled with an intense heat. Their walls are melted and their contents fused. The stones of the innermost cell of a long barrow near Maughold on the Isle of Man have been fused together like the mysterious vitrified towers of Scotland and elsewhere." Many Egyptian tombs and the interiors of pyramids are scarred by intense heat. Caliche ($CaCO3$) adhering to bones and rock undersides in a California burial cairn provide radiometric dates of 19,000 to 21,000 years, whereas archaeological estimates of the many such cairns give 5,000 B. P. or less.

Caves are said to be formed by the percolation of water through weak stone, cracked stone, or interstices of layers of stone. The filtering drops become trickles, and then streams. The cavity is enlarged. The river deviates or dries up and the interior is prepared for occupancy. Time elapsed may be "millions of years."

Although water may quickly hollow out caves, the role of electricity is not to be ignored. Electric fields, as Asakawa has demonstrated experimentally, enhances heat transfer in nearby gases (up to 1.5 times); liquids (up to 2.0 times) and solids (up to 1.6 times), depending upon the positioning of electrodes and the strength of the applied fields. Perhaps caves are ancient hotspots, electrical calderas, where creation time is shortened by the blasting impatience of electrical arc currents.

CHAPTER SIX

COSMIC & TERRESTRIAL LIGHTNING

A powerful; highly developed lightning of ancient Italy, the Etruscans, believed in the strictest set of relationships between the small Earth and the great and divine Universe. They planned their cities astronomically, as did all early peoples, but, more specifically, worshiped lightning and gave "the thunderbolting god" Jupiter to the Romans. They founded a College of Lightning Arts (ars fulminum) at Visul. When a bolt of lightning struck, the ground became at that instant hallowed; no one might disturb it until priests made a site inspection and had concluded which of thirty types of lightning it was and what should be done about it.

They dug wells to receive lightning and marked the wells with the bidental symbol of Jupiter (Zeus), a two-pronged spear. Zeus has been variously portrayed as the hurler of cosmic lightning, with a two or three-pronged spear, and even hurling a bolt whose shape was not forked lightning but like an American football, a plasmoid perhaps, a kind of lightning bomb.

All mountains were sacred to thunderbolting Jupiter. Seneca, the Roman stoic and dramatist, has him dissolving mountain ranges with his bolts. The Bible says the same of Yahweh, all this and more.

Psalm 97 gives us :

"His lightnings lighten the world;
the earth sees and trembles.
The mountains melt like wax before the Lord,
before the Lord of all the earth."

The Babylonians speak so of Marduk, the Indians of Shiva, the Persians of Mazda.

Archaeologist Nicola Rilli dug in one location at Prato (near Florence) and found three distinct heavy ash layers defining three distinct periods of prehistory. He found a small silo of grain, intact but carbonized, a fact that he ascribed to a great fire that had been suffocated. Lightning fires may have played a role in the burnings.

Electrical phenomena akin to lightning are associated with volcanism, earthquakes, and meteoritic phenomena, including atmospheric pass-through and impact explosions. They may also be an independent "instrument of the gods," as strong or stronger than gravitation in their effects when two dense bodies approach one another closely. Further, cosmic electricity may traverse a whole star system or planetary system.

C. E. R. Bruce of the British Electrical Association for many years sought recognition of the place of electricity and lightning in the creation and destruction of whole galaxies of the universe. He described lightning discharges of $6x10^{11}$ miles in width and ten times as long generating temperatures of $5x10^{8}$ degrees Celsius and lasting for 10^{6} years or more. The discharges occur amidst accumulations of cosmic dust.

Bruce's colleague, Eric Crew, who shares his views, has given more attention to cosmic lightning within the solar system and particularly in encounters involving earth. How he handles electrical problems of large-body encounters can be exemplified in the following passages:

If a charged body B (such as a large comet) approaches a planet A which has an atmosphere, opposite charges are induced and the atmosphere will be pulled out towards B. This increases the voltage gradient between B and the extended atmosphere very rapidly and violent discharges may take place even though the

> two bodies are separated by a considerable distance. The effect is intensified if both A and B have atmospheres, and even more so if they have opposite charges.

The effect... is to cause jet of compressed material to form and for the substance to be ejected on to the negatively charged body, or the induced negative charge.

> Charges induced in the solid surface of A as B approaches will cause a ground current to flow and the resistance of its path will cause the induced charge to lag behind the line joining A and B. The electrical force will produce a turning moment on A and B and the resultant motion will depend on the direction of the force in relation to the axis of rotation of A and B. The displacement may be increased if B has a crust floating on a molten interior, as the moment of inertia of this would be much smaller than that of a completely rigid sphere, even if the possible tilting of the axis is ignored .

That is, in the case of the several large body encounters of the Earth, which we think may have occurred, strong lightning exchanges took place, atmospheres exchanged in varying proportions, debris flew into space, powerful ground currents of electricity followed the point of closest contact, and these currents assisted inertial forces to push crustal sections of the Earth over its plastic mantle.

Ralph Juergens' theories of cosmic electricity have been close to the historical events proposed by quantavolutionary theorists. Intimately acquainted with the experiences and ideas of Velikovsky, he worked for many years upon the basic astrophysical problems posed by the Venus-Mars-Earth scenario, specializing in the application of electrical theory.

His primary theory deals with the source of solar energy. It is in one sense non-catastrophic. It is also quite new and unaccepted; yet, as he says "the modern astrophysical concept that ascribes the Sun's energy to thermonuclear reactions deep in the solar interior is contradicted by nearly every observable aspect of the Sun." Whereas the conventional theory is that the Sun derives its energy from a hydrogen-fusion nuclear reaction continuing over millions of years, Juergen's theory is that the Sun's surface bears a negative heavy electrical charge, which it has gathered mostly from galactic winds and from a great many bodies brighter than the Sun, and which discharges itself upon the solar system bodies. The solar radiance that strikes Earth and causes heat is as nothing compared with the galactic radiance that strikes the Sun. The Sun's bloated atmosphere is the anode; its highest levels are of the highest temperatures, which go down, rather than up, as the surface of the giant gas bag of the Sun is approached. Hence, elaborate attempts to catch neutrinos from the Sun's "solar furnaces" as they traverse the Earth must fail; if no nuclear fusion, then no neutrinos.

The Sun's radiance, varying only slightly as its total charge varies, penetrates the electrically neutral plasma of interplanetary space, passes through the positively charged outer magnetosphere, enters a neutral zone and then a negatively charged inner zone, and finally strikes the Earth's atmosphere with warming and radioactive effects. (Jupiter does not "need" the Sun's heat; it radiates several times as much energy as it receives from the Sun.)

A great proportion of all the craters and many fissures of the Moon and Mars, and, though less visible, of the Earth, are explained by Juergens as the effects of cosmic lightning, occurring during the holocene period that we are studying.

The "plasmoids" which I referred to earlier are a type of lightning conducted to Earth as "pieces of plasma." These balanced "things" of positive ions and electrons retain their identity and appear as luminous objects of missile-like proportions. They would cause impact craters or above-ground explosions that leave little trace. A second type of primeval lightning, like that known best to us, would give clear evidence of electric scarring, whether as a crater or as a jagged crack in the ground.

Juergen's theory of Moon and Mars belongs to Earth as well. The Earth must have lunar rilles in large numbers. An unknown but considerable number of craters, "river" valleys, fractures and ravines must owe their origin not to ice, water, volcanos, or meteoroids, but to cosmic lightning. In the absence of well-directed field work, not only are their indications misinterpreted, but usually their very existence remains a surmise. The present level of electrical activity on Earth does not excite research except in imaginative minds, like Ralph Juergens, Nicola Tesla and Frank Dachille.

It is well for geologists to consider meanwhile the promise of such theories. Take, for example, the consequences of the concept that the Earth's global electric potential has not been uniform throughout its history, an idea that I repeat in this book several times; consider its consequences for another insistent idea of these pages, that geological time may be grossly exaggerated.

Juergens argues that the Earth's surface potential is highly negative and low. Suppose that it is lowered further. Rampant radioactivity would occur. The half-life of every radioactive atom would be drastically reduced. Radiochronometric time would be largely erased. In the opposite case, if Earth's potential became higher and less negative, polonium, for instance, which has a short life as evidenced in the geological record by the halos it inscribes upon rock, would acquire a much longer half-life and so would other radioactive isotopes.

Chapter Seven

FIRE & ASH

The legendary and early historical record is replete with assertions that global burning has occurred. Writing apparently about historical experiences, Seneca, the Roman stoic philosopher, gives a common ancient view of the holocaust:

> And when the time is come when the world destroys itself to be renewed, then (Earth, seas and life) will destroy themselves by their own strength. Stars will fall upon stars. And when all material things are in flames, everything which now shines according to a planned distribution will rise up into a single fire.

Of course, Seneca does not declare that a stratigraphical record will be thereafter available; the Earth is "renewed," which implies that few marks would have been left upon the rocks and no bed of ashes would have formed and persisted. Where are the ashes of single or multiple events, for that matter? Sometimes they are present, sometimes not. In certain parts of the world, extensive beds of ashes of possibly local type can be found. They are thin. We can find the ashes of Troy, on several levels of destruction, but can the ashes of the countryside around be found? If not found, does that means that Troy alone was burned, or that ruined Troy alone preserves its ashes?

G. R. Carli, writing in 1780, was already asserting that "the idea of a deluge of fire and a deluge of water was present among all peoples... This idea of fire and water... seems to recall tradition of an event of which the memory has endured. It is certainly odd that the indications manifested by a seaflood should have suggested the idea of a deluge of fire. Carli cites Clement of Alexandria for the observations that Stenelas, father of the king of the Ligurians, lived at the time of the fire of Phaeton and the flood of Deucalion. So fire and flood occurred together. Reasoning from effect to cause, Carli then assigns the coal deposits of the world to burning and water acting in quick succession, a theory now coming into prominence again. He argues that only a comet could burn up the world, drop vast amounts of water, and bring great tides at the same time. Probably this line of argument will stand up: a large body encountering Earth, even if it were not dropping water or ice, would bring both conflagration and flood. Whether it crashed or not, the effects would still be similar. Donnelly produces an abundance of legendary accounts of the world in flames: from Druid mythology, Hesiod's Greek account, the Eddas of Scandinavia, Ovid's Roman account of Phaeton, the Meso-American Toltecs' Codex Chimalpopoca, the Persians' Zend-Avesta, the Hindus' myth of Ravana and Sita, and the legends of the Tupi, Aztecs, Tacullies, Ute, Peruvian, Yurucares, Mbocobi, Botocudos, Ojibway, Wayandot, and Dog-Rib Indians, that is from one end of the Americas to the other, and across both continents. He quotes the Gothic Surt of the flaming sword, "He shall give up the universe a prey to the flames," and also the Algonquins, whose god "will stamp his foot upon the ground, and flames will burst forth to consume the habitable land."

Job of the Bible hears from a retainer that "the fire of God is fallen from heaven and hath burned up the sheep [to the number of 7000], and the servants, and consumed them, and I only am escaped alone to tell thee." (In our days cases of a score or more animals being electrocuted by a lightning bolt are recorded.) There begins then the woes of the stubbornly patient Job against frightful divine tests. It is only one of many references to naturally caused combustion in the Bible. The story of Job may be exceedingly old; there Elohim (Heavenly One) is addressed; it happened in full Neolithic times, perhaps at the ending of the age of predominant Saturn worship.

Later in reference to fire is the "flaming sword", east of the Garden of Eden, "which turned every way, to guard the way to the tree of life." This was after the "Fall from Grace." The image of a sword in the sky may refer to the Great Central Fire of early Greek Philosophy and, as we elaborate in *Solaria Binaria,* to a then intermittent arc between Jupiter and the Sun.

The Latin author, Ovid, is the principal source of the Phaeton story. The Babylonian cuneiform expert, Kugler has explained Ovid's as a true history of a comet. Phaeton is the inexperienced son of Phoebus who demands to be let to drive the chariot of the Sun one day. He prevails, but loses control of the steeds and burns up sky and earth. The constellations are disturbed. The flames turn whole nations into ashes. The ground bursts asunder, the rivers dry up. Smoke billows bring darkness to the world. The ocean shrinks. Ashes cover the Earth.

Mother Earth trembles and sinks below here usual place. She pleads with Jupiter. "If the sea, if the earth, if the palace of heaven, perish, we are then jumbled into the old chaos again. Save it from the flames, if aught still survives, and preserve the universe." Jupiter responds by demolishing Phaeton and the chariot; Phaeton, his yellow hair streaming in flames, is hurled to the earth like a falling star.

The Sun, Father of Phaeton, mourns as in an eclipse. The earth was lit only by its own flames. He would not resume his daily journey until all the gods supplicated with him. The days appeared once more, and Jupiter restored order and life to the heavens and earth.

No one disputes the fact that the earth has been badly burned. Provided, of course, that the statement is properly qualified. The ocean basins are of melted rock; they are fashioned almost entirely of basaltic lava. Ocean abyssal sediments are thin and loose, and composed of organic and dust fall-out for the most part, including some products of combustion. Of the continents, part of the surface that is exposed to view is igneous, a product of old or new melting. Another portion is metamorphic, a word meaning rock transformed mostly by heat and pressure, both old and new; this emerges from both sedimentary and igneous rocks. (It is significant that whereas observers are compelled by the sight of volcanism to say that some lava beds are new, they are reluctant to name any metamorphosis of rock that has taken place very recently.) Igneous rock, if not witnessed as it forms, is also invariably given old dates.

From the standpoint of human primevalogy, the uppermost layers of rock and debris are highly important. These are usually termed unconsolidated, or loosely consolidated, or aggregated, or conglomerate. High energy expressions of "earth, air, fire and water" will produce large quantities of this material and their origins, dating, and relation to the biosphere are hard to discern.

Everywhere one is likely to find soil, a catch-all work for any layer from the thinnest film up to a few meters in which life forms take hold or dwell. Fossil soils often rest between layers of the several types of rock.

Besides the soil, too, exist metals, soda ash, peat, various ashes, coal, oil, natural gas, salt, and other deposits. Some of these are thermal products. Billions of tons of glassy microtektites are strewn over the globe; whatever their origin, they may have fallen in as hot rain on land and sea. Layers of ash are found over vast stretches of the oceans bottoms, perhaps everywhere, since the searches have just begun. Ash is fairly distinguishable; it is more difficult to detect whether the much more profuse sedimentary clays are not themselves in part the products of combustion, carried over and dropped upon the sea or drained off the continents onto the slopes and shelves.

On the land, too, ashes mix readily with soils and detritus to form clays. It is not impossible to detect calcination in soils and clays, but the subject has attracted few geo-chemists. Soils and young marine sediments of northeastern and offshore America reveal, under chemical analysis, evidence of a fiery origin in that they contain polycyclic aromatic hydrocarbons. These are carcinogenic and mutagenic. It is possible that their incidence is world-wide. If so, it would indicate that the whole world suffered one or more fall-outs of burning or burnt material. The burning could have been caused by superterrestrial impact explosions or gases. Or the products of atmospheric fire (burning naphthas and brimstone or sulphur, as the Bible would have it) descended. Or both might have happened.

"Extensive ash layers are now recognized in continental areas throughout the geological record," they point out, citing C. S. Ross. They declare too that "ash of similar composition has been logged in boreholes in many of the dry lakes of the western United States." (These dry lakes are all very young, post-glacial.) Wengret and others showed extensive ashes and calcination in the Nile Valley to which they assigned fairly recent ages; one can only wonder, for similar reports simply are not registered generally, how many cuts and profiles around the world reveal such calcination and why, as has been observed, the older rock-strata show almost no calcination - except that metamorphics, granites, igneous rocks, and perhaps limestones themselves are sign of heavy thermal activity.

Until very recently, geologists, like archaeologists, have been incurious about thin beds of ashes. An alerted surveyor, Heladio Agudelo, wrote this author (Oct. 4, 1977) saying, "In my work... while helping build a new street I noticed a black line in the gravel formation." It was a "one inch thick black line in otherwise homogeneous alluvial(?) formation." Within several weeks it became invisible due to erosion "but it will take no bigger a tool than a hand shovel to expose it again. This is in Londonderry, N. H., no more than an hour's drive from Boston Airport." Thin beds of ashes represent enormous fire, the effects of ordinary forest and construction fires disappear quickly.

The Ewing group comments that "Murray and Renard identified volcanic particles in practically all of the Challenger surficial samples of deep sea deposits, demonstrating that volcanic detritus is an important component of modern deep sea deposits throughout the world. They suggested that the abyssal clays are largely the result of alternation of volcanic ash." Later on, the authors themselves conclude: "It is necessary to study the alterations of fine pyroclastics in the sea and to set up criteria for recognition of the alteration products formed under the full range of environmental conditions." (I proposed such procedures for heavy combustion products in many archaeological levels, exemplified in the "Burnt City" of Troy IIg).

The Worzel ash consists of colorless shards of volcanic glass without sorting by particle size. "In all important respects it is similar to material which has been classified as volcanic ash in the deep-sea deposits of the world." Analysis of the surrounding sediment in the Worzel cores indicates that the bottom waters "must have contained some oxygen" and that the sediments "probably represent no more than 100,000 years and conceivably far less." Whatever the date, mankind was very much present and concerned. Certainly years of darkness, disease, distress and terror occurred around the world with this event.

"The ash is entirely unlike material described as meteoritic dust. Only the wide geographic extent of this layer suggests any source other than volcanic eruptions."To this proposition, with which Worzel might differ, given his quoted remarks, one might take exception. "Meteoritic dust" is too imprecise a term to use in argument, considering that we may have to consider lunar material and the 50 to 150 million mile tails of comets. If, as the authors grant, there is a need to examine and re-examine numerous types of sediment, there is also a need to distinguish, if at all possible, "cosmic dust" from "terrestrial dust". If worldwide volcanism can only originate from an externally interrupted motion of the Earth, or from a titanic large-body encounter, then "terrestrial dust" is also an effect of exoterrestrialism.

Heezen and Hollister write that the Krakatoa eruption of 1883 "produced an insignificant sprinkling of ash" by comparison with six great eruptions of the past "million" years that blanketed thickly the ridge and basin of the Java Trench. "Indeed how great must have been the earlier eruptions if the greatest known to man was too small to produce significant record. Powerful eruptions in the Japanese, Kurile, and Aleutian arcs have produced so much ash that these airborne volcanic products dominate the scenery of the Northwest Pacific in a belt almost 1000 km wide. (We note again that human were already present during these great ash storms and presumably coining legends.) Heezen added elsewhere the Mediterranean Sea bottom as a depository of several heavy ash layers.

Walter Sullivan describes " a succession of ash layers" encountered on the edge of the continental slope before striking the lava basalt of the true ocean bottom. Might this not indicate that the continental slope was laid down subaerially, collected its sedimentary and ash layers and was then inundated by the ocean? Drilling in the Atlantic "has begun to paint a picture of the awesome events that accompanied the birth of that ocean."

Fires in prehistory may have been much more extensive than they are today and their part in animal adaptations may have been considerable. He quotes Harris on "Forest Fire in the Mesozoic," where the author describes vast fusain deposits, identifies them as fossil charcoal, and says "the objection usually used against accepting fusain as charcoal produced by fire is that there is too much of it and in too many layers. It would make the past a 'nightmare.'" Fusain is intimately associated with coal beds and thus reinforces the Carli and Velikovsky thesis, seconded by Francis and Cook, that coal is what remains of a bulldozed burning biosphere, buried deeply and tamped down promptly by successive waves of other material.

Animal fossils are sometimes found amidst ashes. "Ancient Ashfall Entombed Prehistoric Animals," heads an article by M. H. Voorhies where a Middle Miocene prodigious ashfall over hundreds of square miles snuffed out over 200 species at one waterhole alone. When geologist Louis Lartel was excavating Cro-Magnon man (fragments of 15 individuals) near Les Eyzies-de-Tayac (Dordogne, France) in 1868, he uncovered five archaeological layers that had been covered by ash. The Upper Paleolithic was an age of ashes too. The glacial ice, where such great sheets existed, must have been covered with ash, if today ice drilling reveals no heavy ash fallouts it must mean that the caps are exceedingly young.

Erich von Fange has come upon many a recent report of burnt sites. He mentions towns whose calcinated ruins resemble strikingly what one can read of Troy IIg, "The Burnt City," when reexamining the extensive records of its excavation. His cases come not only from the Near East but also from Western Europe and Britain, Central Africa, the Gobi Desert of Central Asia, the Mohave Desert of the American Southwest, India, Cusco (Peru), and Cete Cidades (Piaui, Brazil). The "Cities of the Plain," including Sodom and Gomorrah, flourished in an area that became a scene of utter devastation to this day, over four thousand years later.

All that grew in this Dead Sea Rift area, all who lived there, all that was built there, were wiped out by falls of fiery debris and an upheaval of the earth; asphalt, salt and sulphur are abundantly displayed now.

Examples of the latest possible world conflagration can be found in Greece. These would be in the -776 to 687 B. C. period, by Velikovsky's chronology, which I accept; owing to a major shift in time reckoning, most of the great destructions in these areas that has been assigned to around 1200 B. C. is now scaled down to the eighth and seventh centuries. The new great destructive sky god was Mars in many forms. It was now that King Nebuchadnezzar ravaged the Near East believing himself to be the personification of the planet-god Mars-Nergal: " I am Nergal. I destroy, I burn, I demolish, leaving nothing behind me." (He was, of course, not nearly so effective as his model, and was ultimately killed).

The same age began with the downfall of the Mycenean culture. The evidence of the destruction of Mycenean culture by fire has been available for a long time, but put aside for lack of a cause. A. H. Frickenhaus, a German excavating long ago at Tiryns, described how he had located a burnt Mycenean palace with a new Greek-style temple built right over it. At Pylos, not far away and of the same period, fire was manifest everywhere, burnt rooms, burnt oil, fused metallic implements, scorched pots.

In his analysis of the Pylos event and others, Isaacson has substantially proven the correctness of the revised dating. Apparently, the Mycenean (Greek) Age changed into the archaic Greek period amidst general conflagration. But so did age upon age before, both geologic and cultural.

Chapter Eight

EXOTRRESTRIAL DROPS

Reasoning from exoterrestrial exploration and the shapes and chemistry of the Earth's crust, and with a strong assist from early human legends, the origins of some material of the Earth's surface is assigned to fall-outs. They range from invisible gases to giant meteoroids. The geophysical column commonly displays exoterrestrial products and their effects.

We stop short of using exoterrestrial fall-out like a magician pulling everything out of hat. If still we appear extreme, it is well to recall that the physicist Alfven theorized that the Moon was at first a much larger aggregate which broke up, showering upon the Earth the whole of our continental masses.

1. Falling dust and stone.

When Alexander the Great asked some Celtic leaders in 325 B. C. what they most feared, expecting them to reply Alexander himself, they said it was that the skies might fall.

In the most ancient legends it is common to find references to more than comets and deluges of water. Deluges from the sky consist also of dust, loess, stones, glass, tar, oil, salt, gold, iron, ashes, carbohydrates - all of them sometimes hot and sometimes aflame. They are invariably tied to catastrophes.

Doeko Goosen has gone well beyond the ordinary unsatisfying explanations of soil formations commonly employed. "Not so long ago soils were considered to form in materials derived by weathering of the underlying rock. Over several decades there has been a growing recognition that much of the mantle of soil is allochtonous." But where does it come from? Few are the regions where soil can be shown to have aggregated as humus from the vegetation above. The large areas of Europe and Asia covered with loess are now considered all or in part by Russian scientists as non-aeolian.

The origin of loess may be in an immense fall-out of dust from a comet or an explosion of Earth material into the highest atmosphere whence most of it fell back to form loess and clay covering many hills and valleys to this day. Since humans seem to recall such an event, the time might not be far off.

Till is a stiff clay full of stones varying in size up to boulders; conventional science says it was produced by abrasion and carried along by the ice sheet as it moved over the land. So Geikie said in 1863, and the definition is still useful. Donnelly pointed out that this till, which he called drift, is not in all places where the ice was said to be and exists in other areas where no ice was supposed to have been. Till is common "over much of the most important mineral producing terrain of the northern hemisphere. Till occurs ubiquitously in Canada and Scandinavia and is present as well over significant areas of the United States, U. S. S. R. and United Kingdom."

But why, argued Donnelly, was there a "driftless region" in Wisconsin, Iowa, and Minnesota. And why is very little found in Siberia; there exist "the great river-deposits, with their mammalian remains, which tell of a milder climate than now obtains in those high latitudes, still lying undisturbed at the surface." So wrote James Geikie.Why do glaciers today not produce true ancient-type till, that is, striated stones, drift clay, mountain-top till, and how could glaciers form sheets over 30% of the Earth's surface a million years ago, not to mention pushing boulders up thousands of feet in elevation?

Using the ordinary theories of glacial geology, even though he is an exoterrestrial catastrophist, the Soviet geologist Salop has pointed out "that the Precambrian glaciations occurred under very unfavorable physical-geographical conditions. The glacial deposits are interbedded between strata indicating a hot climate, such as red-beds, dolomites, phytolite-bearing limestones (at present only found in warm, usually mineralized waters along the seashore or in tropical lagoons and hot springs), evaporites, kaolinitic sandstones and bauxite." This association of tillites with formation of warm and hot climates is typical of the Paleozoic Ice Ages too.

But Salop also demonstrates that nine ice-age pre-cambrian "intervals vary from 40 to 125 (or 180) MY and no evident periodicity can be observed." He then associates "biologic revolutions with the epochs of excessive climatic cooling usually resulting in glaciation." Tillites are taken as the signal of an ice age; whatever the climate above and below the till, whether cold or hot, the till is supposed to designate cold.

Tektites are jets of fused silica. They range from microscopic size to large chunks. They are strewn around the world in enormous fields. They are found in the waters and soils of Central Europe, West Africa, Australia, Indochina, Thailand, the East Indies, the Philippines, Japan, China, and the Caribbean. Heezen and Hollister estimated an Indian Ocean deposit of a billion tons that they think occurred upon a reversal of the Earth's magnetic field 700,000 years ago. Glass and Heezen differentiated three forms of tektites found in the Far East. One was melted twice, one melted once, and a third little melted. They deduce a massive cosmic body breaking up upon atmosphere entry into two or several pieces. Of these, one would explode in the upper atmosphere, another closer to the ground and a third close to the ground.

I think that legendary streams of cosmic arrows shot by the gods upon hapless but offensive mankind might refer to the glassier kinds of fall-out. Tektites resemble somewhat obsidian, a popular igneous stone for fabricating arrowheads. Tektites may fall like showers of needles, or arrows, or as arrowheads in size, weight and hardness.

Reporting systems on natural phenomena have gradually become more complete, regular, and valid. Nevertheless, the Edinburgh Philosophical Journal in 1819 issued an enchanting list of "meteoric stones, masses of iron, and showers of dust, red snow, and other substances, which have fallen from the heavens, from the earliest period down to 1819." Among the exotic items were: a great fall of black dust at Constantinople on November 5-6, 472 B. C. accompanied by burning heavens; a kind of red matter like coagulated blood in the middle of the 9th century; a burning body that fell into Lake Van, Armenia, turning the waters red and cleaving the Earth in several places (1110 A. D.); gelatinous matter in India with a globe of fire; and a mixture of red rain and snow whose dust contained silica, aluminum, lime, iron, carbon and loess and was coincidental with a shower of meteoritic stone over central and southern Italy in 1813. Red rains, often associated with meteors, were common. William Corliss, in his compilations, has educed much additional literature on peculiar fall-outs. Peter James, Donnelly, Velikovsky and others have demonstrated the frequent occurrence of red falls in protohistory.

Much meteoritic dust falling upon the Earth is invisible and immeasurable. Micrometeorite dust has been estimated by Fred Singer to fall at a median rate of 1250 tons per day or 456,250 tons per year (the rate may actually be 10 times more or less, he estimates). The calculation is from the detection of aluminum 26 abundance ratio in Pacific Ocean bottom cores. This is $4.5x10^{11}$ grams per year today, but Schmidt's estimate is only 400 tons per year today.

If any exponentialism is part of Singer's scheme and it should be, a fairly considerable portion of the Earth's crust should be composed of gathered-in planetary dust, achieved in a fairly short time. If, for example, we had a measure showing this figure to have been 10^{20} grams per year in 500 B. C. and 10 25 in 2500 B. C., the subsequently plotted curve would give us the mass of all of the continental crust except for the basic granite within a few thousand years. We do not have such figures, but if we consider the obsession of ancient voices with days and years of darkness and ascribe half of this to fall-out of dust, the required substantial deposits would be quickly forthcoming.

A study by L. and W. Alvarez, Asaro and Michel describes a fall-out of dust 1000 times that of Krakatoa from a meteoroid crash, which, they claim, darkened the Earth for years. The crash was deduced from the presence in Italian, Danish, and New Zealand limestones of the fossil break between the Cretaceous and Tertiary periods of iridium, 30,160 and 20 times its normal background level in terrestrial rocks but characteristic of meteoroids. Spain and Holland were added by Ganapathy to the locations bearing the tell-tale chemical signals.

Fish-clay analyses by Kyle and others in Denmark agreed with the limestone findings. A number of additional rare elements were also in long supply, 5 to 100 times their normal abundances.

The correlation of a fossil index set with a distinctive chemical element marks an important advance in geological investigation. A sure layer is now presumed to exist worldwide; even were it not to signal an age boundary, it would permit a tightening of identifications of relative and absolute dates of strata and species. We know that we are dealing with a uniform world-wide event, something that is only hoped for when correlating fossils and rocks. We know that the event is limited in time.

We know further that if the event is not denoted in the strata, the reason is not that the event did not occur. That is, some stratum capable of containing the iridium (or other element) must at the stipulated time have existed everywhere. Where not found, conditions for its prompt removal must have existed, or later removal must have occurred. Alternatively, the fall-out was erratic and initially directed only to certain spots by the presumably catastrophic winds and tides of the moment. Despite all this, with a dozen such exoterrestrial chemical markers, historical geology and paleontology would undergo a quantavolution.

A question arises as to what constitutes outer space or exoterrestrialism for dust and stone falls. Under certain conditions of large meteoroid or cometary impact, and heavy multiple volcanism, exploded material can achieve extreme heights and even be lost into space. Such would be the case, for instance, were the Moon to have been exploded from the Pacific Basin. In such a case, a prolonged fall-out period of a great many years, perhaps centuries, might result. Pebbles, dust, loess, tektites and other types of matter might separately collect in orbit and shower down homogeneously, while simultaneously, volcanism would pave large stretches of the globe.

Once more, we find the gradual fall rates of the present and the more credible exponentially higher fall rates of the recent past so productive of mass and volume for the Earth's crust that a young age for the Earth or a very young age for the catastrophized Earth suggests itself. Whatever the properties of fully exoterrestrial falls to explosion and fall-back, the fall-out even will wreak havoc: darkness, lightning, winds, possible interruption of Earth motions, and biosphere destruction, plus excitation of seismism and volcanism; holospheric transactionism, that is.

2. GASES, POISONS AND FOOD.

Since most chemical elements and compounds can be either found beyond the Earth or conjectured to have once formed from the thermal and electrical conditions that occur exoterrestrially, scenarios of past events to explain present processes are becoming as common, prolonged, and disastrous as the "soap operas" of radio and television.

The gases that we will discuss are mainly effective in the biosphere. We address not only their chemical qualities but their behavior in mixtures and their propulsion by winds. The poisons we discuss are cell destroying chemicals. The food consists of the rare occasion of the descent of digestible cell-building chemical compounds. Hydrocarbons are considered here as poisons.

Gibson and Moore, investigating subsoil samples from the Moon, found so many differences in volatile elements between North Ray Crater and other sampled locations that they concluded it to be the site of a cometary impact. They agree with Kopal that "the total amount of gas which can be acquired by the Moon in a catastrophic encounter with a comet is far from negligible." The Earth is a bigger target for comets than the Moon. We would expect the Earth, then, to have also picked up many elements from foreign sources. Traces of gases and hydrocarbons were found some distance from the crater. Gases emitted by an impacting body would probably cause significant surface phenomena on Earth as well.

Were the original mammoths gassed into extinction? Instant death, fractured limbs, destroyed sometimes in herds and sometimes alone, discovered on hills (not in river channels), some found with their skins and innards intact, several found with food in their stomachs, even their mouths, often associated with an incongruous assembly of other species, they lived and died where they were found, several still standing, one with a rooted tree buried with it. The mammoth and almost all other large animals of the same period were extincted between 5,000 and 30,000 years ago over the face of the globe. The extinctions occurred from over practically the whole arctic area and down to the southern part of the United States, Europe and Middle Asia, where their close relatives, the mastodon, now extinct elephants, and modern elephants browsed. It is strange that no human skeletons have yet been found, since we have their drawings of the mammoth.

Obviously if the date of each specimen were to be taken seriously, we would have, as one writer argued, a series of local catastrophes. All over the world, he might have added. Nor were the frozen elephants found encased in ice, but rather in a muck of pebbles and clay, which is the same kind of muck that is widespread over hundreds of thousand of square kilometers in the frozen arctic regions and contains the mangled remains of millions of animals and plants. It is hard to dispute claims of a sudden, widespread, simultaneous, and single catastrophe.

The assigned dates are hardly defensible. In the preliminaries of such a catastrophe, valid carbondating would be extincted along with the large animals. The supplemental dating is provided by the complicated ice age series, of which more later, but which, we can say, is something of a muck itself. With the unreliable dating shunted aside, a global scenario can be provided, an extravaganza, to be sure, but one is driven to it by the facts.

One may speculate that a large body passed by the Earth perhaps 6,000 years ago. It drew up tides of water and air below its path by hundreds of meters. It drew along and up, then, water and atmosphere from the extreme northern and southern latitudes. It tilted the globe at the same time. Most animals were asphyxiated during the hours of the withdrawal of air. Simultaneously they were deep-frozen by temperatures reaching in directly from outer space in the range of -150°F.

The intruding body departed. The columns of air and water collapsed, and rushed up to the north and south. The winds and tides collected most of the dead animals, tore up the ground, and finally deposited the remains in a muck that sometimes reaches to 1,000 feet of depth, even to 4,000 feet in one that the Soviets have excavated. Much of the air never returned; the supply from the larger envelope around Earth was depleted and the immediate atmosphere was thinned. As the legends say, it was now the bitter, cold age of the "God of the Bright Skies", Jupiter. The mammoths, dry frozen in a vacuum, rested in their packages of muck until the present day.

One has to do with human experiences with atmospheric pressure, not only in moments such as asphyxiated the great mammals, but time and time again in primeval history. Sudden electrical events, not encounters alone, must have raised and lowered the air pressures under which humans lived. At times, mankind must have endured miserable headaches. Anthropologist Kennedy once referred briefly to "certain ritual practices like trepanation (which also developed obsessive proportion in Late Neolithic and Beaker time in Western Europe)." The practice extended in North Africa from the Canary Islands through the Berber lands at least as far as Egypt. It was performed in Mesoamerica as well. George Sarton writes in his history of science of prehistoric skulls that have come down to us with evidences of trepanation (trephination) performed upon them in life. The trepan is a saw for cutting holes in or removing pieces from the skull. It is a dangerous operation, hardly on a plane with piercing the nostrils to hold decorative devices. (But why are these devices so near the sinuses, too?) Extreme headaches and fury can thus be relieved. Trepanning, we surmise, was an indication that some considerable part of the population could not cope with a periodical fluctuation or definite change in atmospheric pressure.

Associating ozone layer destruction with the periods of a reversal in the Earth's magnetic field and these with the extinction of a number of species, discoverable in ocean bottom drilling, Reid, Isaksen, Holzer and Cruzen theorize "that current concern about possibly anthropogenic destruction of stratospheric ozone may be well-founded since it is possible that major depletions occurring in the distant past have had profound effect on the development of life as we know it." Anticipating again what is to be developed later, we can give credence to the theory, but would add that the destruction of the ozone layer will have occurred during any catastrophe involving turbulence in the stratosphere, especially with the passage of a large body.

Furthermore, the authors say, "the harmful effects accompanying polarity reversal, whatever they may be, form only one component of the total environmental stress on a given species." Beland and Russell point out that solar flares of extreme power, of a kind never observed and perhaps occurring once in 200,000 years by probability theory, would have to coincide with the reversal of GMF in order to account for a large number of species extinctions.

Vera Rich, reviewing knowledge of the Tunguska (Siberia) meteor of 1908, brings forward evidence of scabrous infection of the local reindeer in that year, a great acceleration of tree ring growth beginning then, and an increase in the radioactivity of surrounding trees. Another report has it that certain plants mutated as well. The event was exoterrestrial in origin and probably is of the category of "gas-bag" explosions, since scarcely a ton of exogenous particles has been recoverable from the immense scene of destruction.

Perhaps the body entered the Earth's atmosphere with great speed, electrically attracted as well as driven by inertial differences, and thoroughly ablated until it became a gas projectile without a casing, that exploded before striking. Or perhaps it was an explosive plasmoid from its inception. Generally speaking, the radiation effect of a single meteor or cometary train passing through the atmosphere would be heavier than many hydrogen bombs (unless these latter are deliberately "dirtied" by cobalt or other chemicals) because of its great heat, its compression of the ambiant air, its wide path of fall-out, and deep and large explosive cratering.

During the disasters of Exodus, several documents give indications of radiation effects. The widespread "leprosy" effect may denote radiation disease, as I have explained in my study of Moses. Eating fallen quail killed many persons, reports Jewish legend. The manna, too, had to be eaten under supervision; to argue that it was" holy" and thus had to be treated ritualistically is a modern sociological notion overlooking that it might have become "holy" for several reasons, one being that priests, the savants, were called upon to distinguish the edible from the poisonous manna. The Egyptian Ipuwer papyrus conveys the impression that women became barren and that people lost their hair. The cattle herds died of scabrous diseases. The most substantial theory of Exodus times regards them as part of a much larger, a global, event, involving the close passage of a comet, so that radiation effects are logically to be expected.

M. G. Reade and Wong Kee Kuong have more recently discoursed theoretically upon methods by which carbohydrates, such as the manna which fed the ancient survivors of the Exodus disaster, could be produced with the aid of cosmic lightning. Formaldehyde (a compound of carbon, hydrogen, and oxygen) is a partially combusted gas, of which "there will be no shortage.. in a burning fiery cloud, almost whatever its origin." In mixtures of free oxygen, carbon, hydrogen and nitrogen, this is the only product. It has to be synthesized into sugar in an alkaline environment (already done) which is not poisonous and can be converted into starch, rolled into "coriander seed" sizes and dropped at dawn. So goes the argument of Reade, himself a confectioner and engineer. The necessary procedures and formulas are presently at the threshold of laboratory chemistry, he asserts.

In Hindu rite, the soma-devi are celebrants of sacrifices using soma. As a libation to Agni, soma is now superseded in India by ghi. Now the deva is a goddess practically identical with Venus, and the devi are her cohort. Venus, east and west, is worshiped at times in the form of a cow, the sacred cow of India, for instance. Ghi is clarified butter. The "golden calf" of the Hebrews in Exodus is the equivalent Baal-Venus image. These few (from a great many) observations are made solely to point out and complete the coincidence of a great celestial presence (a cometary body), a turbulent atmosphere full of dust and lightning, the availability of carbon dioxide, hydrogen, oxygen, methane, formaldehyde, and water in large amounts, the presence too of many enormous laboratory vessels from which would fall not one but several products, and, of course, the desperate survivors who would eat anything (regardless of its nutritional value) and reverence the imagined donor.

At the same time as the Hebrews, Hindus, Mexicans, Greeks and others were munching manna, they were vitally concerned with a certain redness in their environment. The most astonishing and fearful color had fallen out of the skies and penetrated the surface. Again we take leave to quote from Newgrosh:

> Dr. Velikovsky has produced numerous citations from ancient sources to show how falls of a blood-like substance occurred when a "new" comet (later to become the planet Venus) came into catastrophic contact with the Earth: the Manuscript Quiche of the Maya, the so-called Papyrus Ipuwer from Egypt and the Book of Exodus all record the fact that the water in the rivers was turned into "blood". In addition to these examples, Dr. Velikovsky refers to the Greek myth of Zeus and Typhon, the Finnish epic Kalevala and the lore of the Altai Tartars. However, a more exhaustive survey of such legends would include the Sumerian myth of Inanna (a Venus goddess) who filled the wells of Sumer with "blood", the Egyptian story of the goddess Hathor (also Venus) whose visits to Earth were associated with the covering of the land with a blood-like "beer", and the Norse legends of the "raining of blood" associated with the Valkyries. These myths are widespread and all tell the same story. There can be little doubt that something looking like blood fell from Venus during its close contacts with Earth.

To conclude, possibly electric discharges between the intruder and Earth synthesized organic compounds in the cometary gases, including an edible component and an inedible red silicate that showered down to color the Earth and water a turbid red. Recently, Claus, Nagy, and others have discovered inherent organic compounds, carbonaceous chrondrites, in meteoritic material.

Hoyle and Wickramasinghe have tackled the problem vigorously over the past few years and emerged with two relevant hypotheses: one that life forms originated in space and a second that plagues also descend from space. Comets carry the appropriate chemicals and can carry on the necessary varying experiments naturally, over millions of years, until "photosynthetic bacteria, able to oxidize hydrogen sulfide anaerobically," emerged.

The authors propose a perpetual vigil and a screening of stratospheric contents for microbes. If their theory is correct, one might expect veritable plagues to have had a hand in the great extinctions of species that have marked geological history. The causes of death would not only be mechanical flooding, wind, hailstones etc. - and radiation, but also should include "biological warfare" against the Earth. Actually there is yet another dread possibility, chemical poisons, such as cyanide.

There is normally more in the soil than the erosion of terrestrial rocks: this has become apparent. Equally, new elements are discoverable that convey surprise, mostly unpleasant. The Dow Chemical company of Midland, Michigan, has been for several years in a quarrel with local authorities and environmentalists. The latter claim that Dow has manufactured chemicals that deposit dioxins, a carcinogen, in the soils. Dow says " we now think dioxins have been with us since the advent of (fire). It is perhaps uninformed to discount the company's research, that is apparently discovering dioxins everywhere." Adding more dioxins to the ground, of course, makes matters worse.

A parallel can be cited from the research into "Polycyclic Aromatic Hydrocarbons is Soils and Recent Sediments," conducted by Blumer and Youngblood, on behalf of the Woods Hole Oceanographic Institution. Samples were drawn from "depositional and chemical environments ranging from continental and coastal soils to marsh and subtidal marine deposits, and from high to low oxidation-reduction potentials." The PAH component is significant; PAH is carcinogenic; ancient burning may be producing some of today's cancers; it would be well to perform statistical correlations on populations, cancer incidences, and "background PAH" of soils. PAH are formed at elevated temperatures by incomplete combustion.

Our interpretation would imply that carcinogenic and mutagenic hydrocarbons

> occurred on the earth's surface during geological times spans. This raises the question whether these compounds might have contributed significantly to the processes of natural selection of mutation, and to the evolution of species.

The scientists assess the possible origins of the PAH deposits. They exclude weathering, seepage and spills, they exclude biosynthesis; they doubt early diagenesis in process of formation; they settle upon pyrolysis. This burning might be thought to occur on the site, but "the consistency in the PAH distribution among our samples suggests a predominant single mode of origin;" the sites are distant from one another. The chemistry does not permit regarding the PAH as "urban air particulates." Forest fires are "possible but unproven:" low temperature burning could provide the homology among the samples and air transport of PAH carbon ash from a great central fire somewhere might preserve the similarity. The ash layers are not noticeable, however.

In sum catastrophes, especially if exoterrestrially invoked, display much chemical creativity. Great typhonic explosions on Earth, probably exoterrestrially induced, will behave more modestly, but similarly. Numerous gases, poisons, and foods have fallen out in natural history, and very recently. Precarious life situations have been widely and abruptly generated. Multiple reports of gaseous and fall-out processes in space and atmosphere challenge the credibility of radioactivity rates that have been established under guidelines consistent with presently observable rates.

3. METALS, SALT AND OIL

Iron-working is siderurgy, a word out of ancient Greece and Rome. It translates properly as the working of star-iron. The Greek word for anvil, on which iron was worked, was close to the word for a meteoric stone. The Eygptians called iron "the bones of Typhon" and "a gift from Seth," both names corresponding to bodies crashing into the Earth, devil-monster and devil-god. Meteoritic iron was known to the early dynasties. "The Jews called iron ore nechoshet, which literally means the “droppings of the (cosmic) serpent, “ a nonsensical term unless our interpretation of it is allowed." The Jews forbade the use of iron in chiseling stones for the construction of an altar. "A similar taboo was observed in Greek and Roman cults, it was and still is widespread." Iron, the ancients believed, was meteoritic in origin. What would they have believed if they had seen the now exposed great iron mountains of Minnesota or Venezuela? Could such mountains have fallen from the sky? Unquestionably. Asteroids exist in the size of iron mines and contain as much iron. Would they not have exploded and dissipated into dust upon landing? Some would and some not.

A not-well-understood feature of meteoroid falls is that they can accomplish soft landings as well as hard crashes. In hard crashes, such as at Campo del Cielo (Argentina) where a number of meteoroids fell, "large masses of meteoritic iron and shale have been found in its vicinity." Heide writes, "the 60- ton meteorite from the Hoba farm near Grootfontain, South West Africa, the heaviest of all known meteorites, imbedded itself in friable limestone at a depth of only 1.5 meters. The iron meteorites of Cape York in Greenland, weighing up to 30.875 tons, lay on solid gneiss rock, or were barely imbedded in moraine rubble, without any trace of an impact. Here we may guess that they fell on a thick layer of ice or snow and sank to their final location as the snow or ice melted.

However, as the Mass and Velocity of the meteoroid increase, its Energy of impact increases, according to the formula $E = 1/2\,mv^2$. The atmosphere cannot brake the body in time. Therefore, no iron masses of over 100 tons have been deemed to be of exoterrestrial origin; where such have actually fallen, and few doubt this, they have been vaporized by the impact.

In the face of this formula and the visible facts of meteoritic iron, it would appear that the large iron ore masses on Earth cannot have originated exoterrestrially. The negation, if any, depends upon variable velocity. If the falling iron mass is electrically charged, or gathers a charge, so as to render it less attractive to the Earth, its velocity would diminish. Theoretically, it could waft down in a soft landing in one piece. If it crashed upon landing, it would possibly assemble itself into the form of an iron ore deposit as deluges of water and dust would fill the interstices. Strange objects have been found in the midst of iron ores being mined, such as wood of recent date.

We can imagine, as have several scientists, that the meteoroids fallen upon Earth are those of a late planet explosion in the region of the belt of asteroids and therefore we have been sampling a planet composed as the Earth is supposedly composed, with iron and nickel core, sima mantle, and sial crust. Calculations, given this simple idea, are complex but not enough. There is too much evidence of exoterrestrial dumping upon Earth by other bodies, more of the nature of Jupiter and Saturn, to carry out this algebra of ratios with confidence.

If an alternative to an electrically-assisted soft landing were sought, one might better conceive of a welding process; gigantic lightning strokes from iron bodies in space lasting for a minute would cast molten iron ore down their path to where they now rest in heaps.

Cores drilled from Antarctic sediments of pleistocene age contained iridium and gold in anomalously high proportions. "A sizeable fraction of the noble metals is contained in vesicular, millimeter-sized poly-mineralistic grains that closely resemble ablation debris from chondritic meteorites, and there is little doubt that the noble metals resulted from the accretion of a large extra-terrestrial object."

Flint is found that has undergone controlled heat treatment, with pressure retouching as revealed by spectroscopic experiments; this is at least Solutrean in age, 22,000 B.P. by conventional dating. The skill is as complex as and less enjoyable than metalworking by heat; why then did man wait another 15,000 years to begin his work with copper, tin, lead, gold, silver, and iron? Perhaps they were not available. Or, perhaps the dating is too long and, soon after the flint-working, metalworking began, which is one logic for preserving the conventional origin of metals by casting aside the conventional chronology.

Mankind was ready to work and even to melt and purify iron, it seems, long before it was available. If only in order to supply the type of hypothesis that may lead usefully to historical research on the subject, I would suggest that most metals occurred around the period of the great Deluge and in the transition from Saturn to Jupiter worship, about 4200 B. C., and may be connected to a cosmic explosion that I have in *Chaos and Creation* assigned to a planet with the legendary traits of Apollo. It is noteworthy that the ancient metal mines of Attica had two favorite names, Artemisiakon and Hermaikon, both siblings of Apollo. R. S. Diez is cited by Pauwels for arguing the origination of the immense Sudbury (Canada) nickel mines from a meteoroidal impact of pre-Cambrian times.

Walter Sullivan has presented in the New York Times of Nov. 2, 1966 a map of the world's most productive gold field below Johannesburg, which shows a large primary "bulls-eye formation" rimmed by gold-bearing formations and a much larger 200-mile-diameter, secondary, cratered, rim-like area, also bearing gold, and asks "Did a comet create a South African gold field?" Unless the gold was alchemized on the spot, it might have been part of the meteoroid that crashed.

Most metals, in conclusion, may originate exoterrestrially. If an alternative must be found, it may be suggested, although hardly discussed directly here, that special thermo-electric events might produce the metals. This would constitute electrolysis on a huge scale, in a dense catastrophically formed atmospheric plasma, before or after striking.

The metal, manganese, is exceptionally terrestrial in origin. Its growth out of underseas volcanos is particularly explosive and rapid. Pure manganese is found in cones near the Mid-Atlantic Ridge. Hot water and steam percolate through lava segregating the metal and depositing it in molten pools where it cools shortly. The French-American Mid-Ocean study, "Project Famous," found manganese geysers along the Ridge in the 1970's.

Manganese is also found in nodules on the ocean floors. These, by contrast with the geyser type, are supposed to have required much time to grow. Scott and his colleagues estimated that nodules grow at rates of 1 to 10 mm/million years. They are supported by Ku, Burnett and Morgenstein, using both radiometric and nonradiometric techniques of dating. But Goldberg and Arrhenius reported finding a 50-year old naval shell with a ferromagnesium oxide coating 30 mm thick, indicating a rate of 60,000 mm/m years. Heezen and Hollister point out that the rate of accumulation of manganese is a function of its concentration in water and the availability of a nucleus in the water. Conventional gradualist theory cannot explain the "mystery" so well as quantavolution.

Nodules abundantly litter the deep abyssal hills. They form around a particle, tephra, a pebble, an animal tooth, a bone, or on the surfaces of volcanic or drifted rocks. The nodules should require a very short time to form, if supplied with nucleus, warm water and a manganese rich soup emerging from fast flowing and erupting volcanos. The manganese adheres to any object and rafts to its ultimate destination far from its birth place with fast spreading lava, which also boils out manganese accumulations as it spreads, and by swirling currents of newly forming seas around it, the same currents that hold the nuclear objects in suspension for a time. The process proves itself as turbulent and swift by the nuclei, which would otherwise sink in the abyssal muck if there were such and by the availability of manganese only at the hot spots of the ridge. Thus, contrary to the long-time theory of manganese formation, the very presence of the manganese nodules goes to demonstrate how rapid was the paving of the ocean basins, a topic to be treated later on.

Sodium chloride is of course a mineral compound, and not a metal. The salt domes of the world, averaging 30 cubic miles each, may carry 100,000 cubic miles of salt, about three-tenths of all the salt of the seas. Salt is not found in pre-cambrian rocks, which are said to embrace most of the time since the Earth was created. Basalt of the ocean bottoms contains no salt and salt could not have been precipitated from the melting of mantle rock. Granite is also deficient in salt.

There is no apparent earthly source for salt. A Head Curator of Geology at the U. S. National Museum, George P. Merrill, long ago wrote that sodium chloride (at least the latter) must have come like meteorites from outer space and been caught up first in the atmosphere and then dumped in the oceans.

The salt may have descended both as a solid and in acqueous solution. Salt domes exist beneath the sea floor as well as below the land. Salt domes containing oil have been discovered beneath the floor of the Gulf of Mexico at 12,000 feet of depth (2000 fathoms). Great salt domes have been discovered below the Mediterranean floor as well, giving rise to an idea that the Mediterranean once, 12 million years ago, became a dry basin. Why salt should not then be evaporated and laid in even layers of sediments rather than in intrusive pockets is unanswered.

In South and East Texas many cylinders of salt (with nearby anhydrite, gypsum, oil and sulfur deposits) penetrate the Earth to depths of a thousand meters and more. The Gulf of Mexico does seem to have characteristics of a gigantic meteoroid impact. Since other salt domes have been also discovered beneath the gulf itself, one may wonder whether the meteoric body itself may not have been composed largely of salt and injected its own salt tubes into its crater basin. The fact that salt is so free from contaminants (less than 0.4%) argues for the solid integrity of the salt from its initial appearance on Earth.

Legends imply my theory here. Saturn was the first Lord of the Mill, a grindstone round like the revolving vault of the sky. It ground salt into the sea and was sunk in the ocean during the great maelstrom and deluge that brought the golden age of Saturn to an end. In Hindu myth, the gods churned the celestial ocean and the mill ground out salt into the sea. Norse myth has the heavenly mill churning out gold, then salt, then, sunk in the sea, sand and stones. The unhinging and failing of the Mill implies, too, a tilting of the axis of the globe, a likely accompaniment of the cataclysm.

A South American legend supplies significant detail. "The Arawak of Guyana call the Galaxy 'the Tapir's Way. ' This is confirmed in a tale of the Chirignano and some groups of theTupi-Guarani of South America." According to Cuna tradition, "the Tapir chopped down the 'Saltwater Tree', at the roots of which is God's whirlpool, and when the tree fell, saltwater gushed out to form the oceans of the world." The Cuna cosmology thus unites the idea of the tree-of life found in many places, including *Genesis,* with a Tapir-god, Saturnian-Elohim divinity, and, as the tree of life is destroyed (the old order ends), saltwaters deluge the Earth. (In *Solaria Binaria,* Earl R. Milton and the present author identify this tree of life with the legendary and philosophical axis of fire and this with the presence, until a nova of Saturn, of an electric arc-current flashing between a then-larger Saturn and the Sun, and visible to man.) In sum, various legends independently agree that the salt of the oceans came with an aquatic cataclysm in a time when mankind was an intelligent witness.

That salt came down upon the doomed "Cities of the Plain" at a later time as well is argued by Dwardu Cardona. Yahweh threatens his people with "sulphur and salt and burning, so that its whole land will not be sown... like the overthrow of Sodom and Gomorrah, Admah and Zeboiim, which Yahweh overthrew in his anger and in his wrath;...."

In the same work, Milton and I propose that the Noachian Deluge occurred in cyclonic form, with the salty waters hosing or jetting down at thousands of locations. If this were correct, some of the characteristics of salt deposits would be explained, such as their common cylindrical shapes and great depth below the surface of land and sea bottom. The saltwater would bore through the surface rocks under great pressure and with enough time to penetrate deeply. The water would vaporize promptly in the ambient heat and what was left of it would leak through a multitude of fractures on the margins of the deposits.

Velikovsky's theory of petroleum origins introduces a frightful deluge of oil. He cites references in legends and scriptures to the fall of naphtha, sometimes blazing, and of brimstone, often rendered otherwise as a rain of hail. The Abkasian, a people famous for their long life-spans, convey a story about a fall-out of cotton, which caught fire and burned the Earth; perhaps it was "cotton-candy" mixed with hydrocarbon. The ancient bible of Mesoamerica, the Popul Vuh, tells of the fate of the people of that age:

And so they were killed;
They were overwhelmed.
There came a great rain of glue
Down from the sky.

The "glue" is still found in the land of the Olmecs. William Mullen comments on the work of the pioneer excavators: radiocarbon samples are contaminated by asphalt. "Much of the Early Tres Zapotes level was sealed with volcanic ash. Coe reports that lumps of asphalt were found everywhere at the San Lorenzo excavation." I consulted with an expert on the area. As expected, he said that the area practically floats on oil. I then visited the area. He spoke truth. But the question is: Which came first, the culture or the oil? Here, as throughout the world, the ancient voices give precedence to the people.

Velikovsky's concept can be summarized to a degree in his own words:

The tails of comets are composed mainly of carbon and hydrogen gases. Lacking oxygen, they do not burn in flight, but the inflammable gases, passing through an atmosphere containing oxygen, will be set on fire. If carbon and hydrogen gases, or vapor of a composition of these two elements, enter the atmosphere in huge masses, a part of them will burn, binding all the oxygen available at the moment; the rest will escape combustion, but in swift transition will become liquid. Falling on the ground, the substance, if liquid, would sink into the pores of the sand and into clefts between the rocks; falling on water, it would remain floating if the fire in the air is extinguished before new supplies of oxygen arrive from other regions...

The descent of a sticky fluid which came earthward and blazed with heavy smoke is recalled in the oral and written traditions of the inhabitants of both hemispheres... All the countries whose traditions of fire-rain I have cited actually have deposits of oil: Mexico, the East Indies, Siberia, Iraq, and Egypt ....

Definite legendary, archaeological, and geological evidence of a holospheric catastrophe in Mesopotamia was provided by J. V. K. Wilson for a period tightly connected with Inanna (identifiable as Venus). Large-scale mesolithic rock displacements are displayed, and accounts of rains of oil, the poisoning of the land, and falling sheets of fire are described in the ancient documents. Lion-headed pillars are associated symbolically with mushroom-shaped clouds (our typhonic cyclones) in the legend and architecture of the times.

The Soviet geologist Levin asserts that the hydrocarbons in cometary heads "must have played a part in forming petroleum and in the origin of life." Blumer, in a path-breaking article on organic paleochemistry, pauses to reflect that "man has long been curious about the origin of these materials," coals and oil. "On occasion, early speculations approached the truth in a colorful way; thus, the Triassic Tyrolian oil shales, which are rich in vertebrate fossils as well as in chlorophyll and haemin derivatives, were thought to have resulted from an impregnation of the local rock with the blood of a slain dragon."

We shall certainly not be contradicting him, if we suggest that the chemical transformations producing oil are as likely to occur in space as below ground, probably more likely, if we wished to argue the point. Further, we do not see how it can be asserted either that organic biomass capable of forming oil does not exist in exoterrestrial bodies or, if it does not, that its absence precludes space gases constituting or contributing to the constitution of the oils that are present on Earth.

Most metals, salt, and oil, we conclude, are more likely than not to have originated exoterrestrially or in exoterrestrially precipitated transactions at the Earth's surface.

CHAPTER NINE

HYDROLOGY

Everything on earth's surface begins to appear anomalous. Hence we begin to suspect that the abnormal is really the norm. Water is no exception. To begin with, its reversal of the ordinary liquidity-solidity thermal behavior is awesome: it swells when it "should" contract. There seems to be too much of it to have been squeezed from below. Its excessive salinity in the oceans has already been discussed. Its "misbehavior" is subject of countless legends and its symbols are engraved in religious doctrine and ritual. It has apparently been highly energized in the past in the form of tides, deluges, and ice. That the exoterrestrial connections of Earth's water are multiform should then be ordinary knowledge in the earth sciences. In fact, the scientific literature hardly touches upon it.

1. WATER

The world's oceans contain 1.4 X 10^{18} tons of salted water. Its surface fresh waters -streams, rivers, lakes- come to 5.1 X 10^{14} tons, 50,000 times less, a drop in the bucket. The ice of the continents amounts to a menacing 2.3 X 10^{16} tons. And water vapors constitute 0 to 7% of the atmosphere up to 50 miles high, enough to lay a cloud cover over half the globe at any given time. And there are groundwaters, more voluminous than those of the surface. The fresh waters amount in all to three percent of all waters, and three-fourths of the fresh waters are bound up in ice.

The omnipresence of water in large amounts in all life forms grants it a large role in biological and atmospheric activities. Its employment and bulk make its lithospheric transactions important shapers of the Earth's surface. Where do the ocean waters come from? Since she sees the streams and puddles after a rain, a child reasons that all water comes from the sky, that is, unless a geologist gets to her quickly to tell her that the oceans have always been here from the time the Earth was formed, or almost as long. An eccentric geologist might say that, over the ages, hydrogen atoms descend from the Sun and space upon Earth, unite with oxygen in the atmosphere and then over billions of years drop to form the waters of the oceans.

The conventional myth - by which I intend no slight - is set forth by E. Bullard who assumes "the obvious things.... one of them is the constancy of the total volume of water through the ages." Water is "obviously" in "equilibrium," but "the mechanisms of the equilibrium are unknown."

Thus "it looks as if the water must have been tied up in compounds, perhaps hydrated silicates, until the earth had formed and the neon had escaped." (This last is needed to explain why neon, so abundant in the Sun and stars, is so rare on Earth.) "Water must then have been released as a liquid sometime during the first billion years of the earth's history, for which we have no geological record." Bullard follows this with further apologies for the myth but says that the past decades have revolutionized oceanography and have "unlocked the history of the oceans."

The door may be unlocked, but few have entered. The billions of years of equilibrium can no longer be accepted: new theory has the ocean floors being scraped and relaid by the continental plates at least over the past two hundred million years or less; no longer can the myth hold that the most ancient sediments must rest on the ocean floors, hence no evidence is thereby offered of what the waters may have been like.

Surely there has been water so long as life has existed, but not necessarily salted water nor much water. One may assume little water to begin with and little for long after. Swamps and shallow seas are best for evolution and quantavolution of species; thick atmospheric soup might be even better, at least in the beginning. Even now, life seems to reject the oceanic abyss. This is a sign of youth, for the abyss is not without nutrients, and forms of life exist that require little or no sunlight.

The oceans do not carry all the uranium that they should possess after long eons of riverine deposits. Their salt is excessive and its sources are not organic. One calculation emerges with only 2.6% of the present chlorine of the oceans as conceivably of continental origins.

The sea bottoms seem never to have been compressed and folded, so this indirect evidence of the age of the water is lacking. Sediments are thin, and mostly accorded under 80 million years of existence. That is only one-fiftieth of the conventional age of the world oceans. Have there been fifty world-girdling oceans?

If the ocean basins were filled late in time, deluges from the skies have to be assumed. There is no other source, nor any more apt source, than the waterlogged comets and great planets. One is compelled to seek water there, and bring it here. Hence the need to invoke explosions of water from Saturn et al., and passing cometary encounters.

Once the theory of a deluge(s) is given, the search for the source of the water is by no means ended. The Earth's water may have been injected, boiled off the imagined primordial melt, stayed up in the skies, and then fallen when a crust had formed and cooled below 212°. This may have happened, but then again it may not have. It would seem that if vapors rose high, they would stay there and rotate with the Earth, descending only when terrestrial electrical conditions permitted or were "seeded" by exoterrestrial fall out (which is also an electrical phenomenon).

Professional courtesy grants geologists not only their huge oceans but also the basins to hold the waters. "God" must have made the basins to hold the water, and even if gods are dispensed with, the basins must stay. So just as some communists stuff their religion into the mummies of Lenin and Mao, some geologists stuff their religion into the "nature" that wisely provided ocean basins to hold the great waters.

The waters are too great for the basins to contain; they cover much of the "true" continents. The fact that the basins occur and the waters occur does not mean that they were made for each other. Nor have they corresponded. Yet the presence of the basins is essential to the preservation of the greater part of the continents. If all the earth's present crust were a uniform level, the waters would cover the globe to a depth of a kilometer and more.

There is not enough water in the earth's granite or basalts to fill the oceans. Granite, the rock that underlies the continental sediments, is notably lacking in porosity. Porosity is the ratio of void space to the bulk volume of a rock, and therefore a measure of the water or gas contained in the rock at the time of its emergence from a molten state. Its porosity ranges from 0.3 to 1.5%. That granite could not be generated from the deeper basalts of the mantle is argued by Y. N. Lyustikh, a soviet geophysicist; four times the present water mass of the earth would be needed for the job. Nor can the process by imagined.

The crystalline, glassy, volcanic basalt, which lines the ocean floors, can have a porosity of anywhere from 1% to 30%. Generally the porosity declines with the depth of the sea, a phenomenon attributable to pressures more lately applied than to original pressures, since this rock was often melted and extruded in unfilled basins that is, at less depth that it is presently discovered. Rocks of the same chemical constitution that differ in porosity will have had different histories in at least one significant regard: the rock of lower porosity had larger infusions of water and/or vapors during its last melting and reforming. An expansion of the earth could be facilitated by the incorporation of water and vapors in heated rock. Water could recycle itself time and time again: it would flood a hot chasm, be incorporated in the rock, be extruded, expelled, and again enter a hot chasm.

Water exists exoterrestrially. Only in 1970 were the first observations of comets in the ultraviolet spectral region made. Cometary atmospheres (comas), in which dust and minor molecular components had been hitherto alone observable, now revealed indicators of a large component of water, "confirming the Whipple hypothesis of comets being 'dirty' ice conglomerates." By 1980, other comets had disclosed similar compositions.

The outer planets contain great amounts of water. The rings of Saturn contain about 377 billion km^3 of non-conglomerated swarms of ice particles, by one reckoning. It has been dropping rings in the past. Saturn is 95 times the size of Earth; if Earth carried the same amount of ring ice relative to its size, it would have had 4 billion km 3 of ice particles to fill the ocean basins. The ocean basins contain 1.37 billion km 3 of water. True the density of Saturn's rings is much less than Earth's waters; still, the necessary relation of sky waters to ocean waters can be premised, especially if Saturn were to have shed most of its waters in times past. Moreover, Saturn is only one of many water-bearers in space. Jupiter and the other planets carry water, as do numberless comets.

Ancient wise men of Palestine, Mexico and India are known to have attributed the deluging of the earth to planet Saturn. Thus, the Hebrew Talmud reads in one place. "When the Holy One decided to bring the Deluge on the Earth, He took two stars from Khima and (hurling them against the Earth) brought the Deluge on the Earth." Velikovsky identified Khima as Saturn. In Mexican documents, where ages of the world are called "suns," "the first world age, at the end of which the earth was destroyed by a universal deluge, and presided over by Ce-acatl, or Saturn" The ancient Persians reported the star Tistar appearing in three manifestations to the accompaniment each of a different deluge of rain of ten days' durations.

Long before modern astronomy, Saturn was perceived to have rings and to be watery, never Venus, Mercury, or Mars. How the ancient would associate Saturn with water is a mystery unless the planet had been observed at a distance much closer than it appears to the eye today and seen to blow off some of its rings or gases that ultimately arrived to deluge the Earth. Since Saturn under various names was the ruling god in human cultures at the time of Noah's Flood, the associations begin to appear reasonable. However, the Saturnian deluge followed the Golden age of Saturn, and oceans existed at least to some depth in Saturnian times. They were navigable by Saturnian age peoples.

The ocean waters are geologically young. Granted waters are difficult to date, Melvin Cook has shown that the oceans contain under 100,000 years' accumulation of uranium, even granted a uniformitarian riverine run-off curve (which, of course, would mean much less time on the quantavolutionary exponential curve).

That the basins which hold the water are young, which is yet to be shown, holds significance for the youth of the waters as well. Few evolutionists and quantavolutionists regress in time to a completely water-covered Earth, although the first passage of Biblical Genesis might be construed so: for Elohim separated the chaos by a firmament dividing the waters below from the waters above, and assembled the land out of the waters below. And the primeval legend of the Earth being fished out of the waters is found in the farthest removed cultures of the globe. Also among the first impression and memories of mankind was the image of the vast cloudy universe recurrently pouring water and debris down upon the hapless Earth.

A more correct interpretation is that early man was caught in an increasingly turbulent cloudy world.

Most if not all of the lakes of the world can be thought of as slowly diminishing stagnant floods - the salt lakes like the Great Salt Lake (Utah) and the Dead sea, and the freshwater lakes such as the Great Lakes (USA), and the thousands of Canadian and American "glacial lakes." That these latter are in most cases being fed by rains and streams as fast as they evaporate or drain does not obviate the fact of their origins. They were created under flood conditions. If this is so, it is likely that ground waters and swampland are also behaving as flood waters, that is, everywhere draining at the levels of the ocean basins.

The Caspian Sea has been shrinking rapidly over the past 150 years, not alone because of human diversions, and becoming more saline. According to the idea that this sea may be a remnant of a recent and westwards dumping of the contents of the vast Gobi Sea, now Desert, carried on over thousands of kilometers, ending in the Mediterranean, the desiccation is to be expected. But, too, the local freshwater replenishment of the Caspian may be inadequate, and may always have been since its quantavolutionary creation.

So, too, can the ocean basins be regarded as flood drains, again to make a logical point, which is otherwise an absurd stretching of language. It can be looked at in this way: the basins of the oceans existed before they contained water; some water flowed or dropped into them, "flooding" them. More craters were added, more 'flooding' took place. Finally they were even 'overflooded, ' that is, land not properly abyssal but belonging to continental sial was flooded up to present shorelines.

By origin, basins may have been created by natural dams accreted gradually or thrown up abruptly by avalanche, by calderas of extinct volcanos, by meteoroid craters, by faults and rifts (as lake Baikal and the Dead Sea), and by the bulldozing done by ice and rock thrusts. The original water may have been groundwater seepage, rainwater, deluges, ice melt, or tides.

No lake is geologically old: this is an impressive datum. It says something about the lately tortured Earth. An undisturbed or slowly changing surface should include a proportionately great number of lakes aged in the millions and tens of millions of years. To object that lakes become filled with sediments must imply that such fossil lakes should exist by tens of thousands in the stratified rocks of the world. They do not. Some seemingly ancient lake beds are evident. These should be placed in the frequency distribution.

The results, even by raw conjecture, would be disappointing. The fossil lakes would be all too few. For, if we multiply the present million lakes, say, of an average age of 10,000 years as a guess, and take the last billion years of the earth "history" as providing similar lakes, we get 100,000 periods, and one hundred billion lakes. With climates changing (and Flint, for one, along with many other geologists had to invent a turbulent rain belt to fill his pluvial lakes), and with continents drifting about, and lands rising and sinking, why should not lakes have visited every place at some point in geological time, and be found in all (or say 10% to 100%) of the geological columns dutifully examined. I fear that *reductio ad absurdum* will once more assail conventional geological theory.

Freshwater springs exist in many places, emerging above their "natural" level, often quietly but sometimes with explosive vigor. The subterranean liquids and gases - water, oils, natural gas, and even compressed air - appear frequently to be pocketed under pressure. Calculations by M. Cook and others allow only a few thousand years for their escape, at most. Their burial must have occurred in some form of thrusting and folding, that is, is no longer occurring; we have accounts of many springs that have died, few that are new. This last fact would arbitrate against conventional theory that underground volatile pockets are fed from descending rock strata and then forced up above their local level at some interstices among the rocks, unless, of course, it is granted that the fresh waters generally are draining away, for the reasons given above.

Once more we turn to oceanography for help. The U. S. Atlantic Ocean shelf was drilled in 1976 at water depths of less than 300 meters and penetrated to depths of from 20 to 300 meters, at 19 widely separated sites. "One of the most significant discoveries... is that fresh ground water occurs beneath much of the Atlantic continental self." These fresh and sometimes brackish waters occupy large lenses in rock strata that are Cretaceous or younger.

The investigators considered whether these expanses of fresh water below the ocean salt waters were remnants that had been trapped in shelf sediments when the Pleistocene ice ages lowered the ocean waters, or were submarine discharges from mainland aquifers. Generally the first solution was preferred, although indications of submarine intrusions were discovered at southerly sites. The investigators did not suggest a third hypothesis, which we offer here, that indeed the freshwater lenses are fossils, but not from a period of withdrawal of waters to make ice. Rather they are both remnants and submarine channels of the age before deluges filled to overflowing the basaltic ocean basins. Fresh waters were trapped in the continental rocks as they made way toward the abyss and are probably trapped in the debris of the continental slope as well. They are extensions of normal aquifers, a circulation and storage system that is being broken into and polluted. We speculate (as do the investigators) that these waters have been suboceanic for only a few thousand years, and will not be with us for long.

2. Deluges

The sources of huge flood waters are limited. They may occur from the sudden collapse of an ice cap such as that of the Pleistocene, which covered, it is said, 30% of the Earth's surface. They can be exoterrestrial - from a comet or exploding body of the planetary system. They can descend from a onetime far-flung vaporous canopy. They can be mobilized as tides from an interruption of the Earth's motion, a tilt of the Earth's axis, or a drag induced by a giant passing body. They can, also as tides, be generated from a heavy meteoroid impact on the ocean, directly and also indirectly as in all cases above, from the winds, rock shifts and seismism accompanying them.

Diderot's *Encyclopedia* (1751- 1765) carried an article on "The Deluge" written by a young French engineer and soldier, Nicholas-Antoine Boulanger. Going beyond Newton's disciple, Whiston, who had explained the Deluge by a comet, he then wrote the first scientific work uniting the four factors; comet, flood, terror, and the origin of religion. G. R. Carli followed in a few years with additional world-wide legends and geological evidence of catastrophe. The ancient reports of universal catastrophe, both men reasoned, bore the stamp of truth.

Donald Patten lists sixty-eight deluge traditions on six continents. He might have named many more. For instance, twenty-five of them come from the Americas; but Marie and Richard Andress, folklorist and geographer respectively, found forty-six in the New World, almost twice as many accounts. In his time, Bellamy estimated 500 deluge myths coming from 250 peoples or tribes. The probability is high that every culture can recite story of a universal flood which practically nobody survived. The deluge is frequently pictured, too, in ancient and modern art. A. Dürer and Leonardo da Vinci painted their images of it, both making it a kind of typhoon. And indeed, in the ancient Chaldean story of the flood of Xisuthros the node of the Deluge is spoken of as a waterspout that "swelled up to heaven "and struck fear into the gods; the god Ea pleaded that any and all disaster be visited upon men, but nothing so terrible as "the waterspout of the Deluge."

The Greeks spoke of three great floods, Deucalion, Ogyges, and Dardanus. The first two have been tied to great floods of Exodus times, the mid-second millennium B. C. According to Philochorus (3rd c. B. C.), "deluge-swept Attica remained without a king for 189 [or 190] years " in the wake of the Ogygian Flood. Sextus Julius Africanus said that "all the former population of Attica was killed in the Ogygian deluge and the country remained uninhabited for 270 years." The Flood of Dardanus was probably of the 8th century B. C.

The most ancient of East Indian gods was Varuna, whose name means the "surrounder" or "concealer." He is the regent for the Sun. The root syllable "var" means water, hence "he who covers the heavens with his water canopy." Ouranos is the Greek equivalent: this Heaven-god, ancient Hesiod's Theogony tells us, came from far away to embrace "Mother Earth," Gaea, and "lay close about her on all sides around." The most archaic deity of the Latins was Coelus, ruler of heaven (Coelum), who like all the other heaven-gods, was ultimately banished. The Kojiki, holy scriptures of Japan, maintains that the gods, in the earliest days, brought the heavens and earth very close together. Two light-gods then ruled the world from their "f1oating bridge of heaven." Later, heaven "began to retire and eventually passed utterly away."

In the Hebrew Genesis, the Elohim (the Most High) created the Heavens and the Earth. The Heavens were a "firmament" placed "in the midst of the waters." The "there-waters" (Shimayim or Heaven) existed with lights but not with the sun and moon, for they are not mentioned in the opening passage of Genesis. The Assyrians said also that the sun, moon and stars came into view only when the monster foes of order were dislodged. When the Scandinavian heaven, Asgard, died with the gods, during Ragnarok, "the Sun and his legions came riding through the gap in shining array."

The name "Yahweh" came later when the skies were opened, just as names of the leading gods changed in all cultures, with the coming of a new age.

I should suggest that (a) self-conscious mythmaking mankind was born beneath a high canopy of rings and clouds, without a visible Sun; (b) deluges began and a visible Uranian Sun and the present Sun appeared; (c) the Uranian Sun went nova, the Earth bore forth the Moon and cleaved, while undergoing further deluges that partially filled the newly formed ocean basins; (d) the heavenly clouds remained to some extent thereafter (during the Golden Age of Saturn when the world lived tropically); and then (e) the second great Deluge came, which was the Noachian deluge.

The ancient Scandinavians called snow the "pus of the gods." In many places, the waters of the deluges and floods or tides were heated. Rains came down in gobs the size of a man's head and were at times boiling hot, according to the Zend Avesta of Persia. Josippon bin-Gorion repeats a Jewish myth: "The fountains of the deep broke up first. Then came the flood from above. Then fire fell also, and rain, boiling hot." Bellamy writes that "quite a number of peoples report not only a Great Flood, but specifically a flood of hot water." American Indians of the West claimed that the waters of the Great Flood were warm. The Voguls of Finland said a great fire raged over the world first and was followed by a deluge of hot water. Then the hot waters raged across the land. Fire mixed with the water- even their rafts caught fire, they said. Amerindians of Brazil said that the Sun was a cauldron of boiling waters that tipped over.

3. Floods and Tides

The legends of tides number in the hundreds, but they are usually hard to allocate to periods of time, particularly in this incipient phase of the science of quantavolutions. When the Biblical Book of Exodus says,

"The waters were a wall unto them on their right hand, and on their left," tidal behavior is suggested at the critical point of the Venusian comet, about 1450 B. C. by Biblical- derived dating. And the Psalms are chanting of the same event when "He made the waters to stand as a heap..." And the Midrashim comment likewise, "The waters were piled up to a height of sixteen hundred miles, and they could be seen by all the nations of the earth." (Though here we are bothered by the height and wonder whether, with the tides, there was a cyclonic tube reaching into the far heavens, the famous column of smoke by day and fire by night, that guided the Hebrews in Exodus). Also, in China, if the time of Emperor Yahou belongs anywhere, it belongs around the time of Exodus; and there the waters "over- topped the great heights, threatening the heavens with their floods."

But when the Lapps recite how the angry god Jubmel raged against the wicked, and, "foaming, dashing, rising sky-high came the sea-wall, crushing all things," we are not sure that this is the time of Exodus or earlier or a combination of later and earlier events. So it goes around the world. The tides are there: immense, overpowering everything, wrecking the surface, launched by the gods, accompanied by fire and wind; still each legend has to be examined carefully before assigning it to a given catastrophe.

Tides may be aquatic, but readily transport denser bodies. The velocity of water is as significant as its volume for carriage. Moving currents carry to the sixth power of their velocity. If a stream of volume "X" were to move at 2 km/ h it would carry 64 times the load it could carry if it moved at 1 km/ h. Tidal transport is scarcely less powerful.

The amplitude of tidal waves will vary greatly. Historical explosions have raised waves of 85 meters, as in the Krakatoan case. Earthquakes, as in Alaska, have done as much too. The Thera volcanic tsunami of circa 1000 B. C., is thought to have raised higher tidal waves than Krakatoa. As we have said, an exoterrestrial body may raise tides kilometers high.

A non-tidal moving flood is caused by the bursting of barriers: a natural dam blocks and collects water and then collapses. Some of the behavior and landscaping to be expected of great tides and floods are exemplified in the Channeled Scablands (Wash., U. S. A). They are 15,000 square miles of effects of a barrier burst flood; they were not made by a tide, not directly at least. Some 100,000 miles of this section of North America are thickly covered with lava, in places more than 10,000 feet thick, which can be ascribed to the immense volcanism incurred when the American continent travelled westwards over the global fracture of the East Pacific area. This might have been around 11,500 years ago, not the 10 to 30 million years conventionally given to the set of events. The whole area was then covered with silt and loess.

The Scablands are a water sculpture of this lava surface. Expert opinion asserts that a barrier of ice corked a mountain pass and caused a Glacial Lake Missoula to form. The Lake was half the volume of present-day Lake Michigan, but pitched high above sea level. The lake, it is thought, was of short duration and finally overflowed. The water cut through the ice cork. (The immediate cause may have been Earth movements.) "Within a very short time - perhaps no more than a day or two - the ice dam was destroyed and the contents of the lake were released." So reads a tourist bulletin on the area. A maximum speed of 45 miles per hour has been assigned to the resulting flood, and a maximum rate of flow ten times the combined flow of all the rivers of the world today. A luxuriant biosphere was wiped out, including large mammals, camels, bison, antelope, and, to my thinking, humans. I add "humans" partly because a doll is claimed to have been found in clay below 150 feet of lava, not far east of the same lava field, at Nampa, Idaho. The flood plucked and transported huge blocks of basalt. It flayed the basalt of its skin of loess. It dug channels in the basalt more than 200 feet in depth, and one of 8 miles in width. It made instant falls and plunge pools and eroded them backwards quickly. When the waters slowed they began to dump debris, some 500 square miles of it, to a depth of over 125 feet. The flood crest lasted a day or so, the main flood 2 to 3 weeks. Today, a satellite photo taken from 569 miles up shows the ramified and interlacing channel-ways of the flood cutting through the loess into the basalt, and then generally the unvegetated region around them.

The barrier-burst flood theory originated with Professor J. H. Bretz of the University of Chicago and was not accepted for many years because it was catastrophic. In fact, the theory can be pressed further in the direction of radical catastrophism.

First there are the reaffirmations of certain catastrophic doctrines. Energy kills time. Buttes, ravines, and river channels can be carved from dense rock in days. A biosphere can be destroyed down to bedrock in a single rush. Broad river channels are sculpted immediately through deep soil and loose rock. Giant gravel ripples are laid down; hills are fashioned;

long steep slopes are fashioned à la minute. Heavy stones are sown far and wide, the famous "eccentrics." Basalt is stripped to form monumental columns.

Across the world from the Scablands are Mesopotamia and India, whose peoples claim great floods as part of their historical experiences. These floods - were they originally from deluges or tides? In all cases, the tradition claims several great floods. Just as the Greeks had at least three floods, the Indians seem to have had their flood of Manu and the flood of the Gariga region, both described in the Puranas.

Both were disastrous, and we need not doubt that, as with the Scablands, other floods occurred from time to time. A similar series seems to have happened in Mesopotamia, where for centuries controversy over the number and extent of floods has raged.

The Indian flood area, whether once devastated or several times over, includes the famous fossil beds of the Siwalik hills. These are foothills of the Himalayas, north of Delhi. They are crammed with hordes of specimens of a great many species. Many of them appear for the first time in these beds and are extinguished in them, so far as paleontologists know.

In the *Geology of India*, D. N. Wadia writes, "This sudden bursting on the stage of such a varied population of herbivores, carnivores, rodents and of primates, the highest order of mammals, must be regarded as a most remarkable instance of rapid evolution of species." Tortoises of over six meters, two dozen species of elephant, pigs, oxen, and apes are scattered about. There are signs of earthquake, folding of the land, perhaps folding and deep burial of animals. Similar deposits are found 1300 miles away in Burma, cut away to view in the valley of the Irawaddy River. Two great zones of fossils are separated by 4000 feet of sand. Petrified trees pervade the fossils in the thousands.

Since large upthrusts of the Himalayan mountains are now being dated to post-glacial times, since even mountains much higher than the Siwalik foothills contain "old" marine fossil beds, since the Siwalik-type beds are so young even when conventionally dated, since evidences occur of huge waves of translation moving from south to north in India and leaving great moraines (including the Siwalik-type hills), since neolithic stones are found in the loess of the Himalayas and since great human cultures were flooded over and probably deluged as well, one is entitled to the quantavolutionary hypothesis: a series of abrupt, intensive, wide-scale changes overwhelmed the Indian subcontinent.

Frantic proliferation and extinction of species occurred, while India broke from Africa and crashed into Asia, while tides moved over the land, ramming, ripping, rising, and drowning, while the land raised up in a great arc into Asia, while hominids, then humans, entered and built cultures that were then destroyed and recreated. It may be that from this part of the world will come the easiest and fullest proofs of revolutionary primevalogy, of a succession of geological and cultural ages coinciding with the successive disruptions of what had been Solaria Binaria.

Dwarfing the Scablands and Indus barrier floods was the Gobi Sea flood, which may have been connected with the complex Noachian Flood. Thomas Huxley wrote the first scenario of the event. Bellamy refurbished the story. The Gobi desert, which the Chinese call "the Sea of Sand," was once a great body of water. Numerous settlements lined its shores. Then suddenly it was emptied in a huge barrier-type flood. Its cultures disappeared along with a great many other settlements along the line of the flood. The western barrier of the Gobi Sea broke between Tian Shan and the Altai mountains, and rushed through where today remain the waters of Telli-nor, Ebi nor, Ala kul, Sasyk kul and Lake Balkhash, much of it now saline and disappearing. The great flood spread out into a "Sea of Turkestan" and then drained down into the depression of the Aral and Caspian Seas.

It then poured out between the Ural mountains and the mountains of northern Iran, descended west through the Manych Depression into the Valley of the Don, the Sea of Azov, and the Black Sea. The areas of today's Romania and Bulgaria were temporarily part of a greater Black Sea. Soon it overflowed at the straits of the Bosphorus and pushed through the Dardanelles into the Mediterranean region. The Aegean and Eastern Mediterranean lands were flooded.

Next the Adriatic River, possibly the legendary River of Eridanus, and nowadays the truncated Po River, was turned into an Adriatic Sea. The Ionian Sea overflowed and the land bridges between Italy and Africa were covered with water. The shelves of the region of Tyrrhenia were submerged, the survivors driven to the high places of the Italian peninsula and islands, and contact was ultimately made with Gibraltar.

The Sahara basin may have been filled with water upon this occasion, to have become the ancient sea of Triton. It was this Tritonian Sea that figured in the mythical birth of Goddess Pallas Athena (the planet

Venus) and I think that it was around 3500 B. P., therefore, when the Tritonian Sea broke out and threw itself into the Atlantic Basin. Ancient Saharan ruins and the art of the Ahaggar mountain caves amply testify to the ancient cultures there between 4000-1500 B. C.

The elapsed time for the 4000-mile journey from China may have been months or years. The drainage of the several temporary basins established en route from East Asia to the Atlantic Ocean occupied centuries.

"Tsunami, ' or 'tidal waves' as they were for long misnamed, have an immense effect on shorelines, both in erosion and in the shifting of great quantities of sediment." But what parcel of land in the world has never experienced a tsunami?

"It is generally accepted that tsunamis are usually triggered by earthquakes or violent volcanic explosions. It is also possible that they can be produced by the slumping of large masses of sediment in water..." Or by meteoroid splashes, we might add, or hurricanes and cyclones. "Though infrequent, there are certainly enough of them for geological purposes. From historical records it can be deduced that there have been more than two hundred notable tsunamis in the last two thousand years; this would allow us more than 100,000 in a million years."

Then move the continents a little here and there, raise and lower shorelines, change climates a few times, and add ten, fifty, a hundred million years. We have millions of great tsunamis to work with. Obviously the whole surface of the Earth will have been worked over a number of times by ordinary, uniformitarian waves. Thereupon add all the other high-energy forms: deluges, exoterrestrial impacts, volcanism, and so on: it is a wonder that the crust of the Earth is not a homogenous finely ground mixture of all past life and surfacing rocks. Now add great catastrophes elaborated in this book and the homogenous mixture should be guaranteed.

That is, stratigraphy is hardly understandable by following uniformitarian principles, if we acknowledge what scientists have all along been discovering, but more recently have become acutely aware of. Even if, as Ager writes, "the changes do not take place gradually but as sporadic bursts, as a series of minor catastrophes," the strata of the Earth do not make sense.

Those who believe in major catastrophes interrupting huge serene tracts of time may be wrong, because they must add to the effects of the great disasters the effects of a multitude of minor ones called for during great stretches of "peaceful" time. The result would be a homogenized crust. The effects of the forces that have operated are such as to suggest for the Earth a short and recently catastrophic history. The Earth's surface still retains its forms and fossils because its tortures have been clustered and have occurred following a short total Earth history.

4. Ice fields of the Earth

Glaciers, formed on mountaintops around the world, take their origin usually in a U-shaped nook of a mountain. Their ice forms and slowly slides downwards through valleys, carrying drift and ending in melting waters. They abrade and pluck the drift as they go along. They are broad, and they terminate in broad curls, from which streams form and run off. Many "extinct" glacier forms exist, indicating that once there may have been much more cold and ice. That is, unless these "fossil" glaciers were pointed towards the sun in a global Earth tilt, and melted, or once were a part of a large crustal lateral avalanche that thrust whole areas away from the polar regions. Or unless exoterrestrial ice were dumped upon higher places and melted away from lower places.

In many settings, such as Cape Cod, Massachusetts, large plains end on the downslope with a number of ponds and layer upon layer of sands, gravel, and clay. In it are scratched stones and finely ground glacial flour. It seems that an ice sheet had once moved downwards on all sides from a northerly direction, acting like a glacier on a grand scale. I have not mentioned climatic changes: a very cold climate, as evidenced by the kinds of fossil flora and fauna discovered in old beds, indicates that a great deal of ice might have been nearby. Nor have I ventured to say when the ice ages happened and how many of them there might have been.

Full justice cannot be done here to the case for the ice ages. The conventional literature does so. But because some of the ice age reasoning falls victim readily to catastrophic claims, it may be time to advance the cause of quantavolution. Here three different positions are held: one is that the Ice Ages did not occur. The second says that they did exist but were sudden events, beginning and ending in disaster. A third admits their slow development but claims that they ended in catastrophe.

Melvin Cook, writing in the nineteen sixties, accepts the evidence for huge ice caps at both poles. Further he seeks no exoterrestrial power. His theory is nonetheless the most perfect of catastrophic models yet advanced. Ignoring the beginnings of the ice ages, but pursuing their end, his story commences with the great ice caps.

These, he says, by their enormous and accumulating weight, bore down upon the crust so heavily as finally to cause a rupture of the rim of the crater. The ice caps avalanched. They scraped the earth as they moved. They acted as gigantic bulldozers that caused mountain ranges to be thrust forward and buckled and folded upwards. Giant floods from the rapid melts swept the earth.

The globe fractured and caused the continents to spread apart rapidly. The Atlantic Ocean and the Arctic Ocean were opened up. In the end the surface of the earth was greatly changed. A great many land and life forms, together with cultural centers, were destroyed in the process. As the huge ice blocks descended, they turned over the biosphere and folded it to create coal and oil deposits in a geological "instant." Waters that were buried deeply are still rising under pressure. Yet the end came quickly, occupying a few years, not millions of years.

Continental drift has been argued as the cause of ice ages: "The ultimate cause of glaciation is thus seen to be movement of continents into appropriate latitudes... And much of the fossil evidence upon which the time-honored concept of Tertiary 'cooling' has been founded could be nothing more than a reflection of drifting of what are now the northern- hemisphere land masses and ocean floors toward the pole and hence into cooler climes." Another theory holds that a huge number of tropical volcanos erupted at once, which threw vast amounts of water into the air, which, because the upper atmosphere was darkened, caused less sunlight to bombard and warm the Earth, which finally caused the vapors to fall at the poles in the form of snow and ice.

If it is consolation to the reader, explanations of "the ice ages" have generally been bizarre and fantastic. Nothing less may be expected of our theory here, unless, of course, the reader is conversant ahead of time with our work. It is not unreasonable, we argue, to postulate a primordial age, as recent as 14,000 years ago, when no ice caps existed. The Earth would have been generally comfortable. It would be also enveloped in the gaseous atmosphere of the binary magnetic tube. This Uranian heaven blocked direct sunlight, but afforded an equable climate to the Earth.

As explained by Milton and myself in *Solaria Binaria,* the binary tube atmosphere would itself have been maintained by the same electrical and inertial forces that kept the Earth in rotation and orbit. Then the solar system as a whole was disturbed by the failure of one of its parts. The part that failed was the counter-solar or Super-Uranian node of the binary solar system. When the electrical current between the Sun and Super Uranus diminished, the magnetic field around the current diminished. All the bodies that circled around the current ceased orbiting around the axis between Super-Uranus and the Sun and descended radially to the plane of the ecliptic. They began to find new individual orbital paths around the Sun. They moved out towards larger orbits.

The atmosphere, a remnant, specially attached to the Earth, of the old plenum atmosphere, drew more closely about the Earth. "Heaven came down to embrace Earth," to paraphrase the Greek myth. The clouds were pierced by material erupted from the disintegrating Super- Uranus and blown down the magnetic tube between the binary partners. Some of it precipitated upon the surface of the Earth. The Earth could not melt much of the ice, most of which fell at the electrically least-guarded poles. The now direct sunlight helped the friction of the fall to vaporize and precipitate some of the ice as rain. Flooding began at the edges of the forming ice caps. The time postulated for these events began about 14,000 years ago.

Within a few centuries the threat to life on Earth became extreme. Great ice blocks covered the extremities and local regions of the globe and threatened ultimately to make contact, erasing practically all life. At the same time flooding spread throughout the world.

If one-third of the globe was covered by ice at the time of maximum advance, according to conventional theory, ice was piled three miles deep at the poles; there was twelve million cubic miles of ice. For a hundred years catastrophists and disbelievers in the ice ages have pointed out that an incredible power (heat and winds) was required to evaporate equatorial water, lift it, and transport it to the polar areas. The world would have burned up at the equator while freezing deeply at the poles. The idea supplies its own contradiction; yet it is the accepted theory, that molecule by molecule the water evaporated, drop by drop it condensed in vapor clouds, ton by ton it fell - all off and on for a million years and more. Then the mechanism was turned off, rather suddenly; much of the ice melted and the oceans rose by several hundred feet several thousand years ago.

Direct exoterrestrial deposition of snow to form the caps follows from the heat requirements to evaporate, lift, transport and condense as snow the contents of the ice caps. The surface heat requirements might have stressed the biosphere life tolerances. Further, in order to raise the required mass of water, the clouds transporting water from tropical to arctic regions would become so dense that heat from the Sun of today would cease to penetrate to the surface with sufficient energy to continue the lifting task. The latent heat of aqueous vapor at the tropics is 1000° F. A pound of water vaporized at the Equator has absorbed 1000 times the quantity of heat that would raise a pound of water in temperature by one degree Fahrenheit.

An exoterrestrial catastrophic solution is called for, from beginning to end. The time to erupt the Moon arrived with a passing great fragment of Super-Uranus. The Earth's crust burst. Lava had to flow in endless streams. Great volumes of sky- borne ice must have fallen and participated in the bursting mechanics. Cook has figured the needed forces, but we should add an initial impetus from the eruption and blow-off of the Pacific crust. A fracture shot to the old north pole and down the Atlantic, thence around the world. The ice avalanched. It fed the boiling sea bottoms to help them settle and expand. Much was then evaporated and precipitated again by the conventional method, but under catastrophic conditions. Finally the new world surface shaped up and stabilized. The precipitous curve of disaster dropped exponentially to the slight level of activity where it could be mistaken for a linear uniformitarianism.

It was thus that the worst and best accident happened. The earth cleaved, lost most of its continental crust, and the ocean basins began to form. This greatest of all catastrophes removed the ice and permitted life to survive; it became the greatest of all blessings. A date of 11,500 B. P. may be ascribed to the event.

The ice caps, as Cook has so well calculated the scene, collapsed and avalanched upon all sides, moving into the great chasms of boiling lava directly or through floods that rushed over the land and plunged down into the new oceanic chasms, carrying debris to form slopes. Hundreds of deep canyons were grooved into the land and slopes around the world, where they remain today, "fossils" from the time of ice age collapse and of the filling of the ocean basins. The ice caves were formed - solid ice from the ice ages sandwiched in between layers of once boiling lava flows, still intact, though hollowed out somewhat, now refrigerating food and supplying age-old spring waters.

Over a mile deep in the Greenland ice field around Dye 3 radar station, Greenland, ice cores are being drilled, extracted and analyzed. From its rock base upwards, the ice is expected to afford 100,000 years of Earth history and the beginning of at least the local ice age (cf other estimates of 1 to 3 million years and our own of 14,000 years). Oxygen ratios in sampled slices of the drilled ice are calculated to determine climatic trends and time scales. The units are "annual" ice varves. As depths increase, the distinctions blur. Dust ratios are used as indicators of heavy volcanic events in the world.

The stratification challenges any quantavolutionary attempt, as here, to explain the ice accumulation as a brief episode. Obviously the ice under examination did not fall as blocks, at least not most of it, or, if it did so fall in the region, the blocks splattered and connected up or flowed afterwards under weight, internal pressure and heat, picking up atmospheric exposure and dust.

Heavy snowfalls, whirled about by heavy winds, would, however, establish the great depth in short order, in dozens or several thousands of years, with present snowfall adding steadily to the basic conserved precipitation. It is noted that at an estimated 10,000 years, the "ice age" deposits of tiny crystals end and the large ice crystals of the present era begin. For those who are disturbed by only 100,000 years for the Greenland ice cap, because of ice age theories of a million years, there is the consolation that the ice beneath relentlessly squeezes out to form icebergs that search out more southern climes.

We speculate that the "ice ages" did happen, first in Uranian, then in Jovian times. Much of the Earth was frozen. The ice was mainly exoterrestrial. Vsekhsviatskii writes of Saturn that "observation of its rings over the past 300 years have shown that during this time the middle line has moved 0.17 of its original distance closer to the surface of the planet. Therefore, one may suppose that in a matter of some 1800 years a large part of the material in Ring B will fall onto the surface." We are here back to visualizing Vail's canopy drops, from primeval sources, or as a way-station. But a Saturnian explosion, not a falling of Saturn's rings, deluged the Earth with ice. Saturn's rings today may be fall-back debris of the same incident, still falling back.

The ice came down with falls of gravel and tillites. The great Ice Age extended from about 14,000 to 11,500 B. P. During this time the Earth was wobbling, the atmosphere turbulent and the deposits of ice were eccentric. Life would have been exterminated by the spread of ice and flooding if the greatest of all catastrophes had not cleaved the Earth and formed the ocean basins. Then ice and waters avalanched or fell into the basins as these grew in size, filling them ultimately over their brims.

The present ice age began in proto-historical times. Saturn's explosion drenched Earth with water and ice and the terrestrial axis tilted as a result of the explosive force. The age of "Jupiter of the Bright Skies," as the Greeks significantly denominated him, began; the skies were clearer and the climate colder because of the tilt; the high canopy was almost quite gone leaving merely the present upper atmospheric levels and magnetosphere of Earth. Ice began to gather around the northern and southern poles, drifting over the cultures of the age of Saturn.

CHAPTER TEN

CRUSTAL TURBULENCES

The continent of Atlantis" myth is still mostly a name for geographical and legendary indications of crustal turbulence; it finally sank in one furious climactic day, so said Plato, after exhibiting earthquakes, volcanism, rising or descending and swelling or shrinking earth masses, plus continental cleavage, floods, pandemonium and extinction - to which, remarkably, Plato assigns as a cause "an instability in the heavens..."

1. EARTHQUAKES

The major source of present-day earthquakes is to be sought along the lines of the global fracture. Taking up first the north-south Atlantic rift and following it from the Arctic to the Antarctic, one observes intense seismism throughout its length but largely in the middle of the Atlantic and little on both sides of the Atlantic Basin. The fracture, like an almost healed wound, throbs, festers and drips a little, pushing the continents left and right almost unnoticeably. Perhaps the rocks of the Atlantic Basin are lagging or stretching behind the Pacific rocks, which are being pushed into the basin of the lunar genesis. No theory is yet adequate to explain the difference in intensity and frequency between the Atlantic and Pacific seismism.

Wherever the fracture moves - into the Indian Ocean, across Asia, and laterally across the Southern Pacific and up the East Pacific - it bears with it seismic strains that develop as earthquakes of shallow focus. Quakes of deeper focus take place along a belt that circles the Pacific area of erupted crust, from New Zealand north and east up to Siberia, across the north Pacific and down the west coast of the Americas all the way to Antarctica. It is famous as "the Ring of Fire."

A second belt of shallow and deep focus earthquakes pursues a route along the old Tethyan equatorial region. It begins in mid-Atlantic, pushes through the Mediterranean and the Near and Middle East, shifts to follow the Himalayans where these break upon the Asian heartland, and swings down and across the south Asian seas. Here, where it overlaps the "Ring of Fire," it is intensively active.

But the Tethyan belt does not appear to cross the Pacific basin. It would, of course, have been erased if the Moon had erupted from the region. There thousands of seamounts stretch up from the ocean bottom, and long transverse faults occur. Rather, it resumes off of Central America where, indeed, there is a meeting of all four great earthquake belts - the globe-girdling rift, the Pacific "Ring of Fire," the westwardly moved Americas, and the old Tethyan belt.

Afterwards it proceeds into the Caribbean which may have been once coupled with the Mediterranean. It ends with the outlying islands some hundreds of miles south of its connecting link, from which I began here to trace its around- the-world movement. Geographers have matched Spain with the West Caribbean region; it is to be expected therefore that the Tethyan fracture of the south would tie into a transverse fracture to the north, thus circumnavigating the globe.

As we move back in time, the earthquakes increase in severity. Velikovsky points out that in the eighth and seventh centuries earthquakes were so numerous that when they occurred they were mentioned in a bare line of the astrological tablets of Ninevah and Babylon. Nevertheless, "reports concerning earthquakes in Mesopotamia in the eighth and seventh centuries are very numerous, and they are dated. Nothing comparable is known in modern times." He quotes from a tablet of Babylonia, "The earth shook; a collapsing catastrophe was all over the country; Nergal [Mars] strangles the country." Further, "references to breaches in houses, large palaces, and small dwellings are very numerous in the [Hebrew] prophets of the eighth century." Neglecting such sources, a historian could claim that "the earthquake held a place in the religious conception of the Israelites quite out of proportion to its slight and relatively rare occurrence in Palestine." Obviously, some literalness has to be restored to the language of the Bible, as well as to many ancient voices, if a better natural history is to be written.

A destroyed city may leave no records of its destruction; a sunken land leaves only an outsider's report and a myth. A lifetime (19371975) of work was dedicated by S. Marinatos before the archaeological and geological world came to realize, perhaps too enthusiastically, what earthquakes and explosions befell the island of Thera in the Aegean Sea some 3100 years ago.

2. Volcanism

"A theory of volcanicity" must not only be "taking into account the whole range of geodynamic processes," as Rittmann says in his classic work on volcanos, but also the whole range of cosmodynamics. Five hundred volcanos of the Massif Central in France, now defunct, were erupting 12,000years ago, or less, during the Magdalenian Upper Paleolithic culture. So maintained Escalon de Fonton of Montpellier University.

Volcanism takes the form of cones and fissures. It is also beneath swellings and bubblings of surface features. Most of the igneous basaltic surface of the world, including the ocean bottoms, was created by fissure volcanism. As occurs still in Iceland, fissure volcanos produce lava copiously. "During recorded history more lava has poured forth above the sea in Iceland than in all the rest of the earth's volcanic belts combined. Yet... Iceland's volcanic belt comprises less than one-half of one per cent of the total length of the world-encircling rift."

When the Earth had to erupt magma on a large scale, from far down, because of a loss of crust and an expansion of crust, fissure volcanism had to be the means. The deep ocean ridges of today still supply lava for paving the abyssal surface; the process has assumed a certain orderliness. On the other hand, viewing the Pacific Basin one must conjecture that a very large surface was once removed and a deep wound was left exposed that repaired itself in situ. The concept of cone and fissure volcanism fails, then and there, and one must speak of sheet volcanism, creating its own hard skin.

Fissure volcanism stands for extensive catastrophic venting; if there is so little of it today, the reason occurs in the general global settling. Cone or tube volcanos represent a moderate 'need to erupt.' Volcanic fields denote an interconnected set of tubes with a number of outlets. Volcanic outlets are spaced apart in relation to the thickness of the lithosphere; thinner rock invites closer spacing.

Today, volcanism of all kinds may be remanent. Fascinating and destructive as it may be, it is as nothing compared with the volcanism of times past. The Soviet geologist, A. P. Pavlov, declared in 1936: "At the present time, only a residual, negligible manifestation of volcanic activity is observed on the earth; formerly, this activity was perhaps the most typical and almost universal phenomenon in the life of the planet." Probably the phenomenon is correct, but the volcanism, like astroblemes, may have happened during only several immense exoterrestrial encounters.

The greatest eruption of modern times, some say (incorrectly) of all history, was the 1883 eruption of Krakatoa. The total volume of erupted material has been estimated at 18 to 21 km^3 . "When compared with prehistoric ignimbrite-forming events ranging in volume up to 10^3 km^3 .... the volume of the Krakatoa eruption was very modest." So declare S. Self and Rampino.

Thera's volcano (Aegean Sea) blew away most of a large, high island and its culture three thousand years ago. Ilopango (El Salvador) destroyed a cultured Mayan area of thousands of square miles in an explosion of 1800 years ago. The volcano of Tamboro on Sumbawa Island in the East Indies emerged from the waters in 1812. Within three years it grew the awesome height of 12000 feet, some three miles tall. Then it exploded. Approximately 100 cubic kilometers of material shot into the atmosphere. About 100,000 people were killed, many more than died in the Anglo-American War of 1812 being fought at the same time across the world.

Hawaii arises eleven miles from the bottom of the sea. It is the world's tallest mountain. It appears to be stable. Yet it ends a long fracture out of Mexico and begins an arc of seamounts that strikes Siberia. The scene of volcanism today is the pallid termination of the scenario of quantavolution. The interpretation of a wide variety of observed facts led us to the conclusion that magmas could originate in two ways, and that we could distinguish between primary magmas having their origin in a subcrustal zone encircling the earth, and secondary magmas formed by the anatexis of sialic rocks within the earth's crust, says Rittmann.

One notes here, besides the requirement of a stretching and tearing of the crust, the origination of volcanic magma from the "subcrustal zone encircling the earth" and anatexis, or regurgitation of surficial rock. This region occurs some 15 to 30 miles below the land surface and about 5 miles below the oceanic bottoms. This layer corresponds not only to the Moho discontinuity, as I have mentioned in connection with the base of seismism, but also with the volume of "missing sial" from the ocean basins, which roughly approximates the volume of the Moon. Volcanism, then, like seismism, reflects the level at which, all over the globe, the still landed crust moved in reaction to the eruption of the Moon. Whether or not the mantle on which this lunar boundary level rides jostling is solid or liquid, in the years of its fast movement it would have heated, liquefied, and expanded. The volcanos are probably still draining the liquid.

Studies of volcanic eruptions arrive at correlations between the moment of major eruption and the tidal forces exerted upon the Earth by the Sun and the Moon. Similar correlations have been detected between tides and seismism. In this regard, volcanism and earthquakes reveal themselves as close relatives.

G. Beccaria (1716- 81) with Stokeley, Franklin and others, set the stage early for a systematic approach to electricity in connection with earthquakes, cyclones, and volcanos, but the promised scientific drama has never been enacted. As early as June 21, 1902, Elmer G. Still published his observations of the volcano-solar-lunar relationship:

> The writer has for several years been observing this relation between the positions of the heavenly bodies and seismic, volcanic, and electrical disturbances, and is forced to the conclusion that the latter are caused in part by the conjunctions,

> oppositions, perihelions (or perigees) and equinoxes of the moon, earth, and seven other planets, especially when several of these occur at once.

A second article in the same year stressed that "the influence of the Moon and planets in causing and intensifying seismic and volcanic disturbances is not altogether tidal action gravitational; it is partly, or mostly, electrical, and seismic and volcanic action is an electrical disturbance."

Once more in 1902, the same author, E. Still, continued his prescient argument, now declaring that gravitational tides of the Moon were quite inadequate as explanations of many terrestrial disturbances. "We know [Still was seventy years ahead of the field] that magnetic earth currents (which interfere with telegraphing), brilliant auroras, severe thunderstorms, violent storms of many kinds, and also earthquakes and volcanic activity accompany sun spots. All these are electrical disturbances, and the eruption of Mount Vesuvius and numerous seismic shocks which occurred at the time of the last large sunspots - about September 15, 1898 - were no doubt electrically caused by them."

Flying high over southern Italy, one may luckily see Vesuvius, Stromboli, and Etna all smoking at the same moment. Arriving in sight of the famous seven hills of Rome, there is a grandeur of culture, not nature. Yet Breislak in 1801 was arguing that the seven hills were debris amidst a large volcanic caldera, and Cuvier for one approved the idea. When the oldest hominids, human in some ways, walked the Earth at Afar (E. Africa), some ten nearby volcanos were active.

A great many dormant volcanos exist and an enormous number of extinct volcanos. If the belts of inactive fissures and the unnumbered thousands of seamounts are added, the Earth has undergone periods of the most intense exoterrestrial stress. Or else, one will have to parcel out these millions of volcanos and 'volcano equivalents' over exceedingly long stretches of time.

The innumerable seamounts are a standing reproach to opponents of quantavolution. I have mentioned their origins as pulled mantle taffy in cosmic encounters. They are an impossibility for tectonic plate theory for there the continents move on plates, not through them, and seamounts appear abundantly around the Moon Basin of the Pacific, with a solitary but impressive chain of hundreds off the New England Coast. If the Moon were erupted from the now Pacific region, the seamounts could be visualized as pulled taffy drop-backs that could not follow the Moon into space. But the Atlantic Ocean off New England would only then have opened its abyss and "New England" would have been retreating westwards. To explain this particular "taffy-pull" we must conjecture a prolonged explosiveness or subsequent passes of an attractive exoterrestrial body in order to assist their generation.

The eruption of the Moon certainly extends beyond the conventional concept of volcanism, although Vsekhsvyatskii claims that planets and comets originated in volcanic episodes, especially involving escapes from Jupiter. Explosion is of course a fission; rocks are transformed; gases and electricity are part of the process, and so on. Also, exoterrestrial influences are connected with volcanism, both as to origins and to triggering activity. These influences are provable in our own time by correlations of volcanism with tides, electricity and seismism. They are provable for ancient times by the patterned system of volcanism in the world and the obvious function of volcanism in relieving stresses according to a pattern highly suggestive of transactions in outer space.

Furthermore, it is well to mention, as a postscript, that should the Moon have erupted from the Earth and all ocean bodies are young, then the eruption must have occurred recently. The basins are dated at under 100 million years. Thus the Moon episode, so incredibly destructive, would have occurred with the full realization of life on Earth, including many thousands of existing species and with most Earth rocks still present. If those species could survive, so even could homo sapiens.

3. Sinking and Rising Lands

Aristotle wrote in his book *Of the Earth,* "Inroads and withdrawals of the sea have often converted dry land into sea and sea into dry land." And Strabo, the most reliable geographer of the classical age, declared that "extensive submergence of the land, as well as minor submergence, has been known."

This is macro-geography, indeed! Geologists of East-Central Asia speak of a quarter of their world rising and part sinking. The events described are dated at the later Ice Ages, maybe less than 10,000 years in our own timetable.

An ancient document, the Oera Linda manuscript, which was written in Frisian with runic characters and whose age and authenticity is much disputed, claims a general Atlantis-type sinking of a prosperous civilization of the Fryas between the North Sea and the Baltic, where frost was rare and fruit trees blossomed. There came a summer of darkness, great earthquakes, a spitting of fire from newly bursting mountains, a general holocaust, an obliteration of rivers, and huge floods that advanced to cover most of the land. Whole islands were newly formed by the bones of dead cows and sand (one is reminded of the Siberian islands formed of mammoth bones). The survivors were subjected by invading Finnish bands (just as the Hyksos invaded Egypt after the Exodus).

The Caribbean peoples talk of an "Antilla," now sunk beneath the ocean. The Pacific Ocean and American peoples of the Southern Hemisphere say that once a continent existed where now stand a few islands amidst a great deep sea. The perplexing books of Churchwarden concern this continent of "Mu." Legends of the Greeks speak of a drowned Aegean Sea, and the ancients believed the Mediterranean Sea was recently arisen.

The East Indian peoples and Indian Ocean people offer legends of the sunken continent of "Lemuria," whence came world civilization. T. Huxley and F. Engels were famous supporters of the theory over a century ago. And the islands of the South Seas, where Indonesia stretches out, are reputed to have been of a single piece before the waters rose or the land sank. The Dutch geologist Bartstva claims that a land bridge connected the Celebes and Philippines until Holocene times. In August 1982, Alan Thorne announced the discovery of Chinese human remains in North Australia with an estimated age of at least 10,000 years. A map in *Chaos and Creation* outlines in the most general way all of these mythical lands that are said to have existed in human times.

These were not the only risings and sinkings, but they were by far the major ones. Kondratov, for example, mentions that Bulgarian researchers have compiled a detailed map of underwater archaeological finds, dating from the eighth to the fifth centuries B. C., discovered along a large section of their country's Black Sea.

If one is to believe legend, every large expanse of ocean once had its land mass. A form of quantavolutionary reasoning could proceed as follows: the ocean basins are new, created in the time of man; before the time of man, there was Pangea, a globe covered by continental crust that carried shallow freshwater seas, especially in the then equatorial area, which area, now greatly tortured, is still recognizable in the fabled Tethyan Sea remnants of the Mediterranean area and the "belt of fire" that girdles the world longitudinally.

Where the continental fragments do not remain to be fitted obviously together, then the intervening land was blasted away or sunk. Continental sial has been extracted on occasion from the deep bottoms in the Atlantic and Pacific Ocean; this is surprising given the major discovery of recent oceanography that the ocean bottoms are covered everywhere with lava. Metamorphic rocks typical of the nearby islands and Italy were found 3000 meters below sea level in the central Tyrrhenian Sea, as reported by Heezen.

John L. Sorenson, citing Kroeber and others, examines 200 basic, defined cultural features of the "Old World Oikoumene." What would be called the "common heritage" of the peoples of the Near East. Of these features one in eight is found in Meso-America definitely. He believes that another tenth would be added to the New World list when checked out through the whole body of information; thus about 18 percent of the Old World basic culture traits are shared with the New World. The statistical probability that this percentage of correspondence would occur by accident is low. It suggests land bridges of past ages.

Perhaps it was around 1500 B. C. as well when Thule vanished into the Faroe Rise. Thule is famous in Northern European myth and is referred to in many books and accounts with tantalizing brevity. Russian geographer N. Zhirov argues this theory, citing evidence that Thule was near Iceland, that many islands were mentioned thereabouts, that it was in a warm oceanic current, and its people grew grain and other crops, (We are reminded of the Oera Linda manuscript.)

Indeed the birth of the North Sea may have come so late as 1500 B. C. The famous amber of the North and Baltic Seas is conventionally dated at seventy million years; it comes from submerged pine forests that are assigned that date. Recently geologists have begun to stress the youngness of the area, prodded by archaeologists. Drowned settlements have been found at the bottom of the Baltic Sea, the North Sea, and off the British Isles. These are not to be confused with the sunken settlements of later time - Slavic Vineta in the Baltic by a tidal wave of

1100 A. D., many places off the mouth of the Rhine in 864 A. D. and so on. We are writing of the whole of these seas. "Europe was inhabited when the North Sea did not exist, when England and Ireland were not islands and Jutland and Scandinavia were not peninsulas but were all parts of a single land mass." Thus writes Roy MacKinnon, who gives us a fix on these great submergences.

4. EXPANSION AND CONTRACTION.

Whatever can explode can expand. Worlds explode. Radio astronomy and even visual observation on rare occasions, confirm this. The asteroid belt between Mars and Jupiter was probably a planet until recently. There are some small indications that it may have been identified with the Greek god Phoebus Apollo, hence be so close in time.

The Earth can explode. Therefore it can expand. It is more difficult to construct a model of expansion than a model of explosion. Both layman and expert can readily conjure up an image of "more than enough" energy to explode any body. In so imagining, they may skip over the crucial problem of how much it takes exactly to explode the body. An explosion can be defined as a rapidly accelerating expansion that has achieved a specified rate where a set of effects occurs that is called "explosion."

The conservation of angular momentum of a rotating body depends upon its retaining the sum of its mass, its velocity and its radius. The radius is the distance from the center of rotation to the direction of its motion along the axis of rotation. Expansion signifies a change in radius.

The concept of radius describes a relationship of objects. It is not itself a force or an entity. Therefore, the expression "increase in radius" must signify a changed spatial relation between things that determine the radius. Once more, the salient question points at electro-mechanics, determinants of mass that might act to increase the radius.

Expansion of a rotating body then must be associated with a change in velocity or mass. In the case of the Earth during a lunar eruption, the loss of mass consists of half the crust and most of a dense atmosphere, altogether no more than 2% of the mass of the Earth. An interruption of rotation imposes an abrupt decline of spin velocity upon the Earth.

This then requires an increase in radius and expansion in order to maintain angular momentum.

At the same time, the conservation of angular momentum does not occur in an isolated system. In the present case, energy representing the angular momentum is transferable to other external systems: the space plasma, the proto-Moon, the Sun and planets, and cometary bodies.

A body such as Earth will expand when it is freed from an external pressure. Possession of a dense atmosphere of the type of Venus would have limited the Earth's figure; if removed, the Earth would have expanded. Its outer surface will even spring back, that is, exhibit an acceleration and a counterpressure that causes it to "take off" from its base. Irregularities found in a number of places around the globe may be fossil expansions, if not fossils of impact explosions and massive eruptions.

There are reasons to believe such events can occur and have occurred. In anticipation of stating them, we may suggest why land has sunk; for the two behaviors of expansion and sinking are not independent, although they may occur at different places and lithospheric levels. Lands have sunk by collapse into new basins, by flooding, and by their contents disappearing in explosive clouds of debris. If the force that explodes the land expands the Earth, then we have sinking and rising in a new formula, one which contains its mechanism, and furthermore may be true.

Carey and Jordan have devoted books to the subject of Earth expansion, and were cited earlier. Both see the process as very gradual. Carey estimates a 20% radial expansion and uses the projected expansion as a mechanism to account for continental drift.

R. H. Dicke and C. H. Brans also predicted a slow drop in the force of gravity, and Dicke estimated that Earth gained from this source 15% in volume over 3.25 billion years. When the Atlantic basin was shown to be young, Dicke ceased to credit its widening as support for his theory, because it apparently had grown 300 times faster than his theoretical rate would allow.

In comparison, my estimate of the radial expansion which accompanied the fission of the Moon is about 9%, less than one tenth of the total radius; the estimate is the result merely of topographical scrutiny. The expansion in volume represented is about 20%, much less than Carey's estimate. Carey's expansion took place over many millions of years; the process here discussed would have occurred in perhaps three thousand years. Again, we rely upon a uniquely great exoterrestrial encounter to compress time, accomplishing in centuries what the aforesaid scientists have allocated as the task of very many millions of years. Any evidence at present of an expanding Earth, we would accredit to the extended uniformitarian tail of the exponential curve of quantavolution.

Of the several attempts at demonstrating expansion, Meservy's appears most clear and valid. He shows that "the separation and movements of the continents in the last 150 million years cannot be explained by continental drift on the surface of the present sized earth." This he does topographically. Following the Bullard and Hurley reconstruction of the supposed original supercontinent before its continental elements drifted apart, he retrojects the present continental map as it must have drifted and shows that the present arrangement could not have emerged from the reconstruction.

In order for the supercontinent of one time to fit the map of the continents of today, the continents of today would have to come from a smaller globe. "It seems highly improbable that the area enclosed by the perimeter [of the Pacific] was ever as large as half the earth's present area in the last 150 million years."

By what means could the Earth have expanded at the time of or subsequent to the breakup of the original super-continent? Six means can be suggested, none of them excluding all others and in fact all six could be simultaneously operative to produce a concurrent breakup of the continental mass and an expansion of the globe. Meservy does not consider a sudden loss of over half the Earth's crust, as by Moon fission, but significantly the occurrence of such a loss, concentrated within the Pacific perimeter, only serves to strengthen his topographical demonstration.

An abruptly slowed rotation of the Earth over days of time, never to be restored, would reduce the centripetal force of the globe and tend to expand its volume. This would be especially prominent if the body causing the slowdown were electrogravitationally attractive. The Lorentz-Fitzgerald (1893) equations assert that all matter contracts in the direction of its motion and the amount of the contraction increases with the rate of motion. The Earth rotates with a kinetic energy of 2.6×10^{36} ergs. If an interruption by an external body depresses its rotation by 35%

and shortly thereafter the rotation assumes the level of a 20% reduction, an energy of some 10^{36} ergs is available, along with a large electrical, gravitational, and axial torque energy, to push the continents and expand the volume of the Earth. This heat of rotational slowdown is sufficient in theory to unleash 50 billion Krakatoa's. That volcanic eruption, one of the worst in history, released about 2 x 10^{25} ergs.

The conditions for expansion of the Earth were probably present, but they approached the conditions for a complete melting of the crust of the Earth. They approached, beyond that, the conditions for the explosion of the Earth. Nevertheless, in the end, the sphericity of the globe was maintained, half of Pangea was preserved, and small numbers of most flora and fauna, including homo sapiens, survived. The fall of cold water on the continents helped to preserve their structures against heat from below while the same waters moving into the oceans and the falling waters there catalyzed the expansion process.

The sudden acquisition of a huge heat presented problems of storage and prompt use, if the Earth were not to explode. The Intruder's pass-by and the forces it exercised upon the globe would begin some days before the moment of maximum impact and continue for several days thereafter. Thus the heat would not be applied all at once; by the time the critical moment arrived, the Earth was committed to partial explosion and expansion. The loosening of the Moon-making crust and the cleaving of the globe would take place quickly; then immediately the heat would be drawn upon for the reconstruction of the Earth.

Also, a great proportion of the heated matter would be exploded into space. The global fracture system would help to handle the venting of enough heat and material to cool, pave and expand the Earth's surface. Moreover, it would develop the capacity to do so within the required time. And the density of the Earth's interior would be originally sufficiently high to provide the material.

A decline in atmospheric pressure by the temporary and permanent removal of atmosphere, especially a heavily vaporized one, would also contribute to an expansion of the Earth. So too, of course, would the actual removal of crustal material of low temperature. It is not necessary that the rising magma be less dense than the escaping crust but only that temporarily it be in a molten state, mixing with gases and water as well, and hence capable of freezing into a solid at a higher level or over a larger expanse of surface. H. J. Binje said once that "the driving force of rising magma lies in change of the nuclear structure of the magma itself."

Water added to a silicate solution reduces its melting point. The lower crust and mantle boundary might melt at as low a temperature as 500 ° C under water saturation. The water itself would be provided by old surface waters and incoming deluges of rain, snow, and ice. The upper crust on which the biosphere and sediments rested would be shielded from the abyssal heat by thousands of volcanic vents penetrating its surface and by the cyclonic venting of heat into space over the immense flayed crater of the Moon. Still the thermal pressures throughout the globe would be heavy and accompanied by rises in temperature that would increase the expansion.

The globe would fracture throughout. Pictured as scraped of its biosphere and surficial sediments, the globe today presents a thoroughly fractured appearance. Nowhere on Earth is one very far from a great fissure that would have been involved in expanding the globe. Perhaps one of the reasons for the discontinuity and absence of expected sediments in so many places is the underlying expansion by igneous intrusions that once occurred. Furthermore, the very 'success' of the globe girdling fractures in producing ocean beds of lava and pushing away the continents is that they were engaged in expanding the volume of the Earth.

The sial continents that remained obviously were not destroyed in the process of partial explosion and expansion. However they were penetrated at many points by expanding lava. The sial could be lifted by less force than would be required to dissolve it. Given over half the surface as a direct outlet, and a huge fracture network for disgorging heat and magma, there would be less occasion to obliterate the many large areas of sial overhang. Willis once wrote that "it is established by observations on rocks that the chemical compounds of which they consist can adjust themselves to changes of pressure or of temperature or of both by changes of volume as well as by alterations of form. Larger volume would result if a mass of rock were heated and at the same time relieved of some of the load resting upon it." He even went so far as to say that erosion can cause underlying rocks to expand their volume. Rock crystals respond to new conditions, not even highly thermal, by reorganization of their structure. "Crystals are almost human in that they always seek the easiest way out... Where crystals grow vertically, continents rise."

Seismic signals experimentally transmitted through the Earth produce more or less sudden changes in velocity, indicating "boundaries" at six radial distances before reaching the center: the Moho discontinuity, and at 400+, 950+, 2900+, 4800+, and 5100+ kilometers of depth. There seems to be little explanation for these seismic transitions unless they represent levels of response to an historical torque. The interruption of a rotational motion of a mass must be perceived by the whole body. At some ratios of density to torque, indications of a phase shift should occur. These indicators would be erased by a huge expansion, but by the same token, will remain vivid under conditions of moderate expansion.

In *Chaos and Creation,* I asserted briefly, and probably in error, that the Earth would lose electrical charge in a grave encounter such as would remove the lunar material: "loss of electrical charge may also have decreased the density of the Earth." This was based on the assumption that piezoelectricity from rock turbulence and electrostatic charges would be lost into space to the larger intruding body; then matter hitherto bonded electrically would be unbonded and take up more volume. However, after discussions with E. R. Milton, I became persuaded that the intruder would have carried a heavier charge, since it was transporting charge from the outer solar system toward the Sun; it was also much larger than the Earth; therefore it would have deposited charge upon Earth. The charge would then be incorporated by the Earth's molecules and cause the stretching of their internal atomic bonding. Hence expansion. But where the charges would accumulate is critical, whether on the continental surfaces, or diffused in the mantle, etc.

Yet another set of phenomena may be connected with Earth expansion, rather than simply the adjustment, unexplained, of coastal margins to which it is otherwise attributed. That is the tendency of continental margins to stretch out over the ocean basins. For instance, "as late as the beginning of the Quaternary period the land of Siberia reached much farther north and at the end of the last glacial epoch was broken up, large areas sinking into the sea."

Elsewhere we read, "while exploring the seismic structure of the continental margin off France, Lucien Montadert, of the Institut Français du Pétrole, noticed that the upper part of the continental crust of margins has been fractured into a remarkable pattern of narrow sedimentary basins bounded by listric faults, that is, faults that 'curve, ' being steep at the surface, becoming more horizontal with depth. He suggested that the continental crust at the margin was extended at the time of rifting by up to 20 per cent."

When the Earth's surface is viewed from a detached intellectual perspective, it begins to appear as a thoroughly disorganized assemblage. Instead of its presenting logical conformities on a grand scale, its every feature becomes an anomaly. All of its real rules seem to have come from violating the rules of the earth sciences. When such a condition is manifest in human organizations, such as the factory, or the hospital, or the government, that is, when what is regularly done contrasts with the way things are supposed to be done, the usual recommendation is to change those rules that are inapplicable to reality. Unfortunately in the present case, as in many cases of social organizations, new rules are not easy to write and, meanwhile, the old stable mixture of reality and pretense that has been managing the enterprise dissolves into fantasy and disorder.

CHAPTER ELEVEN

RIFTS, RAFTS & BASINS

The globe is girdled by lines of fracture, and from these lines stretch perpendicularly hundreds of transverse fractures. The pattern of rifts and cleavages shows that they created basins (as in the Arctic and Atlantic) and exuded material to fill the Pacific Basin which was dug out by exoterrestrial force and assembled or dispersed in space. All of this happened fast. Continental drift is true, but not true enough.

1. THRUSTING AND OROGENY

When nineteenth century geologists departed from their original simplistic uniformitarianism, they found it useful to identify in Earth history several points of great diastrophism ("turnabouts" in Greek) or revolution. Whence came the Laurentian, Algonkian, Killarney, Appalachian, Laramide and Cascadian Revolutions, each marked by profound unconformities in the rocks. Naturally a quantavolutionist will wonder why these never evolved into a new catastrophist geology. First, there was the obstacle of ideology, which a social psychologist can appreciate more than a natural scientist: the social atmosphere of the times, the breakaway from religion, the need of biology to pursue prolonged development periods, and the empirical fascination of studying the processes going on before one's very eyes - these acted to subdue diastrophism and revolutionism.

Long periods of slow changes were supplied until the revolutions themselves appeared as continual skirmishes of the elemental forces. The search for internal forces capable of sculpting the Earth's surface went so far as to conceive of the massive core of the Earth wobbling within the globe so as to push out or pull back crustal features. Some have thought of shrinkage, so that as the Earth aged it wrinkled (apparently not willing to move out upon the seabeds). Nowadays radioactive decay, rising from the rock deeps and engendering heat, has been called upon to push the continents slowly about. And this is said to crumple the colliding edges of continents into mountains and to stretch and reform the landscape.

A second reason why catastrophist geology could not evolve is also related to the ideological: geologists have refused to look into the skies for the forces needed to accomplish the revolutions that they perceived; without a mechanism, they were left with mere names - and questions, such as ones asked by K. Krauskopf; "What are the irresistible forces which can twist and break the strongest rocks?" "Where do the forces originate which can raise and lower continental masses vertically? ... Why have not forces in the crust long since reached an equilibrium? With questions like these we have long since reached an impasse."

Since we have essayed answers to questions of vertical movements for the sake of this chapter, we may add: "How can kilometer-high sediments be pushed over thousands of kilometers of the surface of the Earth?" Every thrust that has occurred or might happen can be described by the same few variables. The permutations resulting in reality may be numerous but are still intelligible.

The continents move; this is a form of thrust. Often it is a thrust through water and basalt bottom; then, again, the Indian subcontinent thrusted upon Asia. Sedimentary rock layers are scraped and dumped over the sides of the awesome abysses; here is thrusting. Coal fields are forests bulldozed and deep-buried: this is a form of thrust. Mountains are piled upon one another, again a thrusting action. Tides and winds lay down field upon field of debris, of vast extent; are these not thrusts, too? It is fruitless to argue over definitions. As with earthquakes, which are moving earth, and shaking, so with thrusting: beyond a certain intensity, the vocabulary is inadequate to the quantity of cleavage, the quantavolutions. Then, too, earthquakes and thrusting can come to a marriage; in a discussion of even the relatively mild seismism of our times, Frank Lane writes that "where an earthquake is concerned there is no such thing as an unmovable object, even mountains are moved." They are thrusted.

Yet it has been a long time since "the mountains skipped like rams," as the Biblical Psalm goes. Although the records of solarian geology are far from complete, we suspect that such a sight has not been seen in the past two millennia. The occasional spectacular rock avalanches and submarine mud avalanches that are presently recorded are not what the Psalmist had in mind. In an age that experienced earthquakes abundantly, he was celebrating and reporting Yahweh at a peak of power, probably in the centuries that remained vivid to him - the seventh to the fifteenth before Christ - reinforced by the cherished accounts stretching back to the breakdown of the pangean surface.

He was speaking for the hapless ones who watched the Alps rise up from the Tethyan geosyncline to be "shoved northwards distances of the order of 100 miles" where now are located Italy and Switzerland. Or the American cordillera, thousands of kilometers long, stretching from Alaska to Tierra del Fuego; there mountain uplifts amounting to thousands of meters have occurred, it is agreed by a range of authors from C. Darwin to I. Velikovsky, in absolutely modern times. The Sierra Nevadas of California are a single block, a thousand kilometers long, thrust up westwards. The Himalayas rose steeply in human times. "The highest mountains in the world are also the youngest," wrote Helm and Gausser. But the Himalayas are also reasonably accredited to the crumpling of the "Indian" subcontinent against Asia with the vast inertial forces initiated in continental rafting. And probably the rising of the Tibetan and African plateaus occurred under lateral and subterranean pressures of the same time.

One after another, explorers and writers have expressed surprise at the youthfulness of the mountain ranges until at least and at last all that are spectacular have been moved up in time to the age of humans. Cook uses huge avalanching ice blocks convincingly as the bulldozer of many thrustal incidents in America and, in one case, in South Africa. The ice sheets push the sedimentary strata for many kilometers, melt and flow around them, crack through them, scatter mounds of debris in their path "like loads of loose, dry snow thrown ahead of a fast moving snow shovel."

Possibly the only alternative to his mechanism would be rapid continental movement toward the south, deceleration of the basal movement of the crust, a swelling of the Earth beneath the rear echelons, inertial continued movement in the same direction by weaker overlaying sedimentary strata, in some cases even overrunning the halted forward elements. Under this scenario, one would have numerous cases of inverted strata, older on top of younger, and hence the fossil inversions sometimes deemed a disproof of evolution. One would also expect then the occurrence of thrusts in regions of the world where no ice sheets were at work; in fact the Alpine overthrusts, the Atlas mountains and other overthrusted areas were not near to overpowering ice masses.

Thrusting played a large role in the formation of coal, lignite, and fusain deposits, which range in depth from the surface down to over a kilometer. The distribution of world coal deposits, Cook shows, follows in significant part the radial avalanching of the ice caps. Coal deposits radiate from cracking and thrust points of the old ice cap and shell-slip. "Most coal deposits are found apparently squeezed by crustal thrusts, between the ice cap depression zones and the concentric, flow-resisting mountain ranges." Ice cap fragments moved outwards upon the biosphere with the scooping and scraping motions of a giant earth-moving machine, depositing it, often smouldering, often slurried with ice and sky waters, into heaps, folded and thrust them over and in-between with thin sands, clay and gravel, and abandoned them in a state of thermal-retaining and heat generating compression laterally and from above. Superhurricanes, fast deep water tides, and typhoons can also scoop and pile up the total biosphere. Since the deep oceans did not exist during much of the quantavolutionary crises, the massive scoopers, scrapers, and in-folders might handle the marine life of shallow seas identically. Coal of different grades, and in thin beds, is interlarded with layers of ash, charcoal (fusain), clay, till, and pebble, that is, with all that goes before the blade of the bulldozer.

Where is the thrusting and folding of the ocean bottoms? There is very little of it, unless, as we said, continental drifting is called thrusting. The seabeds are flat, save for the steep oceanic ridges, the great rises, and the innumerable seamounts.

Geophysicist Edward Bullard marks the contrast:

> The mountains of the oceans are nothing like the Alps or the Rockies, which are largely built from folded sediments. There is a world-encircling mountain range - the mid- ocean ridge - on the sea bottom, but it is built entirely of igneous rocks, of basalts that have emerged from the interior of the Earth. Although the undersea mountains have a covering of sediments in many places, they are not made of sediments, they are not folded and they have not been compressed.

The last sentence points up an impossible predicament for conventional geophysics: a supposed situation in which the continental crust folds and thrusts and compresses into abundant mountains while the oceanic crust slides up and under and around without making mountains, having once and for all and by gradual processes made its igneous ridges and seamounts. That no continental mountains are to be found imbedded in oceanic basalts is remarkable. Considering how recently most of the mountain ranges of the Earth have formed, however, we surmise that the mountains came on the heels of the ocean basin creation or thereafter.

The rhetoric of geology is overpowering in its stress upon time. It rolls along in the cadences of an epic poem, stressing eons of time like the pause at the end of the lines. But today has its poetry of the absurd, and this may drive the incessant echo for a moment from the mind. Consider, then, the absurd: that legitimate arguments can maintain, facing the geological world, an age of 10^9 years and an age of 10^4 years - ten billion against 10 thousand years.

The absurd, of course, is the theory of quantavolution: time is squeezed out of explanations of the Earth until only the minimal amount remains, like forcing the air out of a bottle until a nearly total vacuum is reached. The analogy is not so remote: some say that the Earth is losing its atmosphere, atom by atom, until one day, eons from now, it will move denuded of air in the vacuum of space. The same might be done in hours and days by the near passage of a body sufficiently large and electrically attractive to suck up the atoms of the atmosphere.

The absurd in geology makes statements of a related type. All the igneous rock and its formations of the Earth's crust, could be brewed by sudden heat over 1500 ° C and pressures over 5000 atmospheres within a few years. Igneous rock is the greater part of all rock. All rock that is metamorphic needs less heat and pressure to form, and the same short time. Metamorphic rock is a small percentage of all rock. Sedimentary rock, least common but plentiful nonetheless, by definition never boiled or overheated or intensely pressurized, can be laid and formed as fast as material is provided, this consisting of biosphere products, fall-out, and erosion of other sedimentary, igneous and metamorphic rock.

The absurd idea still has not gone far enough; the Earth's surface and crust are a complicated mixture, of thin and thick pieces, of sliced and hacked out layers, and of dense and light materials, under different pressures and temperatures. How is it to be fashioned to bring order? Dispense promptly with the word "order". The natural order is largely in the mind. The "order" is a wish and illusion. Pursuing the absurd, the mixture of forms and materials of the Earth's crust are but the work of a clumsy chef, who shakes his pot, stirs it erratically, burns the bottom and adds ingredients to his strange tastes. Or, to assign no blame to a divinity, the same effects are achieved by forces born within the Earth and coming from outside of it, but great forces, of the kind that can form the materials. The force that can suddenly slow or change the world's motion can thrust and scatter about the formed materials, and concoct others. What can chop and grind and break the materials can inject all the heat and pressure to make them in the first place, and again and again. What is more, in this absurd scenario of quantavolution, processes occur simultaneously. The chalk cliffs of Dover do not wait to form until the Anatolian chalk cliffs are made; nor does the mutation of species await a sunny "bowr of earthly blisse." While the Earth's crust is reforming into the Moon, a multitude of volcanos blaze, and deluges of water and debris fall upon the world. All the rocks everywhere are in movement, under pressure and exerting pressure; electricity exudes from every pore and catalyzes the already hard-working floods and vapors; radiation and adaptive saltations are differentiating many species and exterminating many more.

How does one argue against the absurd conception of natural history? One would draw books on the Grand Canyon of Colorado from the shelves showing "two billion years of history passing before one's eyes." But the quantavolutionary vision of the Grand Canyon springs readily to mind: the complex can be put together in a short time in uplift and cross-cutting floods, then cleaved, supplied with torrents, and finally quieted down to make it attractive for tourists. Should one appeal to radiochronometry to resolve the vision, it occurs that the radioactive isotopes might have been stopped or raced in the catastrophic maelstrom. We recall again some of the features of the Earth's surface previously discussed. One by one, it would appear, the morphological features of the world succumb to quantavolutionary explanation.

"Long distance overthrusting has occurred (a) for whole continents over the ocean crust where overthrusting has been several thousand miles (continental drift), and (b) for the superficial Cambrian and 'younger' sediments over the continuous, strong basement rock." (Cook) The greatest thrust and rift and the smallest rock-crack can be considered as "faults."

"Shields [the flat barely covered rock of Canada, Scandinavia, and elsewhere] are here interpreted as crustal rocks denuded of sediments by thrusts of their original sediments from beneath the ice caps driven by the hydrostatic pressure and the friction of the ice flow" (quoting Cook).

"Welts" define pre-Cambrian rocks (that is, with slight signs of life), exposed at the surface as a result of uplifts and crustal buckling.

Huge troughs such as the Mississippi Valley are the result of an immense flow of turbid ice-laden waters and tidal flooding, so recent that spectacular anomalies such as the great New Madrid earthquake can occur.

Countless rubble hills are dumped in place by floods and wind from rocks expanded and broken up by earthquake. Most of the rubble orogeny has occurred in times of quantavolution, not by evolution nor uniformly bit by bit.

Uprisings occur through collision of rock masses, undercutting, compression, heat expansion of undercrust, and cooling of quasi-exploded material. Here would be included the igneous mountains of the world, such as Mt. St. Helens or Vesuvius. Here also would be earth that did not escape upon explosion and appears as mounds or hills swollen up (not buckled). Here too would be broad plateaus caused by a heat-expanded crust that cooled in its expanded form at great heights.

Finally, closely related to the previous item, are the submarine ridges around the world and the myriad seamounts (guyots). The ridge mountains, the world's tallest, are igneous productions, still bubbling and bursting along their length. The seamounts, as noted earlier, are the taffy-like pull-back, unexploded lava blisters of the lunarian outbursts.

Quantavolutionary theory, then, holds that any hill and mountain of the Earth can be explained by concepts such as these. All involve energies that erase millions, even hundreds of millions of assigned years of time. In fact, one might imagine a quantavolutionary definition of many other features not before discussed.

Where a gorge, a rift, or a canyon is observed, we are traumatized into seeing faults, fissures and turbulent waters rushing to shape them. Where others see placid lakes, long ago hollowed from rock and fed from melting ice, we see land sinks, quick filling with avalanching waters, now stranded and in all short-lived.

Where deep surface deposits of clay, pebbles, sand, till and their associated rocks occur, we see tidal catastrophes, cyclones, and exoterrestrial fall-out.

In lava fields are seen, not occasional flare-ups after long-prepared mantle heating, but the rivers of boiling rock forced up and out by large earth movements and expansion.

Fan deposits are not gradual accretions at the foot of a flow, but sudden dumps by turbulent currents, and the continental slopes are the largest of fans.

Catastrophic winds, tides, and floods form dunes and peneplains, abetted by seismism.

Basins are formed and erupted by catastrophic uplift, changed Earth motions, or meteoroid impact explosions.

What is left to mention in the lexicon of landforms? Faults, for example, can be classified into tilts, grabens, horsts, and troughs and each of these is divided into sub-categories; these are treated in textbooks and present no unsurmountable obstacle to quantavotutionary theory. Each of these pertains to its parent category – faults - and cannot supply something which the parent lacks. Metamorphic rock is of many kinds - schists, gneiss, limestone, marble mycorites and migmatites - and a natural history museum will present an orderly array of them.

The word "order" occurs again. And again the order is in our minds. The several conditions of heat and pressure and the several minerals that altogether manufactured these rocks were a disordered composition baking inside a faulty oven. One is seduced by the vast quantities observed of each type into imagining orderly production. A tall mountain of sedimentary rocks appears orderly to us, but so does the simple snowflake under a microscope.

One is impressed also by the very many material compositions and forms. But this is an illusion arising from the many different combinations which a few conditions and chemical elements can create; a mere eight separate states of being, described in terms of a temperature, a pressure, and a chemical compound free to combine, can, after all, supply some 2^8 or 256 entities to contemplate. There is order in all things and alongside this order there is chaos in all things; that is, we can look at any event or thing as orderly or chaotic, just as Parmenides looked at the permanence of being and Heraclites at its eternal flux.

Where in the world is the remaining virgin land of Pangea? If one is to believe surveys of the presence around the world of all the conventional geological ages, the answer is "practically nowhere." Perhaps 2% of the world's land can claim a full geological column. The ages are either a fiction, or the victims of quantavolution.

Still, even at this early stage of quantavolutionism, when few minds - and even fewer resources - have been brought to bear on the issues, it appears that by employing only a modest increment of time, quantavolution can move from the absurd to some respectable level of probable validity. One can comfortably and scientifically operate given an Earth age of a million years, with a late resurfacing of the Earth accomplished during the past fourteen thousand years.

It might seem impossible to reconcile the 5000-times-greater time span of conventional geological theory. Actually it is not impossible. The processes reflected in the Grand Canyon profile could be temporarily collapsed by a factor of 5000, making every five million years become a thousand years, without scrambling ordinary explanations. The rules to reduce time are: increase heat; increase pressure; add motions; introduce electric potentials; and look into the skies. Says the sage to the astronomer, writes Friedrich Nietzsche: "As long as you still experience the stars as something 'above you' you lack the eye of knowledge."

2. OCEAN BASINS

The planet Venus, which has been shown to have had its share of astroblemes, lightning activity, melting, volcanos, plateaus, mountain ranges, great valleys, and closed depressions or basins, has a "curious dearth of great basins," whence we surmise that Venus never underwent the trauma of Earth, which resulted in most of the Earth being ocean basins. "The tectonic forces that have shaped the surface of Venus have raised only 5 percent of the surface into 'continental masses' and left only 15 to 20 percent of it as basins... " Nor do we know whether these are "real" basins, that is, distinct from the continental material as they are on Earth.

The ocean's "trackless wastes" may be a nice metaphor for the 71% of the Earth's surface covered by water, but the ocean bottoms are marked by enough signs to revolutionize the earth sciences and natural history. Essentially the ocean basins are three in number, the Pacific, the Indian and the Atlantic. The Pacific Basin was the recent scene of the most awesome event ever to have befallen the Earth since its early times, the outburst of the Moon. The Indian Ocean appears to have been created at the same time by the migration of continental land driven to the scene of the disaster. The Atlantic Ocean was rather obviously originated from a great wedge that helped propel the continents east and west so as to distribute the mass, heat, and electrical charge rather more evenly in the expansion and filling initiated in the evacuated areas.

The prevalent view of sea-floor spreading has molten material exuding from the great oceanic ridge volcanos, pushing into place as a strip and jostling the older strips that compose the floor to move further away from the ridges. Ocean floor chronology and drift theory are based upon observations that from one strip to another, every several "millions" of years, there occurs a magnetic field reversal.

However, besides the other problems, which I have recounted, one core (395 A) from the Atlantic ridge flank shows magnetic differences in depth; the upper 170 meters is normally magnetized, the next 310 meters is reversed: and the following 40 meters is again "normal." This is an unwelcome surprise to chronometry and the theory of convection currents.

Still, pursuant to our theory here, we should expect erratic magnetic effects to accompany the great outpourings of lava; as soon as a batch is dumped off the ridge it hardens with the magnetic orientation of the moment. Very soon, before it has moved away, another batch is dumped on top of it, then another, all occurring before the whole thickness of lava moves far enough to be free of additional burdening. If, as we think, the ocean basins could mostly be paved in a thousand years, during which time the Earth's field would be moving geographically and oscillating, the laminated magnetic structure of the floor must follow.

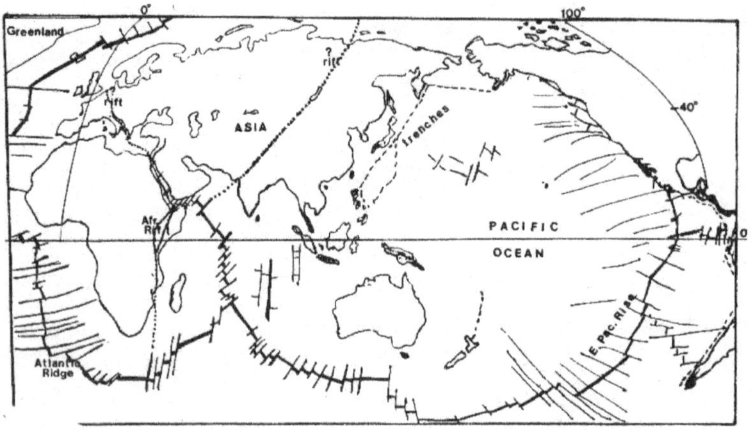

FIGURE A -Sketch of the main ridges and fractures of the Pacific-Indian Ocean bottom(---)with main trenches (----). Possible TransAsian and Trans-Euro-Mediterranean rifts are added (....) to the drawing, which is adapted from O. G. Sorochtin, ed., *Geophysics of the Ocean* (in Russian), vol. II, fig. 17. The lithosphere (crust) is everywhere shallowest beneath the ridge lines. Thousands of seamounts shooting up from, the ocean bottom are not drawn here.

Allen Cox points out that "if sea-floor spreading has occurred at a constant rate, the marine magnetic profiles may be interpreted to yield a reversal time scale going back 75 million years. The apparent average duration of the polarity intervals was greater during the time 10.6<t<75 million years than during the past 10.6 million years, and during the time 45<t<75 million years the average length was still greater."

Cox realizes that this might be an effect of an inconstant rate but dismisses the idea. With our larger theory that negative exponential rates followed a catastrophic opening of the basins, we find this data supportive. The ocean basins opened fast and then ever more slowly, giving the appearance of a magnetic field that used to reverse more slowly than it does now.

FIGURE B: The Arctic Hemisphere, indicating the largely continental (rather than basaltic ocean-type) bottom; and the North Atlantic Ridge passing by the North Pole and proceeding towards Siberia, where possibly it becomes a land rift proceeding to the Indian Ocean via Lake Baikal. (Pages B to F are author's sketches. In all of them, the outlines of the full continents, including shallow shelves, are drawn.)

FIGURE C: The Indian Ocean Hemisphere, noting the African Rift on the extreme left, the East Ninety Degrees Ridge, and the largely continental rock platforms that underlie the vast Asia-Australia area.

FIGURE D: The American Hemisphere, noting how both the Atlantic and mid-Pacific Ridges follow the shape of South America at great distances. A world-circling Tethyan shallow sea belt may once have passed through Central America, the Mediterranean and the South Seas, but can hardly be discerned because the ocean bottom growth and expansion and crustal slippages have largely erased it.

FIGURE E: The Antarctic Hemisphere, showing how ridge-fracture cut the south polar Continent off completely from all land to the North, as by a circular saw. It would appear that the main fractures occurred before the main continental shift, (as in the Arctic Basin to the South), because there still is a semblance of order to their progression around Greenland and into Asia. Furthermore, Greenland adheres in shape to the North American continent and its neighboring western fracture does not seem to descend as deep as the eastern one. And on its East, Greenland seems conformable to the Scandinavian platform.

The total true ocean area created by and in consequence of the explosions and worldwide venting system amounts to 310 millions km2 . Its depth averages 4 km. A floor of 1.24 billion km^3 would have been laid almost entirely in a period of about 2000 years. After that time, the activity of basin-evolution would begin quickly to subside. The basins would continue to evolve at a greatly reduced rate. Most of the vents whave become inoperative. The ridges and fissures are still expanding around the globe but at a scarcely discernible rate; like today, rarely would the oceanic surfaces be troubled by sea-bottom volcanism and spreading.

A full life would have arisen in the warm oceans; the marine species of today originated in the shallow Tethyan waters. Men began navigating the oceanic surfaces now. Whereas ancient fossilized life-forms have been discovered on high mountains, they are absent from the bottom of the sea. It is presumptive, if not incorrect, for geological writings to state that the oceans have covered and uncovered the land on several lengthy occasions. The mountains have arisen from the shallow waters of Pangea, bearing the fossils, or the fossils have been laid down by flooding and tides, or they have been dropped by cyclones. The abysses of the ocean contain only species whose origins in shallower waters are patent. The oceans were born recently, and therefore hold only what has lived in these times. Heezen and Hollister, recounting the scarce record available of life on the ocean bottoms, conjecture that "either there was no abyss then, or the relicts of these ancient seas have been completely destroyed. The deposits of earlier seas are found exclusively on the continents." To us it is clear that these earlier "seas" were the only "ancient seas" and were the shallow Tethyan seas and swamps.

The length of the oceanic fractures and their transverse fissures (transform faults) amounts to some 300,000 linear kilometers. The funnel volcanos number in the tens of thousands. The emission from a volcano cone can be given a value equivalent to 5 kilometers of fissure volcanism, and the number, never counted, can be set at 50,000. Then 550,000 kilometers of venting area was available to produce on the average 2,254.5 km^3 of ocean floor per venting kilometer within 2,000 years, or an average of 1,127 km^3 per year.

In two days in 1902, the Volcano of Santa Maria in Guatamala erupted and emitted 5.5 km^3 of material. A fissure of Laki, Iceland, part of the Atlantic northeast ridge forking, was quite active in 1783 and along a 25 km line emitted 15 km 3 of material in 4 $^1/_2$ months. In the tenth century an Icelandic fissure one year erupted 9 km^3 of lava alone along a 30 km trench. It is evident that if its activity were continuous at its full rate of eruption, the fissure of Laki would eject about 3,000 km^3 in 1000 years, 9000 km^3 in 3000 years, far more than its quota.

The figure used as a base requirement, 1.24 billion km^3 , is twice as large as required. The underside of the ocean floor, comprising half the thickness of the floor, appears to be not a product of lava flow but a melting and cooling of basaltic rock in place. As the gaps widened, and the lava flowed to fill the chasm, the floor of the chasm at first softened from the heat all around it and from the waters, and then quickly hardened beneath the lava flows. This is but the cooled crust of the exposed magma of the mantle. When geologists declare, as does Shelton, that "... we cannot yet explain why magma exists where it does or seeks escape when it does," they are not considering this kind of quantavolutionary and exponential solution.

The continents can be viewed as the rims of the ocean basins. They are steep-sided blocks, whether they plunge directly into the waters or have sea-covered shelves that then plunge down. The continental slopes, on the other hand, are water-covered moraines of continental debris laying on top of ocean abyssal basalt. They have a triangular profile, making nearly a right angle where continental block meets ocean floor; the hypotenuse is a lengthy stretch moving from the top of the shelf at an angle of 5° on the average. The declining rate of expansion of the ocean floor contributed to the profile of the slopes. By moving first rapidly, then ever more slowly, they heighten the illusion that a gradual off-flow of sediments has created the sloping figure. More likely exponentially declining rates of continental debris and seabottom spread worked together to provide the profile. Deep river canyons extend hundreds of kilometers into them. Elephant teeth are found far out on the slopes at great depth; probably the slopes were laid down, occupied by terrestrial life forms, and then lately flooded. Deep turbidity currents, if they were to transport them, would bury them or destroy them. They lay near where the elephant died not long ago.

As told in the previous chapter, the continental slopes are free of continental mountains, as are the true ocean bottoms. The logical implications of this fact have evaded geology. If most great mountain ranges are new, whether by our chronological reckoning or by that of conventional geology, why have none appeared on the continental slopes? The answer suggests itself: the mountains rear up at the edge of the precipices of the continents; they dump their debris into the abysses.

Immense floods and tides traversed the continents and poured off the miles-steep continental blocks into the ocean. The canyons occur where the blocks were fractured, and consequently where the waters poured out most heavily. The canyons, which will be treated soon in more detail, were not submerged beneath the oceans until the ocean basins stopped growing and their waters crept up upon the continental blocks and shelves. The seas do not come in and kidnap the land; they beat back the detritus and even build land. Thus the great slopes could not have formed under uniformitarian conditions or even underwater.

Prolonged, universal run-off of deluge and catastrophic tidal water produced slopes; the blocks were often towering water falls, dropping sheets of slurry into the abyss to form the slopes. The coarse gravel typical of the slopes far out to sea signals the impetuous rush and transporting power of the waters going to fill the basins. The scale would have dwarfed even the scene pictured by K. J. Hsh for the Mediterranean Sea (our dates and events differ, of course), "a giant bathtub, with the Straits of Gibraltar as the faucet. Seawater roared in from the Atlantic in a gigantic waterfall." If the falls delivered 10,000 cubic miles of seawater per year, they would have exceeded Niagara Falls 1000 times, and filled the Mediterranean basin in 100 years. "What a spectacle it must have been for the African ape-men, if any were lured by the thunderous roar." The Mediterranean basin requires in its complexity an analysis that we cannot afford here. It appears to have been primordial, that is Pangean, and shallow. Then it may have suddenly closed and as suddenly opened, dry for a few years, and then overwhelmed by floods of water much greater than at present.

The ocean basins are composed of sima, rich in silicon and magnesium elements. They are of basalt. They are igneous, formed in red heat. They are thin. They are denser than the continental sial. The continents probably sit upon similar material, but much deeper, perhaps directly upon the upper mantle, save where the magma of the mantle may have expanded and intruded upon the continental granites.

The continents and the ocean basins are distinct formations that were produced at different times and by different mechanisms. The sial is old. The sima is new. The fact that the shell of the ocean bottoms is only one-tenth as thick as that of the continents in itself suggests that the ocean crust is the product of a melt, that the seas are new, and that the continents were somehow in a position to resist complete volcanism or explosion. The fact that ocean crust is more basic or less acid than the continental crust indicates that it separated from the primeval melt after the granitic crust; so says M. Cook. The continents were produced by a cooling of the Earth's surface and by their own erosion and debris, and in direct contact with ultra-basic material of a heavier composition. Hence, the igneous marine floor does not cover a former continental surface, and density probes show this to be the case. Nevertheless the floor probably contains continental debris in small amounts. With all the sinking of lands reported in legends, one would expect ocean-bottom drills to collect continental material here and there. Very little appears, leading one to suspect that most sinkings have occurred on the continental slopes or shelves.

The ocean basins are scarcely sedimented; they hold only 1% of all sedimentary materials. Under uniform conditions, this would represent only 16 million years of runoff deposits amounting to 10^{18} tons[3] . Dissolved solids in the ocean waters compose 3.5% of their mass, far from making up the difference, nor can these solids be allocated to detritus removed from the continents.

Often the rocks are bare along the circumglobal ridges. They are 20 meters thick or less. The thickest ocean sediments are not on the basins proper but on the continental shelves and slopes. Further, next to these areas where the abyss begins, sedimentation is thicker and can reach 1000 meters in exceptional areas.

All of these oceanic sediments come either from cataclysmic off-pourings from the flooded continents, or from fall-outs, both volcanic and exoterrestrial. Material lagging at the end stream of the fission of the

Moon might have dropped back to form islands of continental crust in mid-ocean. The time required for such sedimentation is calculable in a couple of thousand years or less under quantavolutionary conditions.

The character of oceanic sediments varies. It differs markedly from much continental sediment that is rock. It is clay and ooze. The shelves carry clay; the polar regions, the slopes, and some of the abyss carries ooze; and the deep abyss carries clay. The polar basins also carry sand and boulders.

Carbonates are heavy on the shelves and bottom oozes, but compose only from 2 to 10% of the clays (since they dissolve in the colder waters). Layers of distinct calcination and ash are interlarded with the oozes and clays in many parts of the world. An unknown proportion of additional ash has been incorporated chemically into the clay and ooze and remains to be distinguished. Much clay is igneous in origin, a product of volcanic tephra, volcanism, and cosmic fall-out. Much manganese has been precipitated onto rocks, pebbles, fish teeth, and bones over many areas, and pure manganese has been found on the bottom near the ridges.

The towering ridges that girdle the world have flanks that descend gradually. They present almost no under-seascape for many hundreds of miles. There is no thickening of the ocean basin crust beneath the ridges, unlike the so-called isostatic thickening beneath the mountains of the continents, much of which is probably due to blunted thrusting. This occurs despite the fact that the ridges rise higher than the continental Alps. Thus they are distinct in origins, as was pointed out in the last chapter. The continental mountains were shaped by horizontal forces, with the intense, sporadic assistance of electro-gravitational forces from outer space. The ridges were formed by vertical forces from within the Earth, with similar assistance; unlike continental mountains, they lack rock roots, evidencing that they were not thrusted.

An impossible predicament is presented to conventional geophysics; how can uniformitarian forces produce this contrast? The continental crust folds and thrusts and compresses into abundant mountains; but the oceanic crust, having made its igneous ridges and seamounts once and for all, slides up and under and around without making mountains, but exudes lava in discrete amounts, and shakes seismically from time to time.

The Pacific Rise conforms generally to what one would expect from an exploded, as contrasted with a cleaved, basin such as the Atlantic.

Worthy of quotation here is a passage from the *Encyclopedia Britannica* (my remarks in brackets): Fast spreading... as is characteristic of the Pacific [because the basin was already blasted out], produces a rise. Slow spreading... results in the formation of a ridge. Sea-floor spreading is a symmetrical process that accretes new ocean floor equally to both flanks of a rift; [The East Pacific basin obviously did not accrete symmetrically.] When a former landmass splits apart, the ridge maintains a median position as the newly created ocean basin increases in size. This phenomenon occurred in the Atlantic and Indian Oceans, but, in contrast, the rise in the Pacific did not rift a landmass when it was formed, and consequently there is no reason for it to be median. [Again, no land mass.]

A slash wound upon already swollen human flesh produces a swelling along the line, but a lower ridge than a single slash wound upon healthy flesh; so the Pacific rise is swollen high off the middle of the ocean bottom, and has a less marked ridge from the slash wound cutting it than the Atlantic basin has from its same slashing.

Although the Pacific Basin is concave, no one can examine a relief map of the Pacific Rise, for example, and say that the volume of the Earth remained unchanged thereby. Since this rise occurred, along with many other bulges, then a considerable expansion might be demonstrated by survey without resorting to theoretical physics. The globe has many slight bulges. Russian geophysicists have recently described its shape as formed by at least two geometric networks of lattices, a many-faceted figure. So there may even be a pattern to the expansion of the global crystal. The latticework can be viewed as expansion joints; the total pattern makes the surface of the globe a set of convex plates rather than a perfect sphere.

Under the conditions imagined here, much of the expansion would be expressed simply in a hurrying of the basin-paving process, accelerated by in-rushing waters. The salt of the dropping canopies would also promote magmatic melting. Molten lava takes up more volume than solidified basalt; wherever the crust was boiling, it would expand the surface of the globe. The tidal pulls of the Intruder, temporarily, and the new Moon, permanently, would draw the surface of the Earth outwards; there the surface would pause, cool, and harden.

The ridge mountain volcanos, and the ridge and transverse fault fissure volcanos, differ from tens of thousands of sea mounts, atolls and guyots that rose tall and slumped back upon the escape of the Moon from the Pacific Ocean Basin and smaller crustal material elements elsewhere. Some of these became instantly created volcanos and continued activity after the others had collapsed back. The Pacific seascape differs from the Atlantic by its incomparably more numerous holdings of seamounts. Morphological examination would indicate that the seamounts do not have the extensive piping systems of continental volcanos.

As the main blow struck and the fracture opened in North America, it drove that continent as a block southwestwards until it overrode the East Pacific Rise (fracture) that had just appeared off what was now its west coast. Much of this western area promptly erupted into volcanism and was covered by huge lava flows and extensive, faulted desert plateaus and plains. The Asian and Australasian coasts and islands do not fit into the North American continent because vast spaces opened up and the whole arc from Alaska to Southern Asia broke away with the explosion of the Moon. A boundary ridge is not easily visible but extends down the Pacific basin on the West from Kamchatka Peninsula to the Campbell Plateau and ties into the Emperor Sea volcanic seamounts and the Line Island Ridge.

The Indian Ocean bottom, unlike the Pacific and Atlantic basins, appears to have been well-traveled. Antarctica has been shoved southward some hundreds of kilometers, and girdled by two great ridges. A newly discovered rift pierces the Waddell Sea, probably a transverse fault from the ridge to its north, and is lost under the great ice plateau hundreds of kilometers inland.

Australia has been ushered eastward by a fork of the same fracture that pushed India north and Antarctica south. Indeed, if one wishes an up-to-date definition of the continents of the world, useful for some purposes, one may say that a continent is a body of land surrounded by an oceanic cleavage. Even in the case of Europe and Asia, some believe the fracture to exist, going up from the Indian Ocean through the Persian Gulf, the Caspian Sea, the Ural mountains and into the Arctic complex earlier described.

Contemporary geological theory has also traced the path of the Indian subcontinent from Southeast Africa to the Tibetan Plateau. "The vast Himalayan range was created when a plate of the earth's crust carrying the landmass of India collided with the plate carrying Asia some 45 million years ago, having travelled 5,000 kilometers nearly due north, across the expanse now occupied by the Indian Ocean." The drift itself took much longer, since it occurred at the rate of 6 to 16 cm/year, if one were to accept the belts of magnetic reversals that mark the stretches of ocean bottom along the line of march and the dates given the lava from one belt to the next.

However, cores drilled into the Indian Ocean bed not far from the observed course produced gaps in dating of sediments by fossils of many millions of years, perhaps fifty millions in some cases. "Why these accumulations are missing," commented the directors of the survey, "is at present a mystery." Fifty from a hundred million or so years is a big proportion. That "there are more gaps than record" is, of course, a familiar complaint among paleontologists on land as well as under the sea. In the other great basins, gaps of twenty million years in the fossil record are common. With sediments so thin, the gaps are not so important, say some - a turbidity current or two, and there you are. (Geophysicists and paleontologists can be catastrophists à la minute, when it is demanded of them.)

Still, when there is a gap in the fossil record of between 50 to 70 millions of years ago, we are speaking of late Cretaceous times and of the disastrous end of the dinosaurs and most marine species. A layer of unfossilized chert tiles the floor just above this zone, "as though some catastrophic development killed off most of or much marine life." One begins to suspect that the Cretaceous boundary may be considered as the primeval age of the ocean beds and that all which is found in the abyss arrived there afterwards; further, the finale of the Cretaceous may have been the end of Pangea and the outburst of the Moon, even if both are to be dated at a few thousand years ago. Almost all of the sea floor assigned to a date is Cretaceous or younger.

An immense part of the Earth's shell is simply missing. It had nowhere to go except into space, for it cannot be decomposed, mixed with plutonic material, or shovelled under the sea bottoms. A psychological fallacy pushes us to believe that the ocean basins were made by and for the primordial waters. That the basins exist is one accident; that waters fill the basins is another accident. The accidents added up to a "miracle" of good fortune for mankind. Better near extinction than a totally frozen or drowned globe. At first, the waters were below the rims of the basins; now they slop over the rims.

The East Atlantic Basin corresponds to the West Atlantic Basin. Their former juncture is plain. The Pacific Ocean is deeper. The East Pacific Basin is sharply marked where the southern, western and northern margins are arranged as a giant set of arcs detached from a blasted area.

All the ocean basins are young, thin, and scarcely sedimented. Oceanographers who recently discovered these facts were amazed; in a few years, the basins became four billion years younger. Only potassium-argon datings, which are vulnerable to catastrophic events, let the bottoms achieve even this young age. The basins are not 200 millions as against 4,500 million years old. They maybe only aged a dozen millennia. The surprise is greater: not one-thirtieth as old, but one-ten-thousandth as old.

CHAPTER TWELVE

FRACTURES & CLEAVAGES

All of the great rifts of the world are connected in time and by cause. They form a system that harnesses the world to the recent fission of the Moon. The individual histories of the sections of the world fracture system are insignificant by comparison with the common historical experience of the whole. The system functioned to balance the world by redistributing the crust and by expansion and to vent gases and heat during the process.

1. CHANNELS AND CANYONS

The model river channel combines the history of an earth fault, a catastrophic torrent, and an erosional runoff bed. Most large rivers, perhaps all of them, are children of Okeanos, whom the Greeks called "the Father of Rivers" who personified the sky waters before the first deluges. His children carried his waters into the new ocean beds.

Many myths appear to conjure rivers where none exist, and, of course, a great many dry river beds of once tremendous rivers are to be found around the world. During the Universal Flood of Deucalion, a small chasm was said to open in Athens into which the waters emptied. According to Lucian the people of Hieropolis (near Aleppo, Syria) "say that a great chasm opened in their country, and all the water of the flood ran away down it." Again, myths warrant hypotheses. In the Volta River Project (West Africa), a onetime shallow river bed was suggested by a deep river bed with a jagged bottom. Local legends spoke of upheavals in the now quiet area, and when the water was lowered prior to constructing a dam, several protuberances became islands, and at a depth of 35 feet revealed carvings whose age was estimated at 3000 years.

The Po River is probably an extension of the African-Rift-Red Sea-Rhine rift valley that connects with a buried rift in the Adriatic Sea. It carried down the immense debris of the sudden uplift of the Alps. It may be the ancient sacred river, the Eridanus, of Greco-Roman legend, long-lost because later a sea. The Po serves in truncated form to water and drain the Po Valley. The Rhine River picks up the graben northwest of the Alps, and moves it far out into the North Sea; not long ago, it shared its burden with a westward flowing river that was then naturally dammed so as to reduce the Loire River of France to more modest proportions. The Colorado River may be a ramification of the East Pacific Ridge, that runs up the Bay of Lower California and strikes through the desert into the raised platforms of the southwestern states, abetting the disintegration of the Rocky Mountain uplift; once its tectonic work was done, it began its present work of erosion.

The great rivers of China flow in the direction they do, says a Chinese myth, because the goddess Niu-Kwa made the waters of the great flood stream off towards the southeast; the whole Earth had tilted and sunk into the sea there. Most great rivers of the world understandably conform to the processes set into motion by the lunarian outburst. Many hasten along courses conveniently provided them and their tributaries by fractures, the Rhine, the Colorado, the Susquehanna, the Indus, the Congo and others.

The Pliocene river beds represent a period of increasing disorder and deluge in the world for about two thousand years prior to the climactic lunar fission. The awesome dead rivers of the Early Quaternary are relics of the phase of mountain thrusting, westward movement of the American continent and the deluges associated with it, which broke down and flushed away the elevated landscape onto the shelves and slopes along the Pacific scarp. The rivers of the American heartland do not exhibit so obviously the recent catastrophic forces. Still, in the late Pleistocene, both the Mississippi and the Ohio rivers changed their courses markedly along an east-west axis, provoked by great seismism, and watched, most probably, by awestruck humans.

Today's third phase finds "pygmy" rivers, many in new channels, watering and draining the country. We group all three phases in the latest of holocene period of the past 14,000 years. "Nothing comparable" with the second phase river action, "ever now breaks the geologic calm," writes King. Then, with prescience of the concept of "collective amnesia," he adds that the idea of "catastrophism is therefore the survival of a terrible impression burned in upon the very substance of human memory."

Scores of impressive submarine canyons extend the courses of rivers around the world. The idea that they were once active as rivers was resisted for a generation. In 1936, geologists Harry H. Hess and Paul MacClintock presented a striking solution. They saw in the canyons evidence of recency, a late Pleistocene age, of suddenness of creation, and of worldwide simultaneity. Here are the three primary tests of quantavolution, all passed by the submarine valleys. Then they are compelled, with reluctance, apologies, and special consultation with H. N. Russell (who advised against it), to advance the quantavolutionary mechanism, exoterrestrial encounter. The passages deserve quotation:

> The valley-cutting conditions resulted from a sudden change in the shape of the hydrosphere, depressing sea-level in low latitudes, raising it in high latitudes; in other words, a change in the ellipticity of the sea surface. At present we can think of no orthodox cause for this change... However, a speculation comes to mind; if a sudden decrease in the rate of rotation of the earth took place, the hydrosphere would respond by being drawn into polar latitudes. The solid body of the earth would less rapidly adjust itself into a new spheroid in equilibrium with the slower rotation, which adjustment, when complete, probably would restore sea level to approximately its present position. But during the adjustment, it is postulated that there would have been time enough to allow rivers to cut valleys on continental slopes. While of course we do not know what could have caused the sudden change in rotation, it is conceivable that a collision with a small extra-terrestrial body would be competent to produce the effect.

The authors then sought for evidence that the depths of the canyons would decrease from the equator to the poles, and, second, that there would be found high marine terraces in the northern latitudes where the shores would have been temporarily flooded. Indications of both were deemed favorable.

The failure of theory to move along such lines is unaccountable, except in terms of the psycho-sociology of science of which we speak in the Velikovsky Affair and The Cosmic Heretics. Many years later one reads in a study by Landes approvingly:

> I claim that the finding of graded clastics and misplaced (shallow-water) faunas deep beneath the sea is not prima facie evidence that they were carried there by turbidity currents: that the finding of cobbles does not prove that they were transported by submarine landslides; and that photographs of ripple marks lying at a depth of 4,500 feet do not necessarily mean that they resulted from current action operating at depth... I likewise believe that deep-seafloor current ripples, like the truncated seamounts, are relics of shallower water.

At this point, Landes should be looking into the ancestral skies. Instead he suggests that the deep ocean basins might once have been over 20,000 feet deeper. Even this idea might lead somewhere, but, instead, the ad hoc argumentation that so often passes for geological theory obtrudes; when in trouble, call upon isostasy, diastrophism, time, lifting, and, as here, sinking, and thus by name-calling the problem is solved and the matter ends; the data are not pushed to their ultimate meaning.

Landes writes: "What manner of logic allows us to accept evidence, such as marine strata, of a sea-level far above present datum of 25,000 feet, but causes us to run from evidence of a sea-level depression of 25,000 feet?... What is so sacrosanct about current sea level?"

He, and others, should be asking the deeper questions: "What are these deluges that humanity has been clamoring about since the dawn of history?" "Must every drop of water bear the holy stamp, 'Made on Earth'?" "How long does it take a pre-designed fracture trough to make a river channel, complete with fractured and non-fractured meanders? .... What is so sacrosanct about the ocean basins having always been filled with water?" I think that we have progressed far enough along in this book to dispose readily of the submarine canyon problem. The canyons were instantly created great river courses that rushed down, first, precipices, then, steep slopes, then gradual slopes, into the ocean basins that were only partly filled with water. Drainage of the water-logged continents and successive deluges filled the ocean basins to overflowing. As the seas encroached upon the rivers, the rivers were also receiving far less water to give to the sea. The underseas box-like, sluice-like channels ended their careers as turbulent rivers within perhaps two thousand years.

Sediments of the trenches are scanty. The same writers say: "It is a general lack of sediment accumulation which is the most notable feature of all the deep-sea trenches. This lack... demands a recent origin of trench topography." Recent must mean holocene or pleistocene, it appears. But now, the plate tectonicists chase in full cry after the trenches as fulfilments of the need of convection cells and subduction of continental and oceanic material. Are trenches barren because they appeared lately or are they barren because they have just digested hearty meals of sial?

2. CONTINENTAL TROPISM AND RAFTING

A Texas association proclaims the slogan "Stop Continental Drift," in its attempts to foil the trend to believe that the Earth's crust has been, and is, in motion. The crust is thin below the ocean bottoms and thick beneath the continents. It is broken up into a dozen major plates whose boundaries are defined by faulting, heat, and turbulence. The plates show signs of having moved great distances over time. Most scientists have been converted to this mobile perspective from a static one during the past generation. "We now have a new, mobilist orthodoxy, as definite and uncompromising as the staticism it replaced." So writes Stephen Jay Gould in *Natural History Magazine.*

Now and then, goes the theory of continental drift - that is, every couple hundreds of million years, the plates renew themselves. The continents do not. They may be split asunder, or bash their fenders, but they move on and on, majestically riding upon the same material, their own mantle magma.

They carry a two-billion-year record of life, while the ocean bottoms have deposited their sediments periodically beneath the sea shores of continents, or beneath other plates which they may be jostling, so that now they carry no more than the last 160 million years of sediments. The plates are pushed around, it is said, by convection currents. These are hot rock moving up to the cooler surface areas and pushing them aside until these bump into other plates which are being also pushed; then they are forced to descend (or force the other plates to descend); they melt once more and become deep mantle material.

We shall concentrate upon the quantavolutionary theory that an exoterrestrial catastrophe brought about the movements of the Earth's crust recently, and we begin with the most obvious fact in topography, namely that the continents of the Earth are concentrated opposite the oceans. "This means that 82.6 percent of the total continental area is antipodal to oceanic area."

So saying, C. G. A. Harrison goes on to describe how, on a computer, he rotated randomly coordinates representing the continents on a sphere, in order to discover how often the actual antipodal percentage would appear. He simplified the continental areas into circles and fed their numbered forms into a computer which then randomly placed them to see how much land would be antipodal to oceanic area. He repeated the random placement 2000 times. "The median percentage of continent opposite ocean to be expected from a random distribution of circular continents is 68.0 percent of the continental area. The observed figure of 82.6 percent is exceeded in... 9.6 percent of all cases. Thus from this evidence alone it would appear that there is a probability of only 0.096 that the present distribution of continents is random over the surface of the earth." He repeated the test with triangular instead of circular simplifications of the proper areas and the results were similar. He concluded that "there is less than 1 chance in 14 that the present antipodal distribution of continents and oceans is the result of a random process."

This is hardly surprising. If the Arctic-Atlantic ocean were closed up, if Australia and Antarctica were fitted to their apparent points of departure from southern South America and Africa - that is, if the continents were rendered into a single mass as they appear to have aggregated before the present age of "drift" began - then all of the existing continental land would be antipodal to oceans.

The implication is strong that before the drift began, the ocean areas were in fact land-covered. The major differences between the Pacific Basin and the other oceanic basins, we have noted, indicate that continental material was blasted out of the former and pushed aside from the latter. The movements of the continents since this time can be interpreted upon the premise of a sudden removal of over half the Earth's crust in what is mostly now southern hemispheric ocean. The land of both the eastern and western hemispheres has travelled towards this vacated area. So have Australia and Antarctica.

The so-called plate movements have not been random, nor can they be interpreted in any other way. The continents all exhibit "lunagenic tropism" and nothing much else. They have moved in the particular direction of the lunar-vacated, now south-central Pacific Basin under the special stimuli of global fracturing, electrogravity slide, earth expansion, hydrostatic equilibration, and isostasy. We define isostasy here (and elsewhere) as the process by which all mutually affected elements in a system, consequent upon any change in one of them from within or without, share the effects of the change by changing themselves in closest accord with their peculiar sites and natures. When a change is introduced to Earth from outside, all possible responses of the Earth's motion and masses are drawn upon to incorporate its effects and to do so in accord with their ranked most possible behaviors. Isostasy has to have a function; in the great post-lunar diastrophism, isostasy functions as a tropism. It moves the continents not randomly, nor to the poles or the equator, nor to gather surviving animals for the Ark, but to repair and redress the lunagenic basin.

If the continents split asunder or were "born separate" and moved several times, there should be abundant evidence of the events. There is little fresh data on this score. The most prominent basin, the Pacific, hardly gives evidence of having been traversed by continents, though they all drift toward it. Mostly, old theories of the lifting and dropping of land during ancient orogeny have been dusted off and varnished to claim the several periods of movement. Fossil ice ages have been claimed, too, as proof of former dislocations of the continents, but these have been embarrassments to ice age theory from the earliest discovery of pertinent evidence; the presence of pebble drift and till, and of glaciers or high mountain freezing may be referred to dense material fall-outs such as were discussed earlier.

"In the whole of geophysics," Defant once wrote, "there is no other law of such clarity and certainty as that there exist two preferred levels in the Earth's crust." Continents are like ships in a frozen sea. "The continents, steep-sided massive blocks surrounded by an enormous world-encircling sea, have deep roots which project 30,000 or 40,000 meters into the earth's mantle while the ocean crust is but a thin

5000-meter-thick film frozen over the earth's massive mantle." Thus report Heezen and Hollister. These continental blocks, I have maintained, are splittable only by a great external force and a responsive expansive force, while the ocean basins are easily producible by fissure volcanism.

The post-catastrophic process is followed by a rapid relocation of the continents and re-encrustment of the globe. The continents were not "just drifting"; they "were going somewhere." A. L. du Toit was veering toward reality when he offered in his early (1937) book, *Our Wandering Continents,* the idea of a "gravity slide," the creeping of continental masses toward rimming geosynclinal depressions.

If the oceanic plates and basins have been completely renewed every 160 my, then they will have been renewed about 35 times since the Earth originated. Each time these plates would have scraped off some of their sediments upon each other. By now the continents of the Earth should be presented in heaps of sialic rock randomly distributed as islands around the globe. Such sediments scarcely exist. Or they are unrecognizable as such.

The thickness of the Earth's crust, as the physicist P. Jordan once said, is a breath of air blown upon a desk globe. The breath should be unevenly blown, for the continental portion is 40 km and the oceanic crust is only 5 kilometers thick. Below the oceanic crust, the Discontinuity zone is less than half a kilometer thick. The Discontinuity seems to be caused by "a difference in chemical composition between crustal rocks and the underlying mantle rocks. "

A minimum viscosity or maximum fluidity would occur at about the 150 km depth, both geochemical and seismic observations being seemingly in agreement on the matter. But if there are no reasonably short gradients of viscosity thereafter, it is hard to visualize a large-scale convection dynamic in operation, much less a host of a dozen giant cells or a pattern of a thousand smaller convection cells working within the mantle. Not unexpectedly, then, Cook and Eardley calculated that to move the continents even in 200 million years would require forces "a billion to a trillion times greater than those that should be generated by the postulated mantle convection currents."

Scientific advance of an important kind occurs when an acceptable interplay of theory and fact occurs. On the issue of the movement of continents, tropism towards the lunagenic basin was suggested as long as a century ago, by Osmond Fisher. But little was known of the ocean basins and the time scheduled for the event was in the dim beginnings of the Earth. W. H. Pickering of the Harvard College Observatory argued the case in 1907, and it was well publicized. Meanwhile H. Baker was evolving his theory. "The separation of the continents by fission," wrote Pickering again in 1923, has for 18 years "been attributed to the great convulsion that occurred at the time of the birth of the Moon, from the side of the Earth. This explanation of the origin of our Moon is at the present time almost universally accepted by astronomers. We see the same phenomenon occurring in many close double stars." He placed the "center of origin" of lunagenesis off the southern tip of New Zealand.

However, Pickering held to the view that, although terrestrial lunagenesis and the Atlantic fission must have occurred late enough so that the continents possessed their modern forms, the time had to be early: "that a catastrophe involving the sudden removal of three-quarters of the Earth's surface could occur without destroying all life, both vegetal and animal, appears impossible." Therefore he disputed Alfred's Wegener's contention that the Atlantic Basin opened up at the end of the Cretaceous period or in the early Tertiary.

Thus an impasse occurred until the present day. Wegener's continental drift theory is accepted but not its cause. Instead geologists cling to their terrestrial ideology and posit convection currents. The effects of ripping some 50 kilometers in depth off of most of the Earth's surface were conjectured to be utterly destructive of the biosphere. Today much new geological and geophysical evidence can be adduced from an examination of the Earth and Moon, tending to support the terrestrial origin of the Moon and the connection between lunagenesis and continental break-up and movement.

Moreover, the Cretaceous-Tertiary boundary is increasingly understood to mark the extermination of most species. Whether this boundary happened at sixty million years or twelve thousand years ago (which I construe to be the case) does not much matter on the issue of biosphere survival. I have pointed out elsewhere that biosphere annihilation was not necessarily predicated during lunagenesis. The immense typhoon that conveyed the crust of the Earth into space would have carried away with it most of the heat generated in the transaction, while at the antipodes of the event, downwards draughts of the then much more voluminous atmosphere would have cooled and re-gassed the land.

3. *Sediments*

About 5% of all crustal rock is composed of sediments that remain in something approaching their state after deposition. They veneer about three-quarters of the continental surface to thicknesses ranging from the merely visible to a dozen kilometers in height, with the average for the globe at over two kilometers. Sediments have been classified by priority of deposition and anywhere from ten to hundreds of major and minor strata have been allocated positions, sometimes only after prolonged controversy, some only among certain believers. Besides containing chemical and mineral traces and distinguishable fossil remains, an estimated 80% of sedimentary rock are shales composed of mud or clay, 10% are of sandstone and 10% are of limestone.

The old problem of sediments missing from the geological column became more worrisome with the discovery that the ocean bottoms do not carry their proportionate burden of sediments, much less the extra quantity to fill the gap in the geological column. The continental slopes are formed of shaken down, wasted down, and blown down debris from the shelves of the continents; but they would constitute only a small part of the supposed accumulation. The composition of the slope deposits is unknown. Perhaps half was carried off the shelves in the continental movements and orogeny following lunagenesis. A fifth may have descended from the sky preceding and accompanying the event. A tenth might have been washed in during the Noachian deluge. The small balance may be divided between river run-off into the oceans and cosmic and volcanic fall-out. Ager reports that "... chaotic deposits and slump topography have now been found at the foot of many present day continental slopes."

The continental shelves and the abysses carry clay. The polar regions and half the remainder of the basins carry ooze. Sand and boulder are confined largely to occasional polar sediments. However, sand composes 10% of the ocean bottoms, too much for long-term sedimentation to have occurred. Carbonates, suggesting organic detritus, are common in the shelf and ooze sediments. Little suggests the continental rock in the oceanic sediments; it is a different world of unconsolidated material.

In a remarkable survey, Woodmorappe has denoted the presence or absence of the ten conventional geological periods on a sample of 967 equal square areas of 406 square kilometers of the continental lands.

Ideally, every square on Earth should exhibit some rocks of all ten periods. Natural history assumes that all areas have undergone similar weathering experiences during any given long period of time; if, as is known, rocks of all ten periods are not found, it is because field surveys have not been competent or complete, or because the weathered debris of given age has been transported as such or as rock later on to somewhere outside the 406 square kilometer area (a journey of a maximum of a dozen kilometers), or because the rock did actually form but was eroded and carried off, or because the rock once formed was later subjected to metamorphosis. Some credence can be given to all these explanations, but, too, it is noteworthy that the "presence" of period rocks in Woodmorappe's study often refers to a minor outcropping within the area and not to full coverage of the area.

The departure of reality from the myth is impressive. In no more than one per cent of this sample of the areas of the world are all ten periods of natural history represented. Some of these widely scattered areas are doubtfully complete (in the Himalayas, Bolivian Andes, Indonesia, South Central Asia, and Cuba). Rarely does one find even three of the ten geological periods in their expected consecutive order. Moreover, "42% of earth's land surface has 3 or less geologic periods present at all; 66% has 5 or less of the 10 present; and only 14% has 8 or more geologic periods represented..."

Individual geologic periods' coverage of the earth's land surface range from a high of just over 51% for Cretaceous ... to a low of only 33% for Triassic. Only 21% of the Lower Paleozoic is represented in 3 or more of its periods; the complete Upper Paleozoic is found in 17% of the areas; the Mesozoic is complete in 16% of the areas. A complete Paleozoic record is found in 5.7% of the areas, and a complete Upper Paleozoic plus Mesozoic in 4.0%. Some percentage of every geologic period rests directly upon Precambrian 'basement', especially high percentages of Ordovician (23.2%) and Devonian (18.6%) doing so.

Derek Ager takes the position of a macrochronic quantavolutionist. "Changes, cyclic or otherwise, within the solar system or within our galaxy, would seem to be the easy and incontrovertible solution for everything that I have found remarkable in the stratigraphical record." The secondary mechanism, which he employs repeatedly but without criticism of its fundamental origins, is plate tectonics. "The theory of plate tectonics now provides us with a modus operandi."

He can use plate tectonics to discover and discuss numerous "periodic" and "episodic" catastrophes around the world. This enables him to be macrochronic: "the history of any one part of the Earth, like the life of a soldier, consists of long periods of boredom and short periods of terror."

He offers a wide range of examples, from numerous eras, of the worldwide distribution of various rock-types and fossils; this leads us to the supposition not only of a Pangea in which sediments and life forms might readily become worldwide but also, and perhaps more important, of species that never reached their potential limits, suggesting forceful interruptions of their spreading. Further, it implies worldwide equal conditions for even very special kinds of sedimentation and rocks to form.

He illustrates the bizarre differences in depth of the deposits of the same age in separate regions both near and distant, pointing out, for example, the one foot of Jurassic sediment in Sicily in contrast to the 15,000 feet of one Jurassic zone's sediment in Oregon. Since they do not form on mountains, sediments, which can fill basins to a depth of up to 20,000 meters, would have been below sea level if the oceans existed when they grew.

CHAPTER THIRTEEN

BIOSPHERICS

Given the increase in studies demonstrating an exoterrestrial connection with general extinctions, can natural history be reodered according to the occurrence, frequency, and type of exoterrestrial disaster? And is large-scale extinction possibly or invariably accompanied by large-scale biological innovation?

1. FOSSIL DEPOSITS

A collapsed time schedule for the creation of the ocean basins demands a reconstruction of how aquatic species developed. Pangea was a world of small waters. Small and shallow lakes and swamps are conducive to the generation of individual variations within species and the prolongation of their careers. Whales and sharks travel great distances, but do not need to do so; they can flourish in a Tethyan sea; so with every other aquatic species. The great deeps are a last resort.

The eels from everywhere descend to breed from their rivers into the salt ocean and there find the Sargasso Sea, the great belt of weed-bearing waters on both sides of the Mid-Atlantic Ridge. They die there and their young swim for thousands of miles and years of time to find the rivers of Europe and America. The American eels have 104 to 111 vertebrae, the European 114 or 115, and n'er the twain shall meet.

Igor Akimushkin conjectures that eels originated or dwelt in the intercontinental fissure when it opened up an asserted 130 million years ago, not far from their fresh waters. Then they expanded their mobility to follow the drifting continents.

Fitting the case to the quantavolutionary theory, it would appear that the Sargasso Sea is a part of the old Tethyan world-girdling shallow freshwater sea; that for breeding the eels found the gulfweed more necessary than the saltwater noxious; that there has been too little time to cast off the habit of traversing great distances, or of adapting to seawater for the long adult life; and that the small differences between American and European eels are an additional indication of a recent common ancestry. The Sargasso Sea seems to be growing, which, since it must precede the eels, indicates that it may not have been in existence long. In sum, eel migrations are as much a proof of continental rafting as continental drift is proof of the reason why eels must be astonishing long-distance travelers.

So also with aquaticized birds: if they migrate today intercontinentally, it is a stretching of their original habits; the irregular geometry, followed by birds that fly away from the arctic directly south and then veer at sharp angles to find their winter grounds, and vice versa to return, may be a function of land-mass migrations; the birds seem to be pursuing their original routes. If so, there may have been little time in which to evolve more efficient habits.

Rocks dredged from the bevelled tops of a number of seamounts carry imbedded fossils of current species that give 8 to12,000 years readings on C14 dating (probably 4000 years old, then). The abyssal floors contain many bones, remarkably preserved. Large shark teeth of unknown species abound. Elephant teeth are found far down the continental slopes of North America. Their preservation for more than several thousand years is unlikely.

A fossil is typically an accident, a disaster, an anomaly. We should not find in Ecuador a mixture of mastodon bones, pottery, and coal. Nor reptiles with full stomachs, pterosaurs swallowing food, a mammoth with buttercups in his teeth, or an ichthyosaur mother in the throes of birthing her infants. The very existence of fossils reflects, says C. B. Hanson, "inefficiency in the natural systems for recycling organic material." He experimented with sending mammal bones down a flume in a laboratory in attempts to replicate natural conditions. M. Coe studied the decomposition of elephants in a Kenyan drought, and concluded that only rapid burial would allow any chance for fossilization. There was no question here of the elephants being assembled to die and then deeply buried away from water and doused with petrifying chemicals so as to produce one of the fossil assemblages so commonly found in natural history. In fact, the best case of a fossil assemblage that geology can afford from historical times is the human population of Pompeii and Herculanum smothered and buried by the gases and ashes of Vesuvius in 79 A. D.

If tides had totally overrun the globe, the fossil record would be much less - all the less because tides dig up old deposits as they move, too. On the other hand, is the fossil record so generally rich as to imply large expanses of peaceful, tideless time when shells could find a quiet home, preserved, until pushed into visibility, there to encounter aeolian forces?

Discovered fossil assemblages number in the hundreds, although they are not nicely inventoried. They occur on every continent, in many countries, in high and low latitudes, whenever land animals, plants and marine life have thrived. A large number remain to be discovered. A list of over fifty is before me as these lines are written, and I realize that they are almost all either late Cretaceous (reptiles) or late Pleistocene (large mammals), and that one must take into account many times this number for the aforesaid periods and then every "rich fossil bed" that graces the boundaries of the total phanerozoic calendar.

An item from *Chemical and Engineering News* comes to mind. Workers "found the fossil skeleton of a baleen whale some 10-12 million years old in... diatomaceous earth quarries in Lompoc, Calif. .... The whale is standing on end in the quarry and is being exposed as the diatomite is mined... The fossil may be close to 80 feet long." A sarcastic reader wrote in (March 21, 1977) that "Everybody knows that diatomaceous earthbeds are built up slowly over millions of years as diatom skeletons slowly settle out on the ocean floor. The baleen whale simply stood on its tail for 100,000 years, its skeleton decomposing, while the diatomaceous snow covered its frame millimeter by millimeter." That is, catastrophes affect the minute as well as the great life forms.

Fossil conglomerates are not partial to genera or to epochs. Many recent studies have been based upon material dredged from marine sediments, and concern minute organisms or creatures. These, too, usually mark boundaries ordinarily termed epochal, or climatic, or even catastrophic, for they involve abrupt terminations of some certain composite of species. Thus, when suddenly a thick band of coccoliths is dredged up from the bottom of the Black Sea, aged perhaps three to five thousand years, a sudden end to a regime becomes apparent: a deluge of strange waters, an abrupt climate change, an electric shock transmitted throughout the body of water, or a sudden break in the food chain occasioned by similar events. We speak more of this when we come to discuss extinctions.

At Bearsden, near Glasgow, a fossil conglomerate termed Carboniferous by age is found. Marine and freshwater strata are interlaced; marine and non-marine life-forms are present, not necessarily tied to their "appropriate" rock strata (land plants and marine animals are mixed); crustacean and shark fossils (rapidly decomposable) are found in high degree of preservation. Though often the material of coal beds, they are not carbonized. A series of tidal thrusts is to be assumed; further, coalification does not occur, it appears, unless an independent heated element is added before or after dumping. The evidence is consistent with the catastrophic theory of coal formation.

Coal deposits are fossil conglomerates of a most impressive kind, and call upon the winds, the tides, and the giant bulldozers of ice and rock.

Quotations from botanist Heribert Nilsson are pertinent:

> Even if our peat-moors grew to a thickness of 2,000 meters, nothing would be similar to the Ruhr Carbon or any other coal district... If the possibility of an autochtonous formation of the seams is judged from the point of view of the amount of material available, the results must be considered as highly improbable. A forest of full-grown beeches gives material only for a seam 2 cm. It is not unusual that they are 10 meters thick, and such a seam would require 500 full-grown beech forests. Whence this immense material? How was it deposited all at once? Why did these masses of living organic material escape decay, why was it not completely decomposed?"

To what degree sediments are "rock fossil assemblages" is unknown. They too, with or without fossils, can be transported by high-energy vehicles. If a tree stands vertically in a sediment does it not demand that its whole depth of burial should be carried throughout its stratum wherever it leads and the whole be considered instantaneous? Should not the vertical great whale referred to above be a measure of a whole stratum's instantaneity? A stratum can only be as thin as its tallest fossil will allow. A poly-strata fossil wipes out practically all the temporal pretensions of the blankets of its bed. Ideally, it should wipe out all identical blankets everywhere.

No part of the world is without fossil deposits. This would indicate that no part of the world has escaped catastrophic experiences. Marine fossils are of shallow seas: the oceans may be too young to have spawned new species, much less to cast them over the continents.

We cannot conclude here from the study of fossil deposits that all major disturbances have been recent. But these conglomerations lend direct and substantial support to the quantavolutionary theory that Earth changes have been sudden, large-scale, and intense, and that most, if not all, have been very recent.

2. GENESIS AND EXTINCTION

The breaking of an age is the occasion for instant creation and instant destruction. The quantavoluting high-energy forces concentrate upon reducing and at the same time increasing the variety of species. Otto Schindewolf, from 1950 on, was tracking what he called faunal discontinuities, for which task D. L. Stepanov called him the "most important and consistent spokesman of the idea of neocatastrophism in contemporary paleontology." In him one finds a more stringent scientific tongue than King's but the same view. "Faunal discontinuities... involve not just the dying out of old, but also the more or less sudden emergence of new phyla. This phenomenon can no longer be successfully accommodated under the term catastrophe in the true meaning of the word: it should rather be described as anastrophe." (that is, 'upturn, ' not 'downturn'). It was partly for this same reason that the term quantavolution was chosen. Probably most species are born or die out at the disastrous junctures of natural history whence the rocks and fossil seas, too, provide evidence of commotion.

Schindewolf dismisses explanations offered for these quantavolutions, none of which he deemed valid, such as gaps in the rock and fossil record, epidemic diseases, climatic changes, ice ages, differing depositional characteristics of species, reduced salinity of seawater, competition and natural selection, mammals eating dinosaur eggs, and changing sea levels.

Then he reaches into the skies. "Since faunal discontinuities are universal phenomena, they must arise from universally active causes. This has compelled me to look for agencies that would (1) have worldwide effects and (2) could extend to the totality of biotopes in the sea, on dry lands, in freshwater and in the air, as well as to stocks of most varied habitats and ways of life."

His explanation lies in radiation storms:

> Since 1950 I have favored the hypothesis that sharp fluctuations in the high-energy cosmic radiation reaching the Earth should be considered among possible causes... I proposed that, on the one hand, the direct impact-effects of ionizing radiation should be considered, and, on the other, especially the increased generation of radioactive isotopes, which would become incorporated in the living organic matter and the molecular compounds of the chromosomes. Here they would unleash a twofold mutagenic activity, through ionizing radiation, on the one hand, and by the liberation of electrons in the decay of the isotopes on the other.

He cites theories of supernovas as the source and media for the transmission of the anastrophic material, and credits E. A. Ivanova with "a connection between the faunal discontinuities and the migration of radioactive elements." Schindewolf points out that the exceptional survival rate of insects compared to other fauna may be due to the fact that "the resistance of insects to radioactive radiation is about ten times greater than that of human beings and other organisms."

There are then the meteoroids and the supernovas as sources of anastrophic radiation. Could "cosmic" radiation come from volcanism? If deep, heavy, and worldwide, radiation closely akin could fall out from volcanism. But such volcanism, as we explained earlier, must look for high- energy excitation from the skies. As for the supernovas, in Solaria Binaria Milton and I attribute heavy radiation to at least three novas - a preliminary outburst of the Sun creating its binary, and two explosions of its binary in subsequent millennia. We also designate several other possibilities of radiation, that would be heavy enough to account for periods of intervening radiation, not from novas but from impact-explosions and crustal removal in passing encounters.

Here and there now reports are issuing of excessive radiation levels in rocks and fossils. Kloosterman was earlier quoted on the subject. Salop speaks of a primary enrichment of uranium in dinosaur bones. Numerous similar findings have been reported since 1956 in Brazil and Argentina. Some bones from an undated red sandstone were radioactive. J. E. Powell summarizes these findings. Fossils from Mongolia also show high levels of radioactivity. Kloosterman located these facts and also discovered that almost none of the world's natural history museums have measured radioactive levels in specimens of their collections.

However, the prevalence of fossil conglomerations around the world implies brief periods of extinction, and forces not alone of radiation, and pre-existing ecologies quite different from those that came after the catastrophic periods. So many rich fossil deposits occur in circumstances that reveal high-energy processes to be at work. In Baja California, fossils were laid down over hundreds of square miles of the desolate terrain, exposed by surface erosion. Living and extinct species mingled in broad confusion. Flint and obsidian artifacts lay also upon the fossil sediments. The bones of mastodons, ancestral horses, a giant tortoise, camels, bison, sharks, whales, sea cows, and fish were plentiful. A shark species found in Mississippi turned up here in the Pacific. Assigned times, prior to investigation on the spot, ranged from 50,000 years (in the case of the artifacts) to 60 million years. Dating aside, the giant, confused, and rich fossil fields signal a catastrophe or a series of catastrophes at short intervals of time, from floodwaters sweeping in from land and sea.

Would the stripping of half the Earth's crust and an associated expansion and cleavage of the Earth, together with a paving of the ocean basins, all occurring within several thousand years and most of it very quickly in a single action complex, exterminate entirely the biosphere? Even the most determined catastrophists have passed over so frightful a concept. If, as has been conjectured, a meteoroid explosion of a few kilometers' diameter would destroy the dinosaurs, the colossal event portrayed here would annihilate all life.

To counter this universal scepticism, there is the fact that life does flourish today despite the event, so that if the event were proved, then the scepticism would have to vanish. However, taken as a problem in its own right, instead of an inference determined by an external logic, we should stress certain possibilities in the event of lunar fission.

1. The atmosphere at the time might have been enormously greater and so extending far into space to permit a reviving reverse flow to replace the escaping atmosphere, and to act at the same time as a great vacuum cleaner against the heavy dust clouds and heated air.

2. Although an enormous number of species may be extincted, only several survivors of a species may guarantee a replenishment of continental scope within centuries.

3. The possibility must be entertained that hitherto unused intraspecies genetic adaptability can permit survivors of modified form under stresses seemingly quite destructive.

4. Holospheric catastrophes by their very complexity can block each other's effects, allowing some life-preserving niches to survive and even fabricating niches where none existed before.

3. Pandemonium

In Ezekiel (XLIII, 2) it is said that the voice of the Lord "was like the sound of many waters." Ancient records and legends are rich mines of electrical allusions from which not only the state of electrical phenomena can be assessed but also the electrical technology of early cultures can be surmised; this field, ignored hitherto, is being researched by J. Ziegler.

The *Books of Moses* carry testimony of great celestial noise that cannot be rationalized as ordinary thunder. And Noah, it is said in Jewish legend, was spoken to by a voice from the sky amidst a great commotion. This followed the failing of things upon the earth and was followed by the Deluge. The story of Job, later on, reads: "Hear ye attentively the terror of his voice, and the sound that cometh out of his mouth." Again: after the ends of the earth are lit up, "a noise shall roar, he shall thunder with the voice of his majesty, and shall not be found out when his voice shall be heard." A circum-global sound.

As the Jews passed from Egypt in the tumult of Exodus, they paused at Sinai. "I am Yahweh," heard the people during the night at the Mountain of the Lawgiving. "And all the people saw the roars, and the torches, and the noise of the trumpet, and the mountain smoking: and when the people saw it, they trembled and stood far off." Ten blasts of the trumpet sounded out the Decalogue, legend tells us also.

Great sounds were reported from around the world: the Babylonian Gilgamish epic: "Loud did the firmament roar, and earth with echo resounded." Hesiod's *Theogony:* the huge Earth groaned when Zeus lashed Typhon with his bolts - "the earth resounded terribly, and the wide heaven above." Velikovsky pursues the name Yahweh elsewhere: he finds Jo, Jove (Jupiter); Yahou, Yao (Chinese emperor of the age); Ju Ju huwe, (an Indonesian invocation to heavenly bodies): Yahou, Yo (in the Hebrew Bible); Yao, Yaotl (ancient Mexico); Yahu (ejaculation of the Puget Sound Indians and other Amer-Indians when they performed the ritual of raising up the fallen sky off the earth).

It is perhaps of some significance that Cohane has found Haue, a Middle-English god-name, in the names of gods, sacred places, rivers, salutations, and objects all over the world into the hundreds of instances. "In the landscape of the Old Testament part of the world is still overflowing with Hawa placenames." All sound alike despite spellings such as oa.., ua.., awa.., huwa.., oua.., wa.., and so on. Provisionally, we may entertain the idea that the sound of great natural events were incorporated in the basic vocabulary of new-born humanity. If so, the popularity of the "awah" sound is at least as ancient as the time of Moses (circa 1500 B. C.), and probably several thousands of years older, and would also then be carried on down to modern times. 'Yow," "wow," and "ow" are everyday American slang exclamations. The divine voices were also heard later on. A Babylonian hymn to Nergal (Mars) is of the first millennium B. C. and reads:

His word makes human beings sick,
It enfeebles them
His word - when he makes his way above
Makes the country sick.

The motions, noise, and gases of a heavenly body of large dimensions seem to be indicated here. The god Mars is referred to in Babylonia as the God of Noise. There is an connection of noise with the planet Mars. The connections between heavenly sounds, sacred events, and the beginnings of music appear to be secure. From Chernikov, in the Ukraine, Soviet scholars reported finding mammoth bones converted into skull drums, shoulder blade kettle drums, and lower-jaw xylophones, at an estimated date of 20,000 years. If the instruments, all of the drum family, are correctly identified, it would mean that the settlement was fully human, with a religion. For nowhere is there any indication of musical instruments or musical sounds that are not connected with the heavenly host.

In earliest China the drums were used to communicate with heaven. The drum comes from K'uei, a green oxlike creature who came out of the sea shining like the sun and moon and making a noise like thunder. He was captured by Huang-ti who made him into a drumskin. But the same K'uei is also the master of music who alone can bring harmony between the six pipes and the seven modes. Without this harmony heaven and earth would lack their essential music. K'uei was also master of the forge, of dance, and of regulating floods.

The sickle with which Kronos (Saturn) castrated Ouranos (Uranus) was also the harp (lyre) of Demeter who had taught the Titans to reap. The strings of the lyre were ultimately five or seven, corresponding to the number of spheres counted as planets. Vail thought that the arch of the harp and sickle came from the opening of the boreal hole of the north when the regime of canopy skies began first to break down; the arch was the sickle; it was also the arch of the lyre, and the strings to be plucked were the beams of light playing down upon the earth.

Here, as in many other cases, an issue is whether the sacred image came before the invention or the invention was made and compared with a later celestial image. As usual I incline towards the position that the sacred example preceded the profane - such seems to be also the lesson of the *I Ching*.

The correspondence between the number of planets and the number of strings on the lyre is an instance in point. It is only one of many. The number of observed planets obviously determines the number of strings on the instrument, not vice versa. An old Chinese text says that "the calendar and the pitch pipes have such a close fit, that you could not slip a hair between them." This seems an odd expression until one realizes that the sacred calendar is replete with a synchronous musical calendar - from Easter music to Christmas music, for example. The pipes are pitched to heavenly sounds and numbers; the calendar is an arrangement of heavenly events.

The Pythagorean philosophy of ancient Greek culture generated the theory of music and the theory of numbers out of the behavior of the heavens. The "harmony of the spheres" of which the ancients spoke was probably first the sounds of heaven of the "better" sort, to which humans might adjust, and which, to them, presaged a tranquil stability, and then later, inferentially, the visual reliable order of the heavenly bodies as noted and welcomed by philosophical astronomers.

Robert Graves has reported an octaval version of the name of Yahweh, Jehuovao. The sacred name can then be pronounced and chanted as a set of vowels running the gamut of a musical scale. We are reminded of the connections between Egyptian and Hebrew culture, when Demetrius wrote: "In Egypt the priests sing hymns to gods by uttering the seven vowels in succession, the sound of which produces as strong a musical impression on their hearers as if flute or lyre were used." The seven vowels were uttered in succession as the divine unspeakable name.

4. SPECTRES

Weeks or months of near global darkness can attend the crash of a meteoroid of 10 kilometers. Scholars studying biosphere extinctions now refer regularly to such effects, as in the study by the Alvarez group referred to earlier. Years of darkness have been claimed in rare cases, the Exodus period being one of these. Heavy winds are reported during the days of Exodus; Talman found the dark days of forest fires to be windless. Perhaps volcanism of a rare kind produced the Exodus dark skies, but more likely is the combination of large-scale volcanism and a prolonged fall-out of cometary dust.

The close of the Cretaceous age with its heavy extinctions saw a darkness of only weeks or months, according to one view, which suggested as the cause an exploding meteoroid of middle size. Nevertheless, most species of animals and plants were extincted, and great physical devastation occurred, so we may suppose that various events combined to worsen the darkness and that they operated holospherically. We suspect much more than the meteoroid was active.

The most impressive of all sights, to judge from many accounts from the earliest records and legends to the most modern of writers is that of a comet approaching the Earth. Unlike the strike of a nuclear missile, the comet gives the fullest visual warning, as well as causing a number of electrical effects from afar. It is "the most provocative apparition of all," in Calder's words, referring to Halley's comet. When the Roman Emperor Nero saw the comet of about 60 A. D., he had many leading Romans murdered to avoid the death he saw for himself in the heavenly portent. "The Incas of Peru regarded comets as intimations of wrath from their Sun-god Inti... In twentieth-century Oklahoma, at the apparition of Halley in 1910, the sheriffs arrived just in time to prevent the sacrifice of a virgin by demented Americans calling themselves Followers." No nation in the world escapes panic upon the sight of a comet's approach, no matter how many scientists their public may include.

That the sight of a comet in itself could so impress people, without ever having caused harm, as so many such as Calder declare, is highly doubtful. Phaeton or Typhon caused several neurotic symptoms everywhere for thousands of years and is probably still working to build up fear over Comet Hale-Bopp or Halley's Comet or all comets that may ever appear. As attested to by the behavior of modern tribes of Amazon jungles, literacy and historiography are not required.

Peoples picture comets in many different forms, none of them impossible. They tie comets into many lessons, symbols, rites, and stories of their religions. Beyond religion, they integrate the comet-complex into sex, work, play, politics, and war, in highly disguised ways. They dread new apparitions and revere substitute portrayals of past comets. Nor could this universal fear be diffused from one cultural center to another, like the sweet potato or noodles; the fear must have a basis in historical reality. As we have demonstrated in so many writings, the comet as an apparition that is followed by catastrophe is a substantially true memory retained of mankind.

Spectres and pandemonium accompany catastrophic events of the earth sciences. In themselves they do not leave vestiges. Still, little by little, research will build up rough measures of the intensity and scale of the events from the visual accounts available in legend and reports. In the case of every important god stretching back before the dawn of classical history, we can elicit and reconstruct from legends of sight and sound the workings of high energy forces that connote catastrophes.

CHAPTER FOURTEEN

THE RECENCY OF THE SURFACE

Conventional chronology today gives about 15,000 years to the Holocene and latest period, and about two million years to the Pleistocene. Then some 35 my go to the Tertiary, with its Pliocene, Miocene and Eocene; 55 my to the Cretaceous; 27 my to the Jurassic; 23 the Triassic; 33 the Permian; 74 the Carboniferous; 72 the Devonian; 22 the Silurian; 57 the Ordovician; 92 the Cambrian; and some 2000 million years (or much more) to the Precambrian era.

By this point, the reader is well-aware of our scandalous departures from the conventional text. We have been arguing, in the whole of our Quantavolutionary Series (and see page 497 below), that all of the preceding ages probably have occurred within a million years, and especially that major elements of the Holocene, Pleistocene, Tertiary, Cretaceous and Carboniferous have occurred within the time usually allotted to the Holocene, namely some 15,000 years or less. The great disparity has occurred, we maintain, owing to the displacement of time by catastrophe. And to denote these catastrophic intervals, we have used certain disruptive episodes that we have tied into astronomical events, bringing a sequence of periods that we begin with the Pangean, and then go on the Uranian, Lunarian, Saturnian, Jovean, Mercurian, Venusian, and Martian, each marked by catastrophe, until the present or Solarian period to which only some 1600 years are allotted.

The stakes in radiochronometry are very high: all of the natural sciences have a stake in the game, plus ancient history, prehistory, anthropology, archaeology, indeed all of the humanities and, in the end, philosophy, theology, cosmology.

As Cook has argued, the early state of the Earth is hardly empirically known or deducible. Yet radiochronometry must proceed as if it were, and, furthermore, somehow, whatever is found now as the result of decay was not present in the beginning but finds its only source in the decay process.

A similar problem is to be seen in the separation of electricity from radioactivity. Ignoring the electrical state (or, better, the electrical history) of the Earth may foreclose alternative life experiences of radioactive materials. But we have intimated earlier that the Earth has had heavy periodic electrical transactions with exoterrestrial bodies and plasmas. Further, the Earth has had electric potentials differing from its potential today.

Sykes placed a standard radioactive cobalt-60 specimen between the poles of a magnet with an estimated flux-density of 0.1 Tesla, positioned a gamma radiation detector in proximity, and took readings of the emissions when the magnet was on and when it was off. The "decay constant," which is supposed to be invariable if it is to be used to clock geological time, speeded up about 2% when the magnetic field was applied. He concluded that "the thesis of decay constancy under all environmental conditions cannot be maintained."

The half-life of radioactive isotopes appears vulnerable to external electromagnetic influences. Since the strength of the Earth's magnetic field has been diminishing, along with that of magnetized rocks, the radio clocks within the rocks will have been slowing down. Further, it is not alone a matter of a long-term trend. In any quantavolution, strong electromagnetic forces are likely to be applied to crustal rocks causing sharp increases in the speed of passage of "radio- time."

"What role," Juergens goes on to say, in passages cited briefly in our chapter on lightning, "might environmental electrification play in setting the rules for nuclear stability, radioactive-decay rates, and energies of particle-emissions in decay processes?" The ambiant electrical stress would be different, whether continuously or for short periods of time. "It would seem to follow that decay rates for radionuclides might well differ radically from today's norms."

Beyond such utilities rest the several advantages that a microchronic model provides in association with the other elements of the theory of quantavolution: such as the negative exponential principle, the holistic principle, and the transactions of exoterrestrial and terrestrial forces. Microchronism, together with its allied quantavolutionary principles, brings all three events together: paleontological ecumenicalism, the moon-eruption, and continental cleavage and rafting.

EPILOGUE

Geology has not been able fully to confront lunar fission because of its notions of time. Nearly all studies favoring the idea have placed the event in the most remote eras, because to place it later would require the reconstruction of later natural history, including that of the biosphere, Furthermore, recent explorations of the ocean bottoms have revealed their astonishing "youth." This finding has been thought to disprove even the earliest fission of the Moon, since lunar fission theory without the Pacific Basin as its point of departure would be unappealing.

But the new evidence piles up in favor of lunar fission from Earth. The physical calculations of mass fit are plausible; the Moon fits its hole. The Indo-Pacific Basin is there; the ocean bottoms are all freshly paved. The land has been cleaved into great and small chunks and directed at the source of the eruption. The cleavages have occurred at a negative exponential rate down to the very present. The only force capable of such large interlocked effects would be the pass-by of a gigantic exoterrestrial body interacting electrogravitationally with the Earth. Such evidence is resisted because it is felt that the atmosphere, lithosphere, hydrosphere, and biosphere would be totally destroyed. This is not a challenge to be met by theory alone. If the facts occur to demonstrate the prior existence of a totally encrusted and thriving world surface and, then, after an epic quantavolution, continuation of the same processes, greatly altered, lunar fission has to be believed.

Yet theoretical logic - call it speculation - has a large role play, not the least in calculating whether the biosphere would survive. A review of all that has been written on this subject allows an affirmative. The extinction of a species is difficult; the extinction of tens of thousands of species is more difficult; the extinction of nearly all species requires the total explosion of the globe. Exponential reproduction over a few years can hide the most drastic reductions of population by fire, flood, thrusting, explosion, fall-out, radiation, de-oxygenating, and de-photosynthesizing conditions.

The very excesses of blast may harbor the secret of survival. Cyclonic action fashions its own boundaries. The cyclonic form is applicable to water, heat, dust, debris, electrical charge, radiation - to all that in a spread-out form would tend to exterminate life. An atmosphere permitting survival, by the theory of solaria binaria, would have been present in a huge plenum or sac surrounding the planets; in addition, atmospheric gases in close encounters can be exchanged, possibly even created under extreme conditions out of water and other compounds.

Surely survival would not be guaranteed. It might even be considered miraculous. Yet there is enough plausibility in survival so that extinction should not be assumed; what is perhaps the most useful and credible theory to explain the tortured Earth should not be passed over. If the Moon was assembled out of a blasted Earth in a highly developed and recent epoch, then the origins and behavior of continental drift are explained, world geography and physiography are explained, the oceans are explained, and the present state and distribution of the biosphere are explained.

It is astonishing and dismaying to consider the huge differences in time allowances between evolutionary and revolutionary morphology. The Grand Canyon has been a showpiece of geology as well as American tourism. Its accepted history is in the range of one to two billion years for the walls and 10 millions and more for the gorge. M. Cook's explanation calls for only 10,000 years to develop the whole complex. The whole world is implicated in such discrepancies, for the types of geological structures of the Earth are limited to a couple of dozens and they are nowhere unique.

Does it not wreck the earth sciences to propose a cut in time by a factor of 200,000? One might as well ask whether it wrecked economics to suffer both Adam Smith and Karl Marx. A quantavolutionary earth scientist can earn a professional livelihood as well as an evolutionary one, so long as his employer is unprejudiced. Further, radical criticism cannot but help any field, if it is properly conducted. Conventional science funds should be tithed to promote tests of the quantavolutionary model.

But beyond these considerations goes the nature of the field. Geology operates upon a few basic concepts, among them superposition, erosion, heat and pressure. And these are commonsense to begin with. When one rock rests upon another, it is younger, unless some force has intervened; erosion is the effects of wind and water upon landscape; heat and pressure can transform and transmute a substance.

Catastrophists do not deny these ideas; in fact, they invented them. Geology also has a large dictionary of names that are given to things large and small, representing infinite combinations of substances, heat, pressure, erosion, and position. The genius of geology is to bring order to this immense variety and to use this knowledge to practical ends like making cement and finding oil. To all of which the quantavolutionist says "amen."

Neither geology, nor any other science in its historical aspect, has to fear the idea of collapsed time, but can derive theoretical benefits from it. Let us speak for a moment of chemical evolution. Should it be as well termed quantavolution? I have here above spoken of the Miller- Urey experiments on the initiation of primitive life processes, and have generally considered the possible derivation of earthly existence from exoterrestrial and atmospheric sources. In *Solaria Binaria* Milton and I went farther into the matter, elaborating the life-creating and sustaining plenum of primeval Earth.

In 1983 C. Ponnamperuna reported the discovery of all five of the so-called "precursors of life" in the Murchison meteorite that fell in Australia in 1959. The compounds are adenine, guanine, cytosine, thymine, and uracil, which are key molecules. He subsequently created all five bases "in one fell swoop" by subjecting a mixture of methane, nitrogen, and water to electrical discharges. This, he said, evidenced that chemical evolution could have been accomplished in a single pool of liquid (or dense atmosphere?) in primitive times. The process might have occurred exoterrestrially as well as on Earth, commented Melvin Calvin who had also studied chemical evolution and won a Nobel prize.

Therefore, collapsing time may boggle the mind but does not destroy geology. Collapsing time introduces the need for high energy forces than can do in weeks what erosion can do in millions of years. The forces - wind, water, heat, pressure - are already present; it is a question of their organization and intensity. The more intense the forces, the more they depart from our experiences, and resemble the catastrophic recitals of the earliest humans.

Also, the more intense the forces, the more likely that they originate exoterrestrially. There appear to be no means whereby the scientific ideology pervading the earth sciences for the past century and a half can continue legitimately to ignore exoterrestrial causes and exoterrestrial effects in explaining our lately tortured Earth.

COLLAPSING TESTS OF TIME

"In one fell swoop:" what, if anything, is this expression but a way of saying collapsing time and quantavolution? Nor can one arrogate to man alone the ability to compress time. Nature may be blind, but she is infinitely large, powerful, and busy.

1. Rapid Sedimentation

Rates of sedimentation are usually estimated on the basis of contemporary rates. Allowances are made for demonstrable past events but these are interpreted on gradualist lines. If the Grand Canyon's age is calculated as an eroded river channel, its age is great. But if it is regarded as a transverse branch of the fissure fracture of the East Pacific followed by deluge and tidal erosion, then it could be of holocene age.

Ocean sedimentation recently examined under conventional premises (with the "help" of potassium-argon techniques), have dated the present ocean basins at nowhere more than 200 million years, incomparably younger than by former calculations. The sediments were found to be astonishingly meager.

Yet, contrary even to this new dating, the ocean sediments could be provided readily from catastrophic sources in a thousand years after the basins formed, as Chapter Seven will show. Furthermore, the ocean bottom, which is under enormous pressure, contains only unconsolidated sediments, a sign of newness. And if the oceans had once been land and the land ocean, then certainly great rock formations should line the bottoms.

In addition, at the rate at which uranium is now flowing into the oceans, the oceans and their sediments have accumulated a supply representing less than 100,000 years of flow, and when the flow off the continents is calculated as a negative exponential curve, the age of the ocean becomes holocene. For most sediments would have been dropped or transported in the earliest years.

Sedimentary rocks are given very great ages in part because the "normal" visible rates of deposit are slow. But a single cometary train might lay down a "hundred million years" of till or detritus clay and gravel-in a day. A coal deposit can be launched by a high-energy "bulldozer" in a matter of hours, covered over the next day by clay and baked until ready; it does not need the "millions" of years of development insisted upon by uniformitarian sedimentary calculations. Petroleum deposits are not proof of long ages, whether terrestrial or extra- terrestrial.

Shelton's marvelous, though uniformitarian, photographic book of geology should be quoted here. After remarking that laminated clay deposits (varves) can permit a time estimate of each layer, he says:

> "For the common sediments... we have no accurate knowledge of how long individual beds took to accumulate or of how much time elapsed between the deposition of each... Some thick beds accumulate in a short time, some thin ones take much longer, and in all probability the period of non-deposition that separate most layers represent far more time than is represented by the strata. As Charles Darwin pointed out over a hundred years ago, with far fewer facts to go on than we have today, from the standpoint of time, the sedimentary record is very incomplete - just an entry now and then with long pauses between."

How did Darwin know the pauses were long? How long is long? Indeed Darwin's idea of "long" is "short" according to today's scientists.

Again I quote Shelton: "Unfortunately most sediments do not contain reliable clues to how fast they were deposited--- or to the duration of intervals between layers... Observed rates of sedimentation range from almost immeasurably small fractions of an inch per century to many feet per hour and make it almost impossible to estimate the average for my large deposit..."

2. *Coral reefs*

Among the most complex challenges to quantavolutional geologists, uniquely related to sediments, would appear to be the coral reefs of the world, both living and fossil. An ordinary statement of the conventional case in the following:

> "Because the coral polyp's existence is tied to that of the algae, coral reefs can grow at depths no greater than around 180 feet -- below this not enough light penetrates to permit algae to carry on the process of photosynthesis. The brittle material we call coral is the polyps' protective external skeleton. The tiny animals absorb calcium salts from the ocean, allowing them to build these calcium carbonate structures around their bodies. New generations of coral polyps attach themselves to the skeletons of dead polyps. In this way the coral reef grow larger - layer upon layer, generation upon generation. Expanding at the rate of only few centimeters a year, some present- day reefs have been developing for 100,000 years and more."

The author does not mention fossil coral found at considerable depths beyond 180 feet. One must suppose a land-sinking or that the water level was rising as the coral grew; the lower coral would die, the higher would grow faster. Suppose the water temperatures were higher; the coral might grow faster; Suppose the amount of calcium salts in the water increased; the polyps would flourish. The opaqueness of sea- water is not an absolute, nor, for that matter, is the radiance at the surface. The algae supply has many variables determining it, including species adaptations and mutations that may cause greater or lesser light requirements. Can coral polyps feed upon bluegreen algae? Do shallow warm lava bottoms and new limestone accelerate coral growth? All those questions can make the coral reef an "anomaly" in short-time reckoning, reminding one of the "anomalies" that are similarly handled by uniformitarians in regard to apparently catastrophic phenomena such as vast "river-formed plains" or the "gradual" erosion of the Grand Canyon.

Even by conventional dating, long-term and carbon-dating-assisted, the seas are supposed to have been over 100 meters lower 20,000 years ago, before the "great ice melt", and, before then, the sea-level was abruptly higher and the coral could not have survived. Hence a continuous coral reef vertical development "for 100,000 years" would be highly improbable. Further, the 180-foot live-depth figure may be more nearly half that -- or 80 feet maximum live depth. The vertical growth rate of coral can be from 1 to 12 meters per thousand years. The lower limit is actually zero, depending upon thermal, chemical, nutritional, wave- energy, and pollution conditions. The highest rate, for all we know. may be limited only by the speed with which the sea-level is rising.

Fossil coral, not heretofore mentioned whether beneath coral growth of the past eleven thousand years, or separately discoverable, as in the Arctic Circle, or at depths of hundreds of meters elsewhere may have originated in the swamps and shallow seas of Pangea, the wholly continental Earth-crust that we postulate in this book. Some of the fossil coral beds may, like the continents, have been displaced and rafted to new locations.

Much of the reasoning employed in the case of coral growth here may also be used to argue the case of limestone caves and their stalactites. That is, subject to discussion in a forthcoming volume, the limestone caves of the world may be taken to be largely new, a product of large-scale electrical discharge of the Earth, water- accelerated. Arguments may be advanced farther, to wit, that the drip-formed stalactites and stalagmites can be grown in short times under non-uniformitarian conditions and yet be strong enough to stand against heavy seismic shock.

3. RADIODATING

William Thomson (Lord Kelvin, 1824-1907) estimated in 1899 that the Earth might be no older than 24 million years if its matter were chemically inert and its heat only the primordial remnant. Other scientists disagreed, opting for longer durations to accomplish evolutionary processes.

How uncertain were the stratigraphic estimates of time that geologists relied upon before new radiometric techniques came into use a generation ago is revealed in their quick surrender to radiometry: it is common joke that the earth has aged a billion years per decade for several decades, all owing to new tests of time by radiochronometry.

Certain elements, such as potassium-40 and uranium-238, which are to be found in rocks of the crust of the Earth, especially at or near surface levels, are radioactive. They are sometimes called "parent elements" insofar as they decay into "daughter" elements by giving up electrons or by other means. They began their decay as soon as they were formed. One calculates their life- span by figuring backwards from today's rate of decay as witnessed in a sample of the element. A rock matrix presumably will contain the parent element and the daughter element in proportion to its age, unless it had undergone some exceptional experience. The dozen or so transformations used for dating purposes include uranium-238 decaying into lead-206, of potassium-40 decaying into argon-40, and of rubidium-87 decaying into strontium-87.

None of these methods is useful directly for the period since 14,000 B. P. because the decay into daughter elements is too slow to detect over the short time. However, radiodating challenges our model of quantavolution indirectly when it produces long-term dates where short-term dates are expected. For example if, by potassium-40 argon-40 dating, the ocean floor appears to be 100 to 200 million years old, then it cannot have been formed between 13,000 B. P. and 9,000 B. P. Also, when igneous rocks associated with hominid bones of the Olduvai gorge, dated by the same technique, produce an age of about 1.75 million years, then the bones cannot be of the holocene epoch.

Major problems occur with radiodating. One is in the setting of a rate of decay and therefore setting a date for "time zero" within a reasonable margin of error. Regarding the "time zero" problem, the radio "clocks" work on vast ages, from one billion to five billion years of age. Adjustments in the so- called decay constant may move all tested rocks up and down the time scale by many millions of years. Although such adjustment never approach a short-term position, they cause doubts as to whether there is in fact a constant rate of decay to be discovered.

A second kind of difficulty deals with high-energy events. Radio-chemical methods of determining pre-historic age are extensions of the uniformitarian premise that the chosen chemical elements have remained unchanged in a closed system, save for the decay process, since the clock started to tick. They assume that nothing would affect the parent or daughter element, apart from the expected normal decay from one to the other; nothing could tamper with the clock. Recent studies cast doubt upon this theory; high forces can break and enter the clock.

The concept of "half-life" is used in radioactive decay time measurements. The half-life of an aggregate of an element is the length of time required for half the atoms of the aggregate to decay into the new element. The half-life of uranium-238 is 4.5 billion years, calculated backwards from presents rates of decay. Can the process of decay be so regular? Decay is the losing of an electron from an atom that is unstable; it therefore amounts to a transmutation. The occasion of the decay is a force. The force is another particle from another statistical aggregate. This force is regularly and randomly applied to the "A" aggregate causing a regular rate of loss. Each "A" atom has an equal chance of being hit in the bombardment. Hence whatever affects the bombarding aggregate will affect the rate of decay of "A".

And all "A's" may not be identical. Some "A's" may be "harder to hit," "resist cleavage," or "repel the projectiles." Still, as an aggregate, "A" might respond uniformly to the force causing is transmutation.

Radiation physicist H. C. Dudley has insisted that the equations describing radioactive decay rates were crudely derived long ago: "Bluntly, they are incorrect; but they nonetheless appear in our latest textbooks... Studies have varied the decay characteristics of 12 other radionuclides [besides 7Be and 90Nb] with changes in the energy state of the orbital electrons; by pressure, temperature, electric and magnetic fields, stress in molecular layers etc.," citing G. T. Emery.

Dudley further asserts that in certain cases, the "decay event A is causally related to decay event B occurring later, such that the time distributions of all decay events were no longer truly random, as required by current theory. There appears to be a chain type reaction operating... similar to that observed in neutron induced sustained nuclear fission," here citing chemists J. L. Anderson and G. W. Spangler.

Dudley asks for the incorporating in decay theory of "the energy state of the entire atom [not just the nucleus] and on parameters of interaction with an energy-rich subquantic medium."

The work of Anderson Spangler and Dudley implies this for

revolutionary primevalogy: decay rates for radioactive elements are dependent upon high- energy forces in the environment, and may be varied little or much. Radioactive decay can be compared with chain reactions in nuclear fission. Hence, at certain points in time, especially when the phenomenon of the catastrophic tube occurred, time pressures (based on today's retrojections) would have been instantly and completely disrupted.

4. Radiation Turbulence

We are conjecturing further, here, that major disturbances in the parent- daughter relationship may occur as a result of radiation storms and typhonic impact explosions. Lesser and more localized in effect, and often inter- connected with radiation storms are jovian bolts, phaetonic atmospheric penetration, titanic large body encounters, and dense material fall-outs. These operate as follows:

Cosmic radiation consists of high-energy particles that bombard the earth sufficiently at the present time to permit the presence in the atmosphere of atoms of all chemical elements. Both the particles striking earth and the transmutations of particles are varied. When, according to quantavolutionary theory, age-making and age-breaking episodes occurred, the earth passed near to heavily radiating bodies and was also subjected to heavy radiation storms from a distance. In fact, every change in the earth's atmosphere lessened or increased the reception of radiation: the cloud canopies, the lowering or dropping of canopies, the rising of exploded vapors, the destruction of biospheres and the loss or gain of atmosphere from comets, meteors and planets.

In all of this, the parent and daughter elements involved in radio clocks have experienced a turbulent history. No pair of elements can be granted to have remained locked in their crystallized rock interior since the beginning of its time. There is no way of commencing the history of potassium and argon at the bottom of the sea. The bottom formed in a turbulent atmosphere and hydrosphere, first wet, then drowned shallowly. then deeply submerged but all the while actively spreading. The waters that poured in came directly from the skies, through skies via the sea and earth evaporation, and through runoffs loaded with detritus.

Under such circumstances the clocks might be deemed invalid. They were set wrongly to begin with. They have maintained a semblance of agreement of very old ages by first of all having had similar recent experiences within their rocks, and through laboratory fudging of tests and samples back and forth.

Yet even "normal" experience of today's solar system presents a severe problem. Nitrogen contained in air and in radioactive mineral undergoes a considerable transmutation of isotropic elements. Lead undergoes the same. The cause is neutronpromoted transmutations. As a result, the decay process of uranium into lead is paralleled by neutron-to-lead activity. When as in certain Katanga and Canadian ore bodies, a neutron-promoted corrective factor is introduced into the uranium- to- lead decay process, the daughter element that is owed to uranium decay is so reduced as to produce a zero age result.

This kind of problem is rendered even more difficult under solarian conditions by problems of selecting and sampling rocks, by the fluxing and painting of the surfaces of rocks where trace elements aggregate, and by the need to transfer (with dubious validity) the findings of a test in one part of the lithosphere to conclusions about tests in other parts.

Problems of leaching and fluxing are severe. Rivers carry an estimated average of 6 x 10 10 grams per year of uranium down to the oceans. If the lead is left behind in the rocks this escaping uranium is effectively turning back the clock. Parents are leaving their daughters, and the remaining parents are being charged with their existence. The amount of uranium in the ocean, moreover, is so small (10 to the 17 the power grams) as to have been produced even under non-exponential solarian conditions within about 10 million years. With quantavolutionary theory, the exponential rate of deposit would eradicate even this time calculation.

Helium in the atmosphere is originated radioactively from the uranium and thorium in the lithosphere and from cosmic rays from the galaxy and beyond. Conventional ages of the lithosphere require that 10^{20} grams of helium should have been released into the atmosphere whereupon some of it would escape into outer space. However, the rate of escape is too slow under solarian conditions to explain why so little helium exists in the atmosphere. Given the amount of helium present there, it has been calculated that the age of the atmosphere must be only 12,000 years. That is, some 12,000 years ago, the atmosphere was reconstituted.

Radioactivity was discovered a century ago but time-measures of radioactivity are largely a post-World War II development. Despite the shortness of its life, changes in the field have been numerous and radical. Its leaders turn quickly in new directions whenever problems are encountered, introducing new half- lives, slicing experimental rocks differently, and giving their favor now to one, and again to another technique.

5. Potassium-Argon Dating

Potassium-argon dating has become highly favored recently, for reasons too byzantine to develop here. For, the criticisms that can be addressed to uranium- lead dating hold also against 40K/40A dating. Indeed, argon (one of the "noble gases" whose exclusiveness or slipperiness gave them their name) is generally to be suspected of vagrancy. Also, the stability of potassium is in question. "Potassium can be made to diverge widely form conventional abundance by countercurrent electromigration."

Argon- 40 will be present in a rock if potassium-40 is present and has had time to decay. Only igneous, and certain types of once-melted metamorphic rock, can be tested. Sediments cannot. The half-life of 40K is so long (1.3 billion years for half the decay to occur) that almost no argon-40 is to be found in a young rock, and therefore tests are not yet considered valid for less than 100,000 years.

Dates produced by related tests are often discordant. Material taken from the Salt Lake Crater on Oahu, Hawaii, dated from 200 to 3,300 million years; the Moon has been dated as older than the universe; and 200-year old lavas, that should show zero Argon, produced enough to allow 12 and 20 millionyear old dates at Kilauea, Hawaii. I shall only mention that, under such circumstances, in other cases, the problems of full and open reporting may become serious in the field of chronometric science; as in public affairs, there arises a temptation to dismiss, "fudge" or even conceal some of the evidence.

Argon, like uranium and radioactive trace elements generally, tends to rise to the surface of the Earth. Hence surface rocks (and these include all that have been measured) will be high in argon content. Argon also can be infused into hot rocks from the air and kept there as the rocks cool. This could have happened to Earth if Mars, thought now to be rich in atmospheric argon, encountered Earth 2,700 years ago; the same Martian argon may be what is making Moon samples, so young in some respects, so old by 40K-40A dating. The U. S. Venus probe of 1978 found astonishing quantities of argon-36 and possibly argon-40 in the burning atmosphere.

Argon, being "exclusive," "slippery," and "noble," leaks. It escapes into the atmosphere; it flows horizontally. It prefers rocks of certain types to other rocks. On the Island of Naxos, Greece, Poul Andriessen found side by side metamorphic rocks which, in tests performed in his Dutch laboratory, produced ages of 5 to 15 million, and of 200 million years (amphibolite ultrabasic rock). Australian tektites have given 700,000 to 860,000 years by the 40K-40A method in 7 to 20 thousand-year-old strata.

Funkhauser and Naughton, faced by the Hawaiian incongruities, speculated that excess argon could be held in crystal irregularities and imperfections such as grain boundaries and dislocations in the rocks. This likely theory would appear to throw the K-A ratio upon the mercy of petrology rather than chronology.

Granted argon is more abundant in rocks nearer the surface, a lava flow will erupt melted surface rock first, than lower rock, then still lower rock. This may falsely date a set of lavas, although the law of superposition is correct. As the law demands, the strata of lava on top will be younger (and hold less argon) than the strata below (with more argon); moreover all will be very old for the reasons given above. As matters stand, it would be a grave risk for geology to rearrange the phanerozoic scale according to 40K-40A dating principles.

6. THE RADIO-HALO PROBLEM

Radio-chronometricians pass restlessly from one measure to another, despite their elaborate equipment, which critics have alleged to be too burdensome to discount and abandon (over 100 laboratories exist today for carbondating alone). While continuously asserting the validity of the great time intervals they have discovered - and indeed imposing this belief upon the geologists and anthropologists - nevertheless they are engaged in a quest for improvements and for new tests that are less vulnerable to complaint.

There are at least a dozen parent-daughter, radioactive decay tests, each with its problems of the type already displayed in the discussion of 40K-40A tests. Discordant time readings within and among individual tests, demonstrable leaking and leaching of elements, and proven possibilities of elements being created under catastrophic heats and pressures are vexing problems, even more than the problems of sampling and contamination.

If, to this time, the restlessness of chronometricians has been excused as a search for technical perfection, that excuse has now worn out its acceptance. The reduction of the uniformitarian ideology is permitting a clear view that elements in varied isotopic forms can and have been engendered by natural and human forces. The implications of various studies, writes Melvin Cook, are that "apparently all the elements are available in cosmic radiation at very high energies as bombarding particles, and that the synthesis of high mass atoms in large decrements of mass increase is possible. It is therefore only necessary for our earth (or its accretion materials) to come close enough to the source of cosmic radiation to effect a complete equilibrium distribution of atoms. At present, the earth itself is too far away form the source of cosmic radiation (owing possibly only to the protecting influence of its atmosphere and magnetic field) to maintain nuclear equilibrium in respect to U, Th, K 40 , Rb 87 , and other radioactive atoms.

These remarks should be taken in connection with the possibilities of catastrophic typhoons or tubes, described in the last chapter, and fluctuations in solar activity recently discovered.

The studies of R. V. Gentry are especially threatening to radiochronometry. He examined over 100,000 radiohalos in the decade just ended. A radiohalo (or pleochroic halo) is a spherical colored ring around a radioactive nucleus denoting the escape of an alpha-particle and its ionizing of a surrounding zone. The ring's size is determined by the speed of escape. When uranium (U238) decays, it does not decay immediately into lead (Pb 206) but produces seven other isotopes *en route,* from thorium, radium, radon and polonium. There occur, then, with decayed U238 eight concentric rings, of which five are distinguishable.

Gentry discovered, however, that many halo systems begin with polonium; they exhibit no uranium or other supposedly preceding halos. And polonium 210, the longest lived of the polonium isotopes, has a half- life of 140 days. If some of the oldest rocks of the world contain this isotope, without a uranium- thorium predecessor, it follows that the host rocks must have been formed in days. Promptly, then, one would have to drop a billion years from the history of the Earth, for the original rocks are supposed to have taken a billion years to crystallize.

Parentless polonium atoms may be primordial, as are uranium238 and thorium-232 atoms, but this would imply that polonium halos "represent evidence only a brief period between 'nucleosynthesis' and crystallization of the host rocks." Incredibly, rocks would form immediately upon the synthesis of the elements in them. Reporting upon a telephone interview, Stephen Talbott says that Gentry "finds compelling reasons to question the entire dating scheme which undergirds our concept of geological time."

Other studies of coalified wood from the Colorado Plateau, buried in rocks of the Jurassic-transition, evidenced such an abundance of uranium and lack of lead that ages of at most 100,000 years had to be assigned to the coal. Then Gentry, in examining the radiohalos, had to report that the coalification required only days, not millions of years.

Sykes has shown by experiment that a magnetic field of the flux density of 0.1 tesla is enough to increase the mean decay count of radioactive cobalt- 60 and to skew the distribution of decay incidents from the normal. The "decay constant" was increased by about 2%; correspondingly, the half-life of cobalt-60 decreased.

7. RADIOCARBON (CARBON-14) DATING

Cosmic rays of the galaxy strike and explode atoms of the atmosphere. These give off neutrons that interact with nitrogen of the air to make carbon- 14 or 14C. This passes into carbon dioxide and then into plants and other living organisms through their food supply. Living organisms also ingest carbon-12 which does not decay. When anything that has lived dies, it ceases to ingest radioactive carbon-14, and the carbon-14 within its cells begins to decay into nitrogen-14. Half of it might decay in 5,730 years, the other half in another 85,000 years, according to conventional theory.

Thus, any once- living organic substance can be tested for the amount of 14C that it now contains in relation to the amount that was originally ingested. The carbon-12 level can be used as the base of measurement. However, not all species ingest 14C in the same amounts, so that specific rates must be calculated for different species. More importantly, "the amount that was originally ingested" may vary.

All that has been said about the effects of high-energy forces upon the atmosphere applies to carbon-14. How much nitrogen was in the primeval atmosphere is unknown and is presumed on today's measure. The carbonization (burning) of the biosphere and the sudden proliferation of flora will directly affect the rate of generation of 14C. Also, if carbon-14 was heavily generated in the atmosphere by electrical phenomena and radio storms, in the times when Uranus, Saturn and Jupiter were worshipped, it was ingested extensively by organisms. Matter of this period would test as "younger" today, provided that several other "constants" remained constant. As disasters diminish in intensity following chaos, relatively less 14C would be created; matter would grow "older." Several disasters involved the desiccation or saline ruination of large areas of the world; this would cause less carbon dioxide to discharge from plants. During short periods of burning, great amounts of non-radioactive carbon are discharged into the air and waters, and therefore contribute to a temporary "aging" of the new life of the time that follows. Whenever both a cosmic brilliancy and a conflagration occurred, today's tests would be contradictory, and averages would mislead.

Libby and Lukens have estimated a "perturbation of about 1%" occurring in the production of radiocarbon of tree rings by lightning bolts. This represents a neutron supply added to the supply produced by cosmic rays. The estimates are based upon present-day assumptions, also upon highly varied conditions and inexact knowledge of the extent of lightning or its effects. It does not consider lightning discharges occurring solely in the atmosphere, and especially the mega-bolts that can be a thousand times more powerful than the average earth-striking bolt and were recently discovered by satellites.

Aside from what is happening in the biosphere, a fixed 14C component of the atmosphere, upon which the test is based, depends upon a constant encounter rate between cosmic particles and nitrogen that produces 14C. Since radiation storms occurred and long- term radiation levels were diminished and increased from time to time, intervals of the 14C scale must have been rendered invalid, except for mere coincidence. Only in the years from about 500 B. C. TO A. D. 1900 might the amount taken in by organisms have acquired some constancy. Even so, strange aberrations of the 14C/12C ratio occur, as with shellfish and coral growths.

8. Radiocarbon datings
as indicators of ecological stress

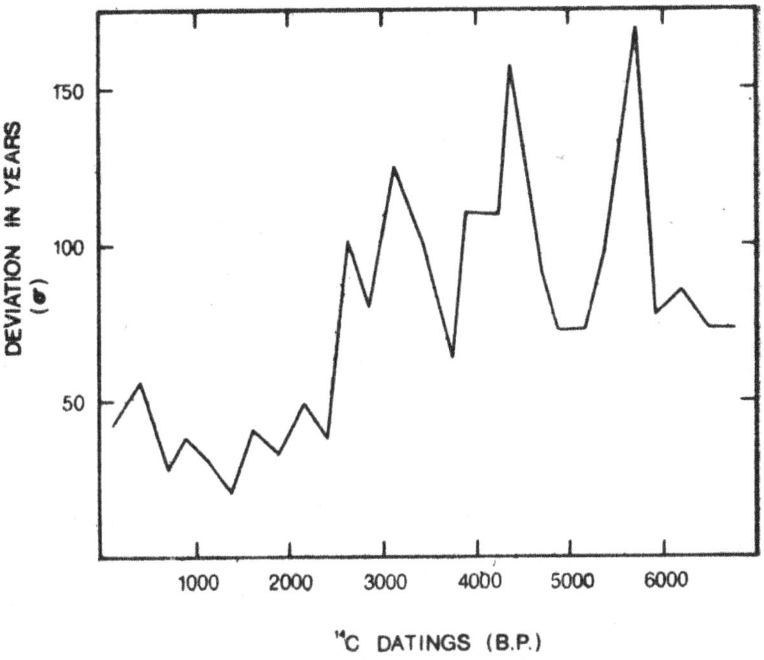

The left hand scale (s) registers the standard deviation of the "true" curve from the trend curve - the number of years by which the radio carbon dates of each 250 year period deviate from the average of the whole group of dates of that period. The bottom scale represents the years before the present (taken as 1950 A. D.). As the chart shows, the dates begin to be erratic increasingly around the time of the Martian encounters (-2687 B. P. by this book's 2000 B. P. standard). The time scale goes to -6750, and thus carries one through the Martian, Venusian, Mercurian, and Jovean ages. However, even the erratic swings shown here do not portray the true extent of atmosphere and ecological disturbance, because, as the text asserts, a succession of quick changes in the atmosphere is possible, from low to high radiocarbon intake therefore by the biosphere, and this phenomenon would cause an evening-out of still a second and possibly much more serious form of deviation. Within a time of several years, an organism could ingest widely varying amounts of 14 C. Hence I suggest that radiocarbon dating may be useless before about 2500 years ago and there may have been a completely different radiocarbon cycle, as M. Cook maintains, before the Lunarian catastrophes.

(Source: Damon et al, extended and applied by G. W. Oosterhout Half- life is 5730 years.)

There are many anomalies in C14 dating, a few of which are mentioned elsewhere in this book. Artifact dating has become quite common, with most of the apparent successes occurring on artifacts and substances of the recent historical past. But it is precisely the problem of C14 dating that, by our theory, it is almost surely wrong in the earlier periods when the tests are most needed.

A group of scientists recently excavated "Little Salt Spring, Florida: A Unique Underwater Site." Among many remains they found in a lower level a tortoise carapace, which provided a date of 13,450 ± 190 B. P., and a wood stake used to pry open the animal, which gave a date of 12,030 ± 200. Some 1400 years of difference. Yet this is not the only problem. The whole range of time may be in question. For a base of a carved oak mortar was discovered and dated to 9080 B. P. and then declared to be similar in style to a piece recovered at Key Marco, 130 km to the South, and dated at about 1200 years ago.

The quantavolutionary hypothesis is disruptive of carbondating, as it has been conceived. An adjusted curve is impossible because the revolutions of the atmosphere in precisely the most critical millennia in primevalogy cannot be positioned and defined sufficiently well for them to be employed in weighing the scale intervals. The 14C method will be useful for dating the past 2,400 years, when allowances are made for short- term atmospheric fluxes owing to extraordinary cosmic, volcanic, solar, industrial, nuclear explosional, or other activity disturbing to the atmosphere.

Mysteriously, corroboration of some of our conclusions comes from a retrogressive calculation by Melvin Cook of the amount of 14C in the ancient atmosphere. Granted the present level of carbon-14 and the fact that it is rising slightly, he found that all the 14C would have had to arrive in the atmosphere within the past ten to twelve thousand years. Far from being constant, prior radio-carbon was at this point wiped out statistically and theoretically a new atmospheric accumulation began. This would appear to be about the time of the climactic Lunarian catastrophe. However, this calculation, as Dr. Cook might grant is more useful as a reductio ad absurdum than as a plotting of the true history of atmospheric carbon.

9. Tree-ring time

Dendrochronology has discovered only one tree whose rings can be used to date associated events back into periods of interest to primevalogy. Such is the hard bristlecone pine, which may achieve 5,000 years of age by ring count. By matching fossil pine with living pine, the ages may be traced back further; perhaps 8000 years B. P. have been claimed by matching . (if conditions of fossilization were uniform millions of years of matching would be theoretically possible!) It is assumed that rings have always grown on an annual basis. Not surprisingly, quantavolutionists have adversely criticized the technique.

"Annual" is a relative standard, presently derived from a revolution of the tilted globe of the Earth around the Sun. Changes in astronomical motions can change the number of rings; if a "year" is shortened, the rings may be increased within the normal lifetime, something that may have caused the Methusalah phenomenon in early reported human ages of the Bible and elsewhere. Also, the rings may increase or decrease if climatic conditions introduce a doubling of seasonal cycles within the same year- time.

The tree has to be matched with human or natural objects of known age, or used to calibrate radiocarbon dating. But tests cannot calibrate each other without reference to a third test. This third test is often a historical date, but such dates rarely exceed 3000 years and even before then are hotly disputed. Furthermore, there occur in the cross-matched trees gaps of rings that may correspond to revolutionary incidents in the arboreal environment. Electron microscopes can find exceedingly thin rings, but cannot explain aberrations among them.

Despite all of this, if bristlecone pines could be calibrated over a span from 5000 to 8000 years, this would mean that the solar system has existed that long in a form not radically different from its present form. Also, no important element of the atmosphere or climate affecting rather similar biological organisms would have changed. Further no major annual motion of the Earth respecting the Sun must have changed (orbital distance; orbital speed; rotational speed of the Earth); or all three motions, if changed, must have added up to the same total solar exposure time.

10. Magnetism

When rocks are near melting, they are stamped with the direction of the magnetic pole. When cooled, they keep this directional stamp. If reheated, they lose it and acquire whatever new stamp is indicated by the current magnetic pole. Also, if a rock changes its position, its magnetism will point away from the location of the magnetic pole towards which it was originally oriented. If also it is heated in a new position, the imprint will be oriented differently upon the rock. Paleomagnetism studies the changed magnetic orientation of rocks. It also judges the ages of rocks, but within severe limits.

Great belts of ocean basin rocks are imprinted with a polarity that is reversed from today's. Moving away from the great hot ocean ridges, alternating belts of reversed polarity occur. It is believed that these reversals occur at intervals, whether a few thousands or millions of years apart. It has been shown that the belts grow older (by fossil record, by inference from land studies, and by 40K- 40A tests), as they move outwards from the ridges. It is believed that many millions of years show up in the magnetic bands.

But magnetic reorientation depends upon the last heating of the rocks that contain the imprint and upon their movement. If the ocean bottom is moving much faster than assumed, then the time between reversals is shortened in proportion. And vice versa, if the reversals occur rapidly, then the ocean bottom must be moving much faster then believed. Probably both have occurred: the ocean bottom moved rapidly and magnetic reversals occurred repeatedly, both within a period of several thousand years, probably between ten and thirteen thousand years ago, or so we shall argue in a later chapter.

Magnetic reversals occur for reasons unknown. Why they should happen at long intervals of time rather than short intervals is also unknown. Short- time intervals between reversals are probably connected with an impulse towards or an actual change of the axial inclination (now 23°+) of the Earth. Impulses were frequent in revolutionary ages. I shall be proposing later, with the support of legendary and geological evidence that the Earth's axis probably tipped on various occasions, both gradually and sharply.

After each abrupt change, the globe may have rocked for a time before stabilizing. The rocking took many years; the multiplex worldwide legends of Hamlet's Mill may reflect this perceived motion. In that case, the belts of differently imprinted rocks would represent rapid growth of ocean basins with a rocks would represent rapid growth of ocean basins, with a slowly wobbling axis of spin and a reversing magnetic field.

A prior period of wobbling of the axis could even produce, in a period of accumulating ice, a succession of seeming advances and retreats (or the illusion of the "ice ages"). But also, pluvial intervals would occur, with melting in- between. The penchant of early man and mammals for living near ice- fields is understandable only because the Earth beyond the ice was not cold (since the ice might come from above). However, it is too early here to take up a position on the "ice ages," which are dealt with in the third volume of this work.

Two terms are used to discuss magnetized rock: natural remnant magnetism and thermal remnant magnetism. Geophysicist T. Nagata of Tokyo has shown that the two are the same. Remnant magnetism, furthermore, will occur and increase with any temperature increase above 200° C.

Magnetism decays. The exact coefficient of decay is unknown. The half- life of paleomagnetism may be only 5,000 to 10,000 years; all magnetism, according to M. Cook, may be less than 70,000 years old. (Nagata guesses 1 million years.) Therefore, paleomagnetic bars of the ocean bottoms or land cannot well be used to measure time. Any considerable intensity must record a young age. A priori paleomagnetic ocean bottom measurements showing millions of years of age must be wrong. The position here taken is that any magnetism of the crust is primordial except where the crust has suffered a melt or welled up as new crust from the interior magma.

11. FOSSIL RECORD AND MUTATING TIME

Organisms that die in a mineralizing setting may become fossils that are recognizable unless subsequently melted or crushed. Fossils are the principal means of dating sedimentary rocks, and, by inference, such igneous and metamorphic rocks as may be connected to them. If two rocks, no matter where they are found, contain the same fossils, the rocks are usually from the same period of time. The more numerous the identical species of the two fossil assemblages, the surer their common age. When the rocks appear to be in superposition, the fossils help to assign them a relative date. Once this is done, if afterwards the same rocks occur in isolation or not in superposition, the fossils which they contain enable their dates to be inferred.

A fossil may be wrongly dated. The record of its period and species may be incomplete. Or the fossil assemblage of various species may have been zoned and then have been transported to another area and placed, say, above a younger assemblage. Or the method of dating may be fallacious. For example, at the Schefferville (Canadian) iron mine, fossil wood specimens, radiocarbon dated at 4,000 years and largely unchanged chemically, were found imbedded (but not intrusively) in iron ore of pre Cambrian age (" over a billion years ago" and before trees evolved) at depths of several hundred feet.

Attempts at correlating results of radiodating with established fossil dating have not helped. They have thrown the phanerozoic scale into disorder. Acceptance of radiodating provides numerous anomalies in traditional fossil successions. Basic difficulties in both methods come out of high- energy processes that devastate the atmosphere, build sediments and transport life forms quickly.

Plant and animal species require time to adapt to environments (life niches), to proliferate and to become extinct. So long as high-energy expressions are absent, it is reasonable to assign long periods to these processes and long life to the species. Originally, evolutionists were composing calendars that were under 100 million years in all. The discovery of natural mutation introduced a dynamic of change, but a successful mutation turned out to be, in theory at least, a most rare event. So more time was needed.

Now a billion years or more is allotted for the evolution of species. But quantavolutionary theory permits short mutation intervals, quick and widespread extinction, the opening up of a great many life niches for pre- existing and new species, and the possibility of less restricted and therefore exponential growth of population. Hence all the time may not be needed to explain evolution, even as evolution is understood by neo-Darwinians today.

12. CYCLES AND ANNIVERSARIES

That the world was created, destroyed, re-created and destroyed again, repeatedly, has never been doubted by any culture anywhere or anytime, except by the modern uniformitarian culture. Five great ages are found in ancient Greece, India, Tibet, Peru and Mexico. Seven ages are put forth in another Hindu source; in Mazda; in Hebraic sources; in the Sybilline oracles; among the Mayas. The Hawaiians and Icelanders count nine; the Chinese reported ten ages up to Confucius. All may be taken as valid relative to localized definitions and experiences. All may be regarded as authentic challenges to the ages set by geochemistry and radiochronometry thus far. There occurs, nevertheless ,an urge to straighten or blend cycles into a helix, even in mythologies obsessed by repetitive chaos of creation.

> "The final step in Aztec speculation, as indicated by their great Stone Calendar, is to assign the four earlier world ages to the four world directions, with the satisfying result that the present age belongs to the center of the world, the place where man likes to think of himself existing... The terror of experiencing a derangement of the cardinal points is transmuted by systematization into the comfort of knowing that all resulted in placing man at the center".

Very recently, however, it has become clear that the competition for chronological veracity is going to be framed in the ancient cyclical - or, as I have termed it here, helical-mode. Natural scientists are becoming "helicalists". Writes Umbgrove, "What creature is this that breathes so heavily every 250 million years? He refers to the Earth and to the cycles of "death and resurrection" that characterize so many earth processes. As we have seen, paleontologists, ice age specialists, solar experts, diastrophists, and electromagneticists - each in their own wayare discerning helices of the ages.

Also York and Farquahar, radiometricists: "Radiometric dates obtained on rocks from a single continent tend to cluster into definite groups. Ages are not uniformly distributed in time."

Furthermore, the timing of the groups seems to be similar over all continents. One can guess from their data that quantavolutions recur and affect the whole Earth.

Every cycle began with a kind of creation or rebirth. There was little of regularity on earth. Life was a continuous commotion. An obsessively fearful race projected itself into the sky. When planet Saturn became the great god, he was king of man and destroyer of man, but also bringer of wisdom and bountiful food. The Saturnalia began and have persisted to this day in jubilee days that follow days of sorrow and fasting. The Jovean anniversaries took over the Saturnalian. The Venusian and Martian came then in the spring near the vernal equinox while the old anniversaries centered around the shortest day of the year (in the northern hemisphere).

From full moon to full moon gave an easy method of counting in the Age of Saturn and it could usually be observed in the often misty nights. Moon calendars, sun calendars, and planetary calendars were often possible in the periods between changes of motion and place. A lunar month can, and does, change its length, without requiring a major social change except to revive terror and encourage religious ritual and related behaviors. Not until the last of the disasters had ended, in 2687 B. P., did a stable moon or sun calendar that was correct by present standards appear. Long afterwards and even until this day in many parts of the world, nothing in the order of skies is taken for granted, and, for calendar anniversaries, for festivals, and for public policy decisions, expert moon-watchers of the priesthood decide precisely when a moon should be termed full or new.

Practically all human constructions that have survived from earliest times are temples, temple- connected, or astronomical. The megaliths, found in many the age of surviving records, that is, the Middle Bronze (Mercurain) and Late Bronze (Venusian) Ages, scientific observations of solar, lunar, stellar, and planetary movements were recorded in several countries; they differ from the observations that scientists today would make of the same movements.

The ancients numbered scientific observers among them, and states were sometimes dominated by astronomer-theocrats. Water-clocks, that measured time by the passage of water, and sun-dials were built; specimens have been found; they mark a time, however, which differs from the present day.

These early observations were made by dedicated, highly disciplined corps of observers and are to be trusted. If they were dedicated and disciplined, it was ultimately because the skies could not be trusted; humans, god- driven, harnessed themselves to the observation of the skies, their pragmatic distrust reinforced by the ever- present subconscious illogic: "To watch is to control."

13. 58 TESTS IN DISPUTE

The quantavolutionist offers his tests of time. They usually lack tubes, needles and gauges and require a general vision of history. The quantavolutionist looks amiably upon tests that mix human evidence with natural evidence, joining an ancient legend or an invention with a change in appearance of the Moon or Mars. To the evolutionist, the quantavolutionist appears fuzzy-minded, gullible, and fanciful. But to the quantavolutionist the evolutionist seems narrow-minded, technocratic and historically lame- brained.

The quantavolutionists say this: Consider all the great natural forces that operate today. Read the ancient myths and accounts to discover how much greater were the expressions of these forces in the beginning. Extrapolate the effects of these forces as known. Then state what must have been the condition of the skies, the earth, and life in the earliest days of human recollection. Then, if interested, go back even farther, to what might have happened before.

The evolutionist offers his tests of time. When these tests are applied, we see time as very long and change as very slow, point-by-point, drop-by-drop. The tests are very many. It would take an encyclopedia to discuss them properly. On the chart of tests , I have displayed four things: the test itself, a brief phrase on its unique quality, the main position of evolutionists in respect to its validity and the contrasting position of the quantavolutionists.

Although it is beyond the capacity of this book to carry explanations and analyses of the fifty-eight listed measures of time, the major objections to their evolutionary interpretation can be set forth. I shall do so, following the categories of the chart, with apologies for the necessary exaggerations and exclusions.

The main objection to accepting the evolutionary explanation of the prominent features of the Earth's surface in Category I is that they are all based upon unproven constancies in the forces working to form the surfaces. High heat and pressures, hurricane winds and tides, or movement of the Earth's crust can form all of these features in short intervals of time. One can move over the surface of the Earth and offer an alternative quantavolutionary explanation of all singular features.

The main objection to the biological measures of evolutionism is again that they may all occur through quantum jumps under high energy impulsion. Once granted that mass extinctions and arrivals of species occur in correlation with catastrophes, then it is only necessary to point out that "successful" mutations themselves are so rare that large numbers of mutations are required, which implies that atmospheric catastrophes are needed. Biological and geological quantavolutions are the basis of the ecological changes that produce the evolution of species.

The third category of radiochronometry almost entirely depends upon a constant radioactivity of certain elements over great stretches of time. Very recent studies have shown, however, that (a) we do not know the original state of the elements and hence the history of their radioactivity, and (b) undecayed and decayed elements have become separated somehow, sometime, and their ratio cannot be now regarded as a measure of time. In the case of item 8, the uranium elements are not found in expected oceanic and atmospheric abundances for a long time record. In the case of item 11, catastrophically produced materials such as water and natural gas are found in an abundance under high pressures that long-term effects should have erased.

Of the astronomical motions, the fourth category, it can be said that (a) proof of constancy of motion is only available for a very short time; (b) even if the laws of motion suggested a history of motion, they do not write the history; (c) some motions are mysterious in origin and best explained as fossil motions from some radically different ancient motion; (d) evolutionary science has been loath to consider the history, presence, and effects of electricity in regards to star systems, solar systems, and the Earth (as to both its external and internal force fields).

	Test (mostly of absolute time)	Idea of the Test	Evolutionary Explanation	Quantavolutionary Explanation
I. Earth Surface Features				
1.	Gross forms and Relations	Estimate of age, based on theory and presuppositions, of lakes, cratering, diastrophism, continental displacement, etc.	Convection-currents from hot mantle, snow and rain, and erosion define features over long times	All forms can be quickly created under high-energy conditions
2.	Rock (or fossils) appearance	Judgments of freshness looseness, etc.	Appearances are deceptive; gradual erosion and cementation	The " eye" , feeling , and taste can distinguish age often
3.	Superposition of strata	Over-layers younger than under-layers, barring intrusions, displacement	Accepted and basic	Accepted and basic
4.	Sedimentation of materials	Rate of deposit, based on observation and/or suppositions of past rates	Generally, present rates are typical of past rates	Revolutions have generally intruded with exponential rates
5.	Erosion of materials	Rate of deposit, based on observation and/or suppositions of past rates	Generally, present rates are typical of past rates	Revolutions have generally intruded with exponential rates
6.	Salt deposits	Same as I-1 plus I-4	Gradual evaporation of stranded salt waters	Dumping of hot waters or sudden burial

7.	Metal deposits	Same as I.1	Old magmatic extrusions from below	Catastrophic column fallout or cosmic dumping
8.	Caves, stalagmites	Present rate of growth, retrocalculated	Water filtration through limestone, etc., long-term	Thermal and electrical hi-energy digging and carving
9.	Metamorphic Rock	Past heat and pressure rates; terminated + I.1-2	Gradual heating and noncastrophic forming	Quasi-explosive and/or electrical heating
10.	Magnetism of Rocks	Magnetometer registers last melt (a) Magnetic N orientation and (b) intensity diminishing with age	Infrequent reversals now believed in with great effects	Successive reversals amidst moving surfaces shortens time
11.	Fluid pressures (gas, oil, water)	Pressure, quantity rate, exit time = retrocalculation.	Entombed fluids finally break through	Continued pressure means very late entrapment
12.	Conglomeration	Same as I.1-4-5	Waters and air aggregate different material	Great forces from long distances agglomerate material

	Test (mostly of absolute time)	Idea of the Test	Evolutionary Explanation	Quantavolutionary Explanation
II. Biological Indicators				
1.	Birth and Extinction of species	Elapsed time for species to develop, ramify, and become extinct	Occurs slowly by II.5-6-7 and individually with slow ecological change	Occurs by radiation and catastrophe. " Missing links" are everywhere
2.	Fossil (includes bones, pollen, etc) superposition	See I.3	Orderly succession of species from primitive to modern	Orderly short-time species succession; sometimes devolution, cross-zoning and confused identification of assemblages
3.	Fossil zoning	Ecological life niches occupied and abandoned, changed and overlaid	Ladder of long-time succession of life niches	Disastrous cross-currents transport and superimpose different fossil beds. Vs I-3
4.	Fossil succession	Superposed beds carry younger, more developed species	Evolution promotes young, developed fossils to higher rocks	Many exceptions, assumptions, anomalies, mixtures
5.	Adaptation time	Time elapsed for accommodating to new surroundings	New species take long breeding for their various life niches	Prompt adaptation occurs with heavy mutation; many vacated niches

6.	Selection time	Time estimates for proliferation of species vs. competition	Requires long time	Rapid diffusion common
7.	Mutation time	Time between successful mutations plus proliferation for species development	With enough time, enough successful mutations for evolutionary branching	Radiation and ecological disasters bunch changes; young species abruptly succeeded. " Missing links" rare
8.	Tree-rings (dendrochronology)	Seasons mark solar years on logs back 5,000 years if matched	Uniformity of recent past evident, earth motions constant	Major gaps and variations in rings; possible substitute cause of rings
9.	Fossil appearance	Youth and age of outcrop or exposed fossils and bones expertly guessed	Experts cannot tell age by visual examination	Experts can see that deposits believed old are really young
10.	Coral reefs	Measure growth; relate to years by present rate	Coral columns of 10^5 years found. Also many fossil corals before	Coral rate varies with minerals, heat. Can grow in shallows of rising waters
11.	Character of assemblages	Variety, numbers, ambiance, and cause of death	Views assemblages as accidents or repeated incidents	Views for cause, dimensions of catastrophe, extinction
12.	Fossil fuel formations (coal, oil)	Time required to form coal and oil	Organic sediments, long-term, special conditions	Bull-dozing, deep burial of heated biosphere and some cosmic oil fallout

	Test (mostly of absolute time)	Idea of the Test	Evolutionary Explanation	Quantavolutionary Explanation
III.Chemo-physical Measures				
1.	Radiocarbon (^{14}C)	Decay of ^{14}C to ^{14}N in dead organism proportionate to time of death to 50,000 years	Accepted and basic, with occasional minor deviations	Atmospheric and decay inconstancy invalidates pre-2500 B.P.
2.	^{40}Potassium – ^{40}Argon ratio	Decay of ^{40}K to ^{40}A in igneous rock proportionate to last melt, $10^5 – 10^9$ years	Can approximate rock age back to pre-Cambrian	Migration, new infusions of ^{40}A invalidate; anomalies
3.	Uranium^{238} – Lead^{206}	Decay of ^{238}U to ^{206}Pb in minerals (non-sedimentary) measured from original crystallization of rock	Can approximate rock age back to pre-Cambrian	Migration, and decay inconstancy of U-Lead chain invalidates
4.	Thorium^{232} – Lead^{208}	Decay of ^{232}Th to ^{208}Pb in same	Same as III.-2	Same as III.-3
5.	Rubidium^{87} – Strontium^{87}	Decay of ^{87}Rb to ^{87}Sr in minerals with ^{87}Sr measured from original rock	Same as III.-2	Migration of Rb invalidates completely
6.	Thermo-Luminescence	Loss and luminescence of rocks and ceramics since last heating beyond 580°C	A promising method not yet applied and reliable	Promising, little applied, still in realm of nuclear uncertainties

7.	Uranium fission tracks dating	Etchings of charged particle collision with micas are counted and related to decay rate	A promising supplement to III.-3	Subject to radionic disasters and cosmic-ray erasures
8.	Oceanic uranium abundance dating	Elapsed time for uranium to reach present level given known input and output	Unmeasurable or incomprehensible	Short time (10^4) to reach present levels of accumulation
9.	Radiocarbon non-equilibrium method of dating atmosphere	Rate used for dating implies decline of ^{14}C to zero in 13,000 years	Anomalous or unverifiable	Atmosphere transformed implying catastrophes
10.	Temperature gradients of planet	Loss of heat is function of known conditions, known event, hence tells elapsed time	Inner dynamics of bodies unknown; radio-activity; earth fractures are hot plate boundaries	Declining heat is function of elapsed time and recent catastrophes; heat is escaping at fracture boundaries
11.	Soil chemical analysis	Micro-constituents of sediments (soil, ash, clay, etc.) show events	Normal fires, air, water deposits produce extreme constituents	Volume, mix, context permit revolutionary period indexing and portrayal

	Test (mostly of absolute time)	Idea of the Test	Evolutionary Explanation	Quantavolutionary Explanation
IV. Astronomical Motions				
1.	Planetary and Satellite Orbits	Regular measureable orbits permit retrocalculations of ancient events	Accepted; universal laws retroactive to solar system genesis	Orbits are empirical; changeable by high energy displacement
2.	Rotation of bodies	Regular rotation permits reliable retrocalculations	Accepted; calculable to original emplacement of body; known regular and minute changes	Origin of rotation from older motions; all motions changeable radically, *cf.* Venus
3.	Electrical fields (charging)	Periods of heavy Earth discharges exist and correlate to human behavior	Speculations; Earth is lowcharged and always has been	A developing analysis of several periods by evidence of cosmic discharges
4.	Solar emission inconstancy	Solar winds and radiance rates can be retro-correlated with major events	Minor duration have occurred but generally constant behavior	Major and minor; patterned electrical behavior occurred
5.	Resonances	Planets Venus and Mercury achieve periodical positions indicating some special events	The events are assignable but of unknown timing	The resonances corresponded to legendary large-body encounters

6.	Librations; oscillations of satellites	Gravitation explains and permits retro-calculation of Moon' s increasing distance and Earth' s slower rotation	Accepted; all irregularities explainable by gravitation and motion laws	Fossil motions detectable; Moon recently emplaced
7.	Gravitational constancy	All retrocalculated practically are measured with laws of gravitation	Laws are absolute and must be consistent with speculation	The constancy is experienced, not absolute; may be in an " unknown" multi-force field
8.	Age of universe	Retrocalculated in billions of years by " big bang" starting assumption	Speed of light basic; no short-time theory conceivable	Irrelevant to most problems and highly speculative and adaptible
9.	Meteorite ages	Correspond through chemophysical measures with age of Earth	Meteorites are solar system and resemble primal Earth material	Meteorites recently exploded and aquire false ages like III
10.	Solar cycles	Periodicities of sun-spots reflected in natural periods on Earth	Periodicities vary but probably not much	Periodicities are reasonable if experienced, but radical departures probable over thousands of years
11.	Star-movement	" Fixed" stars give perfect intervals for retrotiming	Where possible events can be tied to fixed stars	Star movement are anciently implicated in planetary movements

	Test (mostly of absolute time)	Idea of the Test	Evolutionary Explanation	Quantavolutionary Explanation
12.	Planet and cometary movement	A position today gives a reliable position through the ages	No force disturbs much the orbits of planets	Planets and comets have behaved changeably in recent times
13.	Cf. IV with I, III, V	Earth--other planet comparisons and correlations of events and causes	Each planet has its own individual unaffected existence	All planets bear the marks of their experiences with other planets
14.	Succession and behaviors of divinities	Planet-associated divinities are function of time when planets active	Mythological materials are superstitions, gibberish; unproven correlations	Mythology permits increasingly exact and reliable correlations; V.1-2
V. Cultural Measures				
1.	Mythological time	Sublimated and fantastic sequences	Pure fantasy, primitive, and/or indecipherable	Simplified hidden but true messages, sometimes decipherable
2.	Legendary time	Collectively affirmed fixed events and sequences	Same but more intelligible	Same but sometimes clear and accurate
3.	Anniversaries	Asserted or evident temporal fixation of intensive event of type. IV.-14, V.1-2-10	Random celebrations or based upon eternal recurrences	Celebrations of single or physically combined natural events

4.	Calendars	Pre-historical and historical systems of time reckoning	Often confused but OK if in line with present time	Rarely confused; reflect different times; decipherable
5.	Mechanical clocks	Devices; sundials, shadow sticks, and process flows calibrated to time	Unintelligible if not in keeping with present time	Correct " fossils" of different time-periods
6.	Archaeological location and succession	Cultural data arranged by I.3, development stages, and causes of hiatuses	Usually human-caused or normal present forces	Often radical changes and hiatuses catastrophically caused
7.	Timed records.	Reports of experiences, successive events, or astrology	Often inept, primitive, conflicting with true time	Usually correct but mistakenly edited or for extinct time-period
8.	Memorial generations.	Oral long-time transmission plus V.-1, 2, 3 one M.G. = 50 years	Unreliable; to many breaks an much fantasy	Highly disciplined, sometimes valid and reliable
9.	Cycles of Ages.	Theories of successive ages of chaos and creation	Mythical and fantastic; superstitious	Plots main lines of natural-cultural history; fragmentary
10.	Cross-cultural co-experience	Synchronization by common experiences of culture items of catastrophism	Independent invention or diffusion	Traumatic independent invention and coercive diffusion

In the fifth category, evolutionists have wrongly, yet persistently, defied a multitude of ancient voices even when these voices are in consensus on events and time sequences, They have blighted the growth of the science of mythology. Moreover, they have not considered catastrophes in the explanation of discontinuities of excavations, whether strata were disrupted or erased entirely. As Claude Schaeffer declared in his monumental survey of Near East excavations, "Our inquiry has often been rendered difficult by the rarity in most reports of observations on beds of destruction.... Some reporters have regarded these beds as a nuisance or of little interest."

It should be clear, therefore, that the hints given in the table can be expanded into major criticisms of each category of tests. In addition, several general criticisms may be directed at off categories. One may object to the frequent unwarranted claim that the skies, air, waters, rocks and biosphere have changed always at the same rate and under the same conditions as we see them charging today. Inconstancy afflicts most gauges of time. The more that the quantavolutionary hypothesis is insisted upon, the more that the past processes seem to deviate from present ones geological, biological, chemo- physical, astronomical or cultural. The planet Jupiter, for example, has become more and more of an ogre since Velikovsky predicted its radio noises in 1950, and a scientific dragnet is now out to trap all indications of its stormy past, present, and future behavior.

A second reproof is that evolutionists have committed often the same scientific misdemeanors that they accuse the quantavolutionists of. Possessed of two results, each based on a common or different debatable assumption, they claim that the results, since they agree, are certainly true. They have concealed anomalies, allowed the contamination of samples, exaggerated the certainty of their observations, generalized from insufficient data, pleaded their premises as proof, selected the evidence, used special cases as proof, and been thoughtless when it comes to larger theories.

Moreover, observations are often uncertain and unreliable in the tests of time. Significantly, progress in instrumentation many have the effect of disclosing hitherto unobserved phenomena that tend to nullify the aim of the measurement. For example, C14 dating aimed at using a constancy to establish dates, but it has helped greatly in discovering inconstancies. The brilliant and thorough attempts to perfect radiocarbon dating have already given some needed proof of the Martian and Venusian catastrophes" and may paradoxically end up as a most valuable source of information on the ravaging of the atmosphere before Solarian times.

Also, the search for pure samples to test for dates has sometimes shed more light on other problems than upon time. Analysis of Thera (Aegean Sea) explosions ash samples has led to studies distinguishing earlier eruptions of Ischia (Italy) and casts doubt upon various cultural modes of dating for the Eastern Mediterranean. Hydrocarbons from widespread fires have lately been discovered in "normal" land and off- shore cores drilled in the eastern United States.

Frequently a lack of data hampers conclusions about time. Whole realms of nature are missing from the annals of times past. Catastrophic events not only compress time but also destroy the evidence of time. Floods, tides, and hurricanes can erase levels of the biosphere completely; it is permissible to argue that all centers of civilization of the Saturnian age to be described later were completely eliminated, that all "neolithic" discoveries are of survivors, especially of peripheries of cultures - just as the Hebrews, Sumerian, East Indian, and other legends declare. Then, too, the subsequent Bronze Ages chronology for the ancient Near East has lately been shown to be awry, largely because catastrophic premises provoked a re-examination of the domestic and international problems of the dynasties of Egypt.

Finally, the evolutionary theory has had the services of practically all scientists and scholars of all disciplines for 150 years. By contrast, quantavolutionary theory has survived without media or funds and only enough scholars to make rare guerrilla forays into opposition-held country. From lack of focused case studies, the revolutionary time-tables have been excessively imaginative, including that which is to come in the next chapter.

It is fair to say that the five classes of time-tests include practically all techniques of telling prehistoric and ancient time. One should stress that tests on a given site or material or problem are often multiple, as they should be, to see whether the tests support one another. If they do, of course, the probability of validity is increased. It may seem appropriate to annual or ignore the results of one test on particular or general grounds such as contamination or even general theory; but it is hard to knock out several tests on the same grounds. Nevertheless, one should bear in mind the set of general problems confronting tests of time, the special problems inherent in each category, and the particular problems inherent in each testing technique as indicated in the chart. In the case of several areas - sedimentation, potassium-argon tests, radiocarbon dating, tree-ring dating, paleomagnetism and the fossil record - my comments have been sufficiently extended to show that the debate is generally complex and ramified in respect to all types of time-testing techniques.

I have by no means exhausted the range of criticism. For instance, I credit thermoluminesce dating, involving its decay since the last high heat of its matrix, with "promise." Yet the pioneers of the field are commonly frustrated: " There is a gross discordance between the TL age and the radiocarbon age.." of sedimentary samples baked by lava from the Massif Central's Chaîne des Puys (France), the one giving 26,000 years, the other about 11,000 years.

Still I can sympathize with the person who, after all is said and done, consults the conventional time-tables and reasons as follows:

"Thousands of scientists of many fields have worked with one or more of some fifty tests. Even if nobody is an expert in more than a couple of test areas, the scientists all lean on each other. And all agree on the long- range thrust of the many tests. Their agreement should add up to a certainty for either long-range evolutionism or long-range revolutionism. Short-range revolutionism must be wrong."

In reply, I can only stress what has already been said above and elsewhere in the book:

"Every test has its problems of design, administration, reading, and interpretation. Fifty problems do not make a solution. I could readily declare that ancient catastrophes are absolutely proven because not 50, but 150 or 500 cultures unanimously declare that they survived universal disaster. But more than this proof by agreement of sources is needed, in my view."

14. THE DISSOLUTION OF TIME

The idea of long-range time is the bedrock of present-day intellectualism. It is ideological. It performs a great, but fundamentally non- rational, service. By extending time to inconceivable lengths, one makes of it, in effect, a constant, which need no longer be accounted for in factoring the causes of ancient events. Nevertheless, every ideology or "ism" is at best a model, at worst a blinded mule, pacing a circle endlessly.

Of the 58 tests listed, only 1 (one) does not depend upon the empirical experiential proposition that the processes of nature have been proceeding at a constant pace with only minor lapses.

The one exception is the principle of superposition of strata (I. 3). It is a logical principle. It can only come into effect when natural and human material is laid down; it is only valid when the material is not overturned or undermined by igneous or over other intrusions.

The reluctance of "Nature" to tell her true age is perennially a frustration. In a day when even solar time is not accurate enough for some functions and tests, and when even star-time is introduced, the fact that some people must be wrong by hundreds of millions of years in telling historical time cannot help but make one wonder if the minority, at least, is not mad, or whether the whole of science is a sham. Neither is the case.

Knowing how wobbly and weak a grip the human mind has upon time it should come as no surprise that "Nature's" time is disconcerted and disparate. Only by the greatest exertions and mutual discipline and only at the highest peak of group organization are we able to hold a tenuous grip upon a schedule of time; even then, the individual's psychological as well as active deviations from the severely imposed bonds of time are very many and dominant, if one were to be brave enough to count the undisciplined vagaries of time in relation to the ordered ones.

If this temporally disordered mind has difficulty in ordering time in relation to the ordered ones' time in its immediate contexts of group cooperation, it is not to be expected that its farthest expeditions into space, species, and events could establish a nice clockwork. Historians like to tell a story: God, according to Isaac Newton, had set the machinery of the world to move like a clock, but had to intervene upon occasion to make adjustments in its regularity, (an idea that reminds us forcibly of Plato's God at the tiller of the world ship). Whereupon Leibnitz was prompted to remark that Newton had not only made of God a clock-maker, but a poor one at that.

15. OF MAMMOTHS AND AMBER

If, as the preceding pages imply, there may be a general failure and collapse of long-term methods of time-reckoning, a need for a radically alternative chronology arises. But where lies the possibility of such?

Quantavolutionism brings to bear on the problem the abilities of great forces to compress astronomical, geological and biological time. By adding human testimony to anomalous current scientific findings, enriching these with new evidence, especially of an electrical nature, and integrating them within a new hypothetical structure, it can propose a new chronology of the holocene period.

There is little chance that a single technical device. a test, will calibrate the ages. A holistic method must prevail. A thing to be dated must be evaluated by every technique available, in as broad a context as possible; and, even while it is being tested, it is testing the test. For example, Carbon-14 presents us with dates between 30,000 B. P and 21,000 B. P. on three different frozen mammoths; then, for the carbon-14 dates to be acceptable, Siberia must have remained frozen for the duration of the period, else two of the carcasses would have rotted. But then the mammoths would have suffered three catastrophic time-points of sudden death and sudden preservation, by asphyxiation and deep-freezing. A peculiar repetitive kind of disaster would have to characterize this long period of time. If we believe that the species was exterminated at once, then the carbon-14 method cannot be at all valid here. We must still await a definitive study of this long-discussed puzzle. Its solution is important; the utility of carbondating hangs upon it.

Another case involves the fossilized resinous exudation of dead pine forests, amber. The Greeks cherished it for its beauty and its electrical properties; its name was "electron." At Pylos, a Mycenaean city, whose buildings collapsed under intense heat, large stores of amber were found. The substance was in ancient times transported by well-known routes sacred to Apollo from the coastal towns of the Baltic Sea. There it was being washed ashore from vast sunken pine forests. Recent radiocarbon dating of pollen conflicts with conventional belief, according to which the Baltic basin was filled 70 million years ago, and places the flooding of the Baltic Sea in the middle of the second millennium, a catastrophic period described in the chapters on Venus. Presumably, only after several hundred years was the amber fossilized, exuded, washed ashore, evaluated, and incorporated into international trade.

Isaacson has independently established the burning of Pylos in the period of cosmic perturbation involving the newly great god Mars, that is, the eighth and seventh centuries. Fossils themselves tend to be proof of local or general disaster. The abandonment of a precious store of amber also indicates natural disaster, not an aftermath of a battle or accident or ordinary earthquake. Might it not be that no one was left to dig up the treasure?

It would appear that all evidence can be put into a mutually supportive context that is much broader and convincing than a set of dates contributed by single technique. Reasoning from the sacred, the commercial, the behavioral elements, one has grounds for disputing the geological theory that assigns millions of years of age to the Baltic inundation; how could amber have been so abundant that it was still washing ashore in quantities sufficient to support a thriving business? The origin of the mysterious amber was carried in Greek myth: the Heliades, sisters of Phaeton, who drove uncontrollably the solar chariot and was sent crashing to Earth by a bolt of Zeus, wept amber tears in grief for their brother.

16. Schaeffer and Velikovsky

Still another type of reasoning can be shown in relation to Schaeffer's demonstration of widespread concurrent site destructions in the second millennium B. C. Schaeffer follows conventional Egyptian chronology and dates the periods of destruction by the association of Egyptian artifacts with the site level artifacts under scrutiny, whether at the site or elsewhere. That is, the Egyptian chronology was regarded as absolute, just as the radiocarbon dates were once so regarded, and still are given significant shifts and weights.

The revision of Egyptian chronology by Velikovsky, now being completed, shifts whole centuries forward and about, and shifts the whole Greek- Near East chronology with it. For the moment, confusion reigns, and there is bitter resistance. But soon it will become clear that innumerable historical and archaeological problems will be solved simply by switching to the new chronology. Thus, all that Schaeffer "automatically" consigns to the end of the Middle Bronze Age, at around

1750 B. C. I assign to the same time, but dated at about 1450 B. C. The many destructions that he consigns to 1200- 1300 B. C., I assign to 800- 700 B. C., granting special consideration to exceptional cases.

The results are remarkable. Suddenly, the vast "hiatus" between "13th century" destruction and 6th century proto-classical times becomes only a brief hiatus. It is clear that the vast movements of "the peoples of the seas" were a fiction employed by scholars to explain the widespread natural disasters of the 8th and 7th centuries, the Mars disasters of our calendar.

It is tempting to conclude this discussion of current problems of chronology with remarks made lately about Lord Kelvin's three methods of arriving at the age of the Earth in the 19th century. "All three methods employed unproved assumptions and very shaky estimates; nevertheless, they conveniently agreed on the age of the earth." Geologists promptly adjusted their figures to his lead and although "it was not a case of 'fudging', it still took a lot of lively imagination for all those different scientists using different dubious methods to come up with the same erroneous result."

Since Kelvin's day, chronometricians have overlept one another in their eagerness to add time. Even most catastrophists have been catapulted into the race. Long-term catastrophists heap scorn upon short-time catastrophists in order to keep in the running. They may be warned, however, that long- term catastrophism is thus asking for more and more time to do nothing. L. J. Salop has discussed the effects of a 1% increase in the solar constant which causes an increase of ultraviolet, hard radiation by 100,000 times. There would occur one of the many vast destructions that mark the history of the biosphere. A natural catastrophe may not require as rare a combination of events as is believed even by non-uniformitarians. Hence, the greater the success of the long term catastrophists in proving historical disasters, the more implausible is it that these disasters were separated by hundreds of millions of years of time. The catastrophist theory will itself demand a compression of geological and biological time.

Should the moment arrive when the far-flung outposts of time represented by radiochronometry have to be pulled back, they will probably not be able to pause at chronological defenses of the old geology; all the troops of tests would retreat to the confines of short-time chronometry.

With this, I think enough has been said in this chapter of the tests of time to obtain permission to try in this book and its successors a radical calendar that largely disregards radio chronometry; that treats carbondating as exponentially erroneous as it moves backward in time; that subjects geological stratigraphy to catastrophic premises; and that regard human legendary reports to be correct and reliable in the large. Since all long-term measures of time have become suspect, we can proceed by using only as much time as we need for the accomplishment of the studied events. Whereupon 14,000 years delimits our temporal structure

Part Five

THE WAY OF HOMO SCHIZO

Chapter One

THE NORMALLY INSANE

My thesis here comes close to a remark once made by Mark Twain: "The human race consists of those who are dangerously insane and those who aren't." Humans, that is, are naturally somewhat crazy, by all definitions of that term among practicing psychologists.

If everybody, at some time, acts a bit crazy, it is not because they are departing from their normal human state but because they are reaching for their normally insane nature.

Of course, then, the term "insane" should have to be dropped. "Insane" is a deviation from a standard, that of "sanity." If the standard is "insane," then the deviations must be something else— sanity? It is uncomfortable to say so, but, yes, in a way, although and until a better term should be found. For the insane of society are no more fixed and pure representatives of the core of human nature than the sane. All of humanity, sane or insane, normal or abnormal, typical or untypical, forms globally around the core of human nature that we can best describe with the word "schizoid."

Human nature is a set of qualities to be found only among people. Of course we must keep a wary eye on the animal kingdom and its curators, the ethologists, who persist in finding identities between animals and men where once only large differences were thought to exist. We must avoid saying what is human nature, only to find that it is animal nature as well.

But, granted that a quality may be proven distinctive, who is to say that it is important? We must say, partly begging the question, that what is important in human nature is whatever has the greatest effect in producing those human traits and activities that we regard as most important. This leads directly to the human mind; the nub of human nature is in the mind. In the "minds"— because, whatever the propensities of the individual mind, the human species does not exist except in transacting minds.

We have then before us homo sapiens. We declare that we shall make of him more specifically *homo sapiens schizotypus,* homo schizo for short. We would strip from our tunics the noble title of homo sapiens sapiens, which is often now accorded us, reserving it for a species of some future event and time. So drastic an action may not be taken, however, without due process of law, and our book is intended as a hearing on the allegation that homo sapiens sapiens not the "wise wise" man and cannot by nature be so.

What is the nature of homo sapiens that he should be relegated to the status of schizotypicality? According to Pascal, "Men are so necessarily mad, that not to be mad would amount to another form of madness." Mainly the nature of the human is that he is either normally insane or insanely normal, or both. If either or both, he is not the man we thought he was. Whereupon we should analyze his nature more critically than has been the custom, and learn what makes him behave so, and what we can expect of him.

In searching for the roots of human nature, I have to use a number of concepts that are modern, psychiatric, and originally invented for the diagnosis of disease, beginning with the very word schizophrenia. However I also use the terms of old science, like human nature and instinct, and the jargon of the computer age, and of electricity and politics. For the assault upon the problem of the human constitution and its origins levy verbal troops from everywhere. If the assault is successful, there will be time enough to provide these with the linguistic uniform that new science invariably prescribes.

A first step, then, is to show how homo sapiens is insanely normal, implying paradoxically that the concept of normality is quite confused in practice, and ends in contradictions. It cannot then be a helpful idea, if its meaning collapses from one moment to the next. It is like a tall pole without a base, which can stand upright only so long as you steady it; nor does it matter where you stand to steady it.

"Normal" is an interesting word of the 18th century Enlightenment; it is a late arrival to our language. Its function was probably being served before by words such as "good" and "healthy." It comes from geometry hence science, directly from the Latin *norma,* the right angle rule used in drawing. From "rule," came "mould," "unit," a point of comparison. Then a century later "normal" came to be a state of a living being or an organ which is not affected by any pathological modification, as in a "normal human oral temperature." Later it acquired the senses of "devoid of exceptional character," "conforming to the most frequent type" (typical), "occurring according to habit," a "normal person." So we see an object needed by the new applied physical sciences which is expanded abstractly to include a model or verbal rule; already it is forcing its way as an objective concept into the moral sphere; then we watch it attach itself to an undiseased state of a living being (which concerns us here) and ultimately to the statistical type or the ordinary (which we also consider).

The problem when the world deals with human nature becomes apparent: the non- pathological state, and the ordinary or typical, are mingled in the idea of a normal human being. The normal person should be of the normally healthy majority. The trouble is that there is neither a normal standard state, nor a normally healthy majority. On those matters closest to the important code of human nature, we cannot decide on what should be termed "non-pathological." And on even those traits which are conventionally deemed healthy, we cannot find a great concentration of individuals to cluster.

The normal engages the range of the abnormal, even some of its extremes, and the abnormal is a set of improvisations on the normal. We do not deal with a rational person of healthy mind and then someone who is broken down into insanity as with a bad fall off a bicycle. Rather there is the human being whose essential functions are the same, homo schizo, who is always behaving "madly," but as one of his defence mechanisms divides people into the sane and insane according to largely societal canons.

The prudent approach in these circumstances is to locate a foundation for normality. By all tokens, it should be in the idea of sanity. The human species has to be composed of normal sane people; else it is a contradiction in terms. But suppose that we find mostly insane people; then something is wrong with the definition of insanity, or of normality. So we go in search of the normal great majority of sane human beings.

Painstaking investigations of cultures, from the deep forest primitives of the Philippines to the penthouse dwellers of Manhattan, bring to light only cultural forms that are readily analogical, even homological, with all other cultures. They can all become the life-style of whoever happens to become engaged in them from infancy. This transferability, universality, and relativity of culture, is highly important to the definition of normalcy. It enables one to say that whatever may seem to be abnormal behavior in one culture will be found to have a normal place in some other culture. Properly directed hallucinating is a gift, or a symptom of insanity, according to cultural norm; in a doubting and liberal culture, as for instance the United States today, one may discover even psychologists who cultivate hallucinating, whether for religious reasons, adventure, or self-experimentation. Practically every symptom of nervous disease disappears into the tolerant maw of culture.

People are not at all sure of behavior occurring within their own cultures, whether in Singapore or Chicago. "You can't imagine the things that go on!" a tendentious paranoid third of the population will tell you. And they are correct, even if their attitude can lead to some undesirable social distrust that pulls at the weak fabric of social consensus. There are many kinds of abnormal "things that go on," conspiracy, for example, as when gangsters, politicians, businessmen and any other group for that matter plot actions better kept undisclosed, for tactical or moral or legal reasons. There are undisclosed criminal offenses, to which the American people confess in great numbers to priests, psychiatrists, and interviewing strangers. It appears that nearly everyone has committed at least a couple of crimes that he knows about and can recite.

Of the conspiracies and crimes, a great many are moral in nature. Child-abuse, spouse-abuse: these are examples. Millions of such cases occur; are they normal behavior? Are they crime or illness? Or both? Not foreseeing how uninhibited literature would become, a French writer, Jules Barbey d'Aurevilly, addressing himself to famous writers, in his short stories about diabolic women of a century ago, said: "Ask them how much incest is to be found in families, whether of the poorest or highest class, and see if literature, which is accused of being so immoral, has ever dared to tell of them, even to frighten the reader," complaining that one had to go to the ancient examples of Myrrha, Agrippina and Oedipus, although, all around, there were cases to be found.

But then there is the category of abnormalities called madness, or mental illness, which the legions of science strive to segregate from acts of conspiracy, immorality, and crime. Possibly they are moved by an instinct of territoriality— for without a field of study there cannot be a fat herd of scientists. More largely, they are searching for their identity in their object, as a shepherd in his flock, a priest in his parishioners.

We cannot speak individually to a whole people, asking them whether they belong to the psychiatrist's flock. But we have learned to sample a population, assuring that those sampled truly stand for the whole, and we can interview these. Perhaps here the search for the normal person can end. If there is anything that is uniquely human and normal to mankind, it will have emerged from the great factory of the mind to find its way into the communication of ideas, thoughts, and feelings. All other normality can be consigned to our generic kinship with apes, fish, and bacteria.

The result of such surveys of mental health gives scant comfort to expectations of normality. In one study of a sample of householders of a rural Canadian county, as many as 69% were deemed to be psychiatric cases. Most others had troublesome mental problems. Only 17 % were classified as "well." To be "well" then is to be statistically abnormal.

In a middle-class white section of New York City, a different sample survey of mental health was conducted. Here 1660 adult residents were interviewed at length by trained workers and the materials adjudged by psychiatrists. Some 18.5% were deemed mentally "well. " A third had mild symptoms, and the rest, about half, suffered moderate or severe symptoms of mental illness. Again the "well" or "normal" are statistically abnormal.

Other studies can be cited. A recent World Health Organization report gives a figure of 40 millions for the gravely sick of mind in the world; another 200 millions are too ill to function well. One in every six Englishwomen is receiving care for mental disorder, one in every nine men. Deliberate self-therapy must treble these figures. Then, too, people go in and out of treatment, self-administered or not. A third of the American population, where hallucinations are neurotic or psychotic, nevertheless hallucinate from time to time. From one-third to one-half of normal persons aged 12 to 35 years report episodic symptoms of dissociation or depersonalization. A third, not necessarily the same as those, suffer neurotic anxiety or worse. Practically everyone engages in psychosomatic illness from time to time. There appears to be little doubt: the normal is abnormal and the abnormal is normal, statistically, intra-culturally and cross-culturally.

When one adds up the diagnosed ill, the ambulant ill, the suffering normals, the individually destructive, the sexually different, the genetically abnormal, the culturally and criminally diverse and perverse, the infantile and the senile, and proceeds to some type of summation the remainder, if they have a strong sense of normality, can be classified as megalomaniacs.

Feeling bad is the norm, in one or more of a thousand ways. Mental suffering must be on an immense scale throughout the world. The "normal" human being is not the healthy animal he is supposed to be. Throughout history, anxiety has been recognized as an inherent part of man's being. Discussion of the origins of anxiety has become explicit in the 20th century and is a frequent theme in today's literature. The definition of anxiety is as varied as the experience itself, and its biological basis is obscure. While anxiety may be thought of as an unpleasant state, characterized by uneasiness and apprehension, it is also a strong motivating force in many forms of behavior and, like fear, has fundamental adaptive and perhaps evolutionary significance.

Perhaps the very idea of "normality" is a sickness. Psychopaths and neurotics often hate abnormality or atypicality in others, as Hitler hated gypsies and a meticulous drill sergeant may dislike a tall recruit who stands out from the line. One must consider whether the idea of the normal human is not some unrecognized myth, functioning to hold an individuated lot of persons in a tighter society. If so, we should dredge up the myth, for it may be blocking our understanding of human nature.

THE IDEAL PERSON

An obvious relative of such a concealed myth would be the idea of the "noble savage." Something akin to this idea is to be found going back thousands of years.

Upon the Age of Discoveries, from the 16th to the 18th centuries, and despite much evidence to the contrary, Europeans conjured a vision of happy primitive peoples living in a benign state of nature, even foregoing governments, taxes, war, and civil strife, to which the Europeans were subjected. Jean-Jacques Rousseau played a greater role than any other thinker in building up the myth - although it must be said that Rousseau himself insisted that the "noble savage" was for him a mere tool for philosophical reflection, that "he had never existed and would never exist." Rousseau himself felt a compulsion to reveal his own shocking "evil" and errors; his deservedly famous *Confessions* frankly recited his many "abnormal," "neurotic" behaviors. He suffered from paranoia, strong ambivalences, incontinence, phobias, hallucinations, occasional impotence. He enjoyed being spanked on the rump. He believed that his portrait by Van Loo, judged highly by art historians ever since, was part of a conspiracy to make him look like a monster.

Rousseau yet claimed that the human being was born with natural reason and good motives. In his educational writings, he argued that to confine or restrain the pupil was to pervert him. Rusticism, the belief in simple rural existence and its virtues, has been a chronic "neurosis" since times immemorial. One encounters the idea among people of all classes even, or especially, in America today, while only three per cent of the population are farmers; many politicians play upon the rustic theme.

Sigmund Freud, of all people, may be perceived, in his essay on *Civilization and Its Discontents,* to assign mental malignancies to the burden of discipline and complexity attributable to civilized life. He, at least, explains how impossible it is to take up the rustic life again. But he does not doubt that there was and can be a rustic life. And there is little doubt he regards the human being as potentially happier if ever he would return to the "normal" of his mentality. Given his many different writings, this can only be regarded as a contradiction and a minor nuisance, still it is capable of distracting him in his search for the origins and condition of human nature.

Psychologists, disobeying the first principle of science— factuality— have been loath to lay their cards on the table. "The mentally ill person, helped to discover the origins of his illness, will use the knowledge to cure himself, to become normal." So goes an ordinary principle by which many psychologists operate. Finding oneself, coming to terms with oneself, and similar slogans amount to the premise that there is a normal human self that is within us all, a core that, when struck, will resonate natural goodness.

Attempts to elaborate scientifically the syndrome of normality scarcely produce an integrated core of rationality, goodness, or creativity. They elicit conventional syndromes. They tend to bring out diffuse characteristics that are tolerable. It seems almost as if, in striking parallelism with the myths of the noble savage and the rustic, they are turning back towards a docile and agreeable hominid. Johnson remarks that "tests show that so-called normal individuals have little imagination, limited interests and social activities, limited aspirations, and no ambition."

A schoolmaster or legislator might define the normal person as one who imitates well the norm, who obeys the authorities, who eats moderately, does not take drugs, sleeps well, fulfills sexual desires within suggested limits, accepts responsibilities when charged with them, has only appropriate fears and associates these with the real source of danger, is hygienic, loves one's family, is careful in dealing with strangers, feels gratitude, and believes in gods.

Not only are such persons unusual, but they do not constitute an integral psychological type wherein contradictions are absent. They, too, must be the results of high test-scoring on separate items of inquiry. More than this, however, is a fact which will be looming up as crucial to this book, that "normal" qualities, such as moderation, hygiene, responsibility, belief in gods and others as well, are qualities that are sculpted out of a basic natural "insanity."

Rarely, a scientific writer, (Harold Lasswell is one of them) will so much as frankly acknowledge that he is interested in advancing a certain kind of person in society, an "ideal man and woman," an "ideal citizen," in effect, whose etiological dynamics psychiatrists (and statesmen) should explore, understand, and propagate. The "normal" is waived in favor of the "ideal." Indeed, one suspects that it is precisely in order to control the true normal population that the ideal norm is set up. He is what he is, not because he is a natural man, but because he might be artificially created to go against nature!

Lasswell speaks of the values for which humans strive. These are a type of instincts; they are generalized appetitive urges that crop out in many ways. The values are power, respect, rectitude, safety, wealth, well-being, enlightenment, and affection. The objective of public policy should be to develop the sharing of these among the people of the world.

The means to the end is the creation of democratic characters who are willing and ready to share. "Failure to develop democratic character is a function of interpersonal relations in which low estimates of the self are permitted to develop," says Lasswell.

Reminiscent of Alfred Adler's "inferiority complex," a prevalent low estimate of oneself leads people to wish to deprive others, thence producing misunderstanding and aggression, among a host of other neuroses and psychoses.

The democratic man, then, has an intact ego, open to thought and impression; is possessed of many values and disposed to share them with all. He is free from disabling anxieties and increasingly in command of the flow of energies from his unconscious self. Lasswell grants the difficulty of creating a dominating psychic type: "The task is nothing less then the drastic and continuing reconstruction of our own civilization, and most of the cultures of which we have any knowledge."

Lasswell frees himself from the rustic fallacy: a new kind of man is to be created, whose life is to be supported by especially designed institutions, a utopia, to be sure, but one unencumbered by dreams of normalcy and myths of a golden age.

Our arguments here against the prevalence of "normal people" are not intended merely to "broaden our minds" regarding normality; many writers have done this job well. Nor are we aiming to set up an ideal type, which has an elite that can create artificial normalities. We assert rather that the very logic, the very substructure, the very physiology of the concept of normality sought for as a base for judging abnormality is not present.

CHAPTER TWO

THE SEARCH FOR LOST INSTINCT

Most babies cry when they are born. If they do not, they are liable to receive their first spanking. This, say their attendants, will clear their lungs; it will circulate their blood; it will exercise their reflexes. It is good for them. This crying may be uniquely human. Calves, for example, do not cry when they are born. They lay stunned for a moment, and then gradually pick themselves up, pull themselves together, receive some licks, and crawl or stumble around. The mothers of apes generally drop their young almost disdainfully and hardly attend to them at first; the orangutan mother is more considerate, more human.

Perhaps babies cry because they are already more frightened than animals. A famous psychoanalyst, Otto Rank, found the mental state of the baby deplorable, and traced the major behaviors of later life to the trauma of birth. Some conscientious mothers even followed his line of reasoning to the point of giving birth by caesarean operation, thus assisting the baby's birth and relieving its pain of passage from womb to open air. I do not know that the effects of their altruism have ever been reported.

Dr. Rank, passing over the possibly similar plight of calves, babes of monkeys, and puppies, ascribed the most marvellous effects to the human birthing experience. He sees "in the birth trauma the ultimate biological basis of the psychical." Even the myths of the creation of the world, that are told in many cultures, are regarded by him as a sublime attempt to undo the birth trauma and to deny the separation of the infant from the mother. Although everyone has undergone and many have later witnessed the radical experience of parturition, Rank interpreted myths and fantasies of the end of the world as wishes and efforts of the human individual to be reabsorbed into the great All and Oneness.

Further, "the Flood which initiates a new world period is nothing but a 'universal' reaction to the birth trauma, as the myths of the origin of the earth or the sea also show." Religion, creativity, humanness — all are to be attributed to the tragedy of parturition. No one is exempted. I think that he is reversing the order of nature.

Attacking the perinatal problem by another method, Stanislas Grof employed psychedelic drugs, particularly LSD, in seeking to learn of people's re-experiences of the earliest events of their lives. He could identify four matrices of recollections. One expressed feelings of unbounded ease; another, frightful threat, confinement and torture; a third appeared as a struggle for survival and an ecstatic release; the fourth, a separation that seemed a kind of death followed by resurrection. These would connect with the physical sequence of natal events and incorporate analogous later effects. The unbearable situation of the foetus, sensing uterine contractions while confronting a still-closed cervix, would and did engender visions of fear, plague, and natural catastrophe.

Lacking evidence that historical experiences can affect the germ plasma, we must regard these as visions of other life experiences, as secondary or derivative suggestions, and attached to the perinatal process by mental association. We do not deny that they originally occurred. Rather, when they did occur and were experienced and remembered, they reinforced the analogous perinatal feelings. And, if what we have said concerning Rank's theory is correct, the perinatal experience is a reinforcement of the pre-existing genetic fear of oneself that already begins with the foetus. Hence, a double reinforcement may operate upon the original fear. Every species, indeed every individual, has its own pre-existing structure for experiencing; experience is species-specific and organism-specific.

Otto Rank and Stanislas Grof are asking too much of the nasty surprise of birth. Rather, I should say, the baby is crying because, unlike the animals, he already possesses a kind of fear that they do not, and cannot, know. This is his existential fear, an anxiety sensed upon the realization of the existence of himself. He is reacting to a harsh accident, true, but knows already this existential fear and is demanding immediately that he be relieved of it. He already has a frustrated Voluntarism, a will that he expects to appease by action. He is terribly frightened because he is already trying to put his head together and to find himself in himself, whereas the animals have merely to compose their circulation and limbs.

The infant already wants more than to fix upon comfort, although this, and food later on, will usually quiet him. Surely, given the option, the baby would prefer to return to the womb. But there, too, he may well have been frightened, not only by jostling and growth pains, but by the sense of the greatest problem of existence, how to form his identity. For his brain has begun to operate in the peculiar human way.

1. INSTINCT-DELAY

The baby cannot realize his problem. He does not know that his instinctive mechanisms are blunted, blocked and delayed, and that he will pass his life in a mammalian vehicle, a jalopy that he must tinker with and fix up at every turn of the road. Luckily, he is to be trained immediately as a mechanic. Not for him are the joys of a long life of instinctive behavior, consisting mainly of speedy, replicating responses to specific stimuli.

The delays of instinct, touching upon the whole gamut of behavior, provide the human with a continuous fear. The instincts cause tension in their persistent efforts to complete themselves. At the same time, the person finds himself inundated by the tongues of instinct, lapping upon areas of behavior often little related to the original direction or object of the stimulus-response mechanism. The sexual instinct, for instance may emerge as Oedipal, romantic, homosexual, fetishistic, sadistic, aesthetic and/or aggressive behavior; rarely is it merely the highly relevant "display; hop on; hop off; go away" sequence that even higher animals perform.

The delay of instincts by a possibly genetic blockage is all-important: it permits the chaotic creational flood to rush into all the crevices of the forming human nature. The retardation of his animal instincts is viewed as flexibility in the human's behavior. Human flexibility is both cause and consequence, therefore, because both the "decisions" and the follow-up activity are subject to delays in the central nervous system.

The animal world, by contrast, exists by instinct: events emit stimuli; instinctive reactions succeed or fail; the residuum reassembles amidst the continuing events. Almost never is the process punctuated by pauses to consider a dilemma. If its spasms of calculation fail, the animal surrenders to the inertial process. It falls back upon its collective line of defence, it breeds. The animate world can depend upon exponential reproducibility to render individual choice unnecessary for species survival.

Meanwhile the human creature depends upon what he calls mind or intelligence. He has not chosen to do so. He cannot do anything else. Would that he could, for his lot is fearful. His reactions blocked at every turn, if only for an instant, he lives in anxiety over the last turn, the present turn, the next turn, and all the ones he can remember or imagine or foresee. Stripped of ready instinct, he confronts an instinct-ready world. He is innately, individually, and culturally anxious, and the world he encounters is too large to cope with non-anxiously.

Terror drives him to act quickly, but not instinctively, and instinct is the quickest action. Why must he forever fearfully reflect? If, unlike animals, man has to make up his mind, there must be some unique quality in the mind. If two or more options rush into the open question raised by the blocked instinctual response, there is a conflict, an anxiety. Even if there is

but a single and obvious solution, a pause to determine so can cause anxiety.

2. Self-Fear and Self-Control

Because our selves share nearly identical anatomical housing and have highly privileged access to each other, for all practical purposes they are considered as "one in body and soul." In fact, the compulsion to be oneself is so suspiciously strong that no matter what the proof to the contrary, the self will always be the irreducible unit of human existence. Like non-Euclidean geometry and warped space, the poly-self will remain a theoretical construct, out of the realm of the common sense. It will help to understand human behavior, however, and have powerful applications in psychological therapy and law. It will, among other things, clarify an idea that is to be found in a many "primitive" and "advanced" cultures, that a person "possessed" is not to be treated as himself. So now we have the human creature, living in a house of fear, governed by a committee of selves, and acting accordingly in non-instinctual ways.

What is the agenda of this committee of egos? The agenda seems infinitely varied; it can contain anything in the whole world, internal, external, from a microbe to the stars, unlimited, too, in time or space. But the preamble to every item on the agenda is always the same: "This is a bill to control fear by..." So the human seeks control, first of himself, that is, of the committee: "This house must put itself in order." Then he must seek to control others; he must at the same time, just like a government, govern himself while he governs others. And, besides the others, he must seek to control the world, for he senses that not only these others, but also the whole world, threatens him and needs to be ordered. Something of this conduct of operations seems implied in Anna Freud's idea about the ego's "tendency to synthesis," as opposed to the fear of ego destruction, which she recognizes as an instinctual anxiety. (I say, of course, that the ego was hardly there in the first place.)

Control implies power, the determination of the wills or behaviors of people and events. "Power is an ingredient in the transactions which take place within all object relationships and is thus an ingredient in the interlocking forces which determine personality," so declares Arieti.

Control is said to occur when someone determines the behavior of persons or things. The drive to control is a reciprocal of fear. All humans, possessing existential fear and self-fears, must seek control, primarily of the self. Very quickly in all situations the physical self becomes the arena of only a portion of the struggle for control. External objects and beings are also incorporated into the struggle. The self, others, and the natural world are the triple object of efforts at control.

The total of objects, both inner and outer, operate subjectively without discrimination. Whether a person feels he controls his temper or an empire, by physiological indicators of stress, or by any other tests, he is calmer and more relaxed. If he is insatiable in his wishes to control, the diminution of tension with success will be brief and shallow; his fears will cause him to move immediately to assume control in other spheres. Under no circumstances can the urge to control be satisfied.

The absolute renunciation of control-needs and efforts is sometimes attempted. The cosmic indifference of Buddha lets nothing matter save finding nothingness. Once filled with nothingness, one will be at peace. All identifications and attachments are renounced in order to concentrate upon control of the self. With an elaborate strict regimen of diet and exercises, and by abstemious human relations, bolstered by a sophisticated rationale appeasing conventional philosophical demands, a mental balance is achieved that is distinguishable from selfishness, catatonism, or bestiality. Resemblances to autistic trances are present. The discipline remains severe; the drive for control of the world is not abolished.

A more than casual resemblance to the Buddhist outlook is to be perceived in Teilhard de Chardin's attempt to extricate mankind from its dilemma. He sees the remedy as man's reflectiveness, enlarged greatly until the world is co-reflected in his mind in a universe of ultra-reflexion. This is a vague formula, but it takes on greater meaning when we ask what is Chardin's human dilemma. This, it seems, is a state of fear, which he describes abstractly: the great fear of the human species is to be closed in and lost in an unfriendly world, condemned to live with himself as a fixed species.

3. The sense of "I am"

Identity and identification begin with the question of the self or ego. "Everyone is to himself that which he calls self," wrote John Locke, in discussing the idea of a person. The self is "an object to itself," said G. H. Mead. The reflexive form reveals "that which can be both subject and object." This is what distinguishes man from animal, he argued, rather than the alleged possession of a mysteriously endowed soul.

Building one's self is then every person's lifelong occupation. As we have said, he is driven by the fear of not being oneself to begin with. The self is a predisposition, but not a bequest, of nature. Indeed, it is never fully achieved. Man is always an infant in this regard. While apes grow quickly and soon act "self-possessedly", the human can grow in every respect but this, that he never achieves a single self. "The mature person is self-confident," by which is signified that his existential fear is under control and that he egoistically regards himself as "one." A total lack of self-confidence results in a kind of vegetative existence, a sickness as grave as any; even the most elementary kinds of self-control disappear into incontinence and catatonism.

Both "identification" and "role-playing" are in the area of the dispersed self. A "role" is behavior according to a social subtype, which is employed to escape insecurity by virtue of a more secure status. A role may be manifested as casually as a costume for the mardi-gras once a year, or as intensely as a permanent switch in identity accompanied by amnesia. Role changes are common in modern society; they are rare in simple communities, where a fisherman is son of a fisherman, but even there the person goes through life-roles such as adolescence or grandparentage, has a role in a church, and so on. Roles are culturally defined, often assigned, and when fully developed and effective, encapsulate the dispersed self, guarding and maintaining it against dissolution.

Identification can be attested in the assumption of a role, as the boy who identifies with his father, the fisherman, but can more broadly extend to all manner of being and abstraction. Thus one may detach some part of himself and affix it to an identification with the working-class movement, or with the Virgin Mary, or with his family and neighbors, or with a bird. Identification is associated with the wish to control its object; this may be difficult, for frequently ambivalence arises out of an obviously uncontrollable identification.

The self, though it may appear so, is not a social creation, as G. H. Mead and others would have it be. Man would never have a self, a poly- ego, if he were not structured genetically to engage in the search for self by a mind that has to be pulled together. Mead's work is completely intelligible and useful, except on this crucial point. It is significant that he does not seek to go beyond society and culture as the determinants. Meanwhile, he provides us with precisely those kinds of observations which we need, as, for example: "The phenomenon of dissociation of personality is caused by a breaking up of the complete, unitary self into the component selves of which it is composed, and which respectively correspond to different aspects of the social process in which the person is involved."

He advanced and stressed the concept of "social roles," those social housings for the individual selves, and showed how small children could play games with their selves, as well as others, being now one kind of person, now then another— father, mother, evil one, good one, police and bandit, and so on. My own grandchild was raised bilingually in Athens, and when playing with a toy car and policeman, would speak as the policeman in Greek, then reply as the car-driver in English.

The source of the phenomena of self-awareness is the dispersed selves. One would not know oneself unless there were at least two of one, the observer and the observed, the knower and the known, or, better, two mutually perceptive observers. In Hilgard's experiment above, he was able to elicit two speaking selves with contrasting points of view regarding a painful stimulus.

Hilgard's studies of divided consciousness by means of hypnosis expose a "hidden observer" or "co-conscious" as an ordinary concomitant of existence. This self among selves is not a monster, a "beast of the unconscious." "The concealed part sometimes turns out to be healthier than the openly presented self." Expert though he was in hypnosis, Sigmund Freud fashioned his theory of id-ego-superego from classical social psychological theory, from Plato's Republic (I argued in a paper of 1949) rather than from experiential materials readily available to him. He thus may have posed the wrong parties in psychic conflicts. The poly-ego concept is structurally and biologically manifest; it can be the subject of experiment; it can be operationally described.

The origins of the poly-ego, the core of human nature, must be in neurological transformations at some time in the past, as we have strived to show previously. Sufficient for the moment is the hypothesis that when this transformation occurred, a number of critical innovations occurred with it, enough so that we can assume a quantavolution of creation, a Hologenesis. Human nature came all at once. As the first humans experienced for the first time a poly-ego and have until now repeated the experience with every new person, we look for a massive effect upon the human being and find it in the eternal fear that possesses mankind.

4. EXISTENTIAL FEAR

Students of fear in humans and animals are rarely satisfied by obvious causes; human fear is not a pie to be cut up and assigned to wild animals, bad dreams, strict parents, and the like. It is well to make tallies, thus a third of the population fears snakes, most of these fear them intensely; a great many fear heights or being alone in public places; many fear injury and illness; and nearly everyone fears an assault at the time that it occurs. But, perhaps because they are difficult to study and even to conceive of, "little systematic research has been applied to the nature of what are sometimes called existential fears."

As with the concepts of human nature and instinct, many psychologists would like to rid themselves of the concept of "fear," believing it to be vague and operationally undefinable. But, as Jeffrey Gray puts it, "Experimental psychology— as well as common sense— has been forced to invent the hypothesis of a complex psychological state, 'fear,' precisely in order to make sense out of the otherwise shifting and imprecise relationship observed between stimuli and responses."

We do not distinguish here between fear and anxiety. "Anxiety, the psychological equivalent of pain, is characterized by a feeling of dread, a vague fear.. not related to specific situations or objects.. part of the human condition." So says Mendel, abstracting from a lifetime of administering intensive psychotherapy. Physiologically, insofar as anxiety can be detected, it exhibits the chemistry and muscular tensions of fear. And fear, when slight, is indistinguishable from anxiety. And anxiety can become terror and panic. The common use of the term "anxiety" has to be attributed to the need to allay people's fear that they may be suffering from fear.

Fear is part of the human and of all that he creates. The role of fear in religion is large, so that a working out of fears often has taken place in the arena of the sacred. Religion approached by faith, says Rudolf Otto, cannot be the same as religion approached through reason. Central to faith is numen, the specific non-rational religious apprehension and its object, at all its levels, from the primitive stirrings to exalted spiritualism. And central to numen is dread, for it is the sacred, holy, awful confrontation of man with god or the divine essence.

Fear can be both immediate and existential. Immediate fear erupts upon the encountering of threat to the poly-ego system, a learned and/or sensed emotion that sends an ad hoc electrochemical alarm through the central nervous system. Existential fear, also an electrochemical effect, is normally at a constant level which we posit to be above some pre-human level.

What evidence is there for a continuous higher level of existential fear in human nature? That man is an anxious animal has been a byword in psychology. This means ordinarily that the human is never at ease with himself. Rare cases of such are a subject of marveling comment, probably misplaced and incorrect.

It is often argued that humans are culturally indoctrinated in fear, and therefore generally exhibit that continuous anxiety which has every conceivable object as its trigger or focus. Cultures are discoverable that train their children not to possess or display fear. Many mothers of modern western culture earnestly try to preserve their children from the sense of fear. The mothers are reinforced by cultural institutions that have special needs. These, where successful, invariably train only an ignorance of or resistance to fear in some respects deemed crucial by the society, such as facing up to an enemy in battle.

Here, in the first place, there is reason to regard the training as retraining, that is, the acquisition of one set of habits to overwhelm a contrary set. The partial training underscores the practically limitless outlets to existential fear; global courage is not hoped for. The brave Spartans were obsessively fearful of their Helotic slaves; fearful of alliances; fearful of women; fearful of their gods; and would turn tail for home even from a battle if an earthquake occurred. The display of fear is culturally determined; the fear itself is universal.

Comparing the occasions for fear it is doubtful that the human lot is beset by more fearful stimuli than engage the attention of animals. Yet we see in man a variety of psycho-pathological tendencies and behaviors — such as merciless aggression and global attentiveness — not present in mammals and apes. Hence we must seek the source of existential fear in a logical and real condition, which we say is the poly-ego. Selfawareness, inevitable in mankind, produces continual anxiety over his inevitably and profusely invented fears.

In 1933, Freud laid down the theme "...that the ego is the only seat of anxiety, and that only the ego can produce anxiety," and "that the three main varieties of anxiety — objective anxiety, neurotic anxiety and moral anxiety— can so easily be related to the three directions in which the ego is dependent, on the external world, on the id and on the super-ego."

This reads, in our terms: on nature, on others, and on the variegated selves. The self is too complex to be divided into id, ego, and superego. There is a pragmatic instinctive principle involved, but there is no reality principle. The self is never a real self, either.

In systematizing psychology, Freud might better have dispensed with external objectivity and relied upon a phenomenological theory of the world as a wholly subjective creation of the mind. The clutch of components of the ego engage themselves in anxiety-reduction operations. The so-called id, ego and superego elements are ancient and misleading ideas of how the mind works, even though they are conventionally handy for political, moral, and hence therapeutic disputation. Certainly, though, Freudian psychology is erected upon the presumption of ever-present anxiety.

Not until we learn how this continuous drizzle of fear and anxiety is precipitated in human life by the delayed instinct and the split self will we understand existential fear. For the moment, we should counsel alertness against assigning to any experience the accountability for generalized fear. This means to avoid any commitment to sweeping theories such as that of Rank's birth trauma, or to presuppositions like Otto's that dread is validated by its divine associations. Or such large categorical explanations as "castration fear," which is undoubtedly of diagnostic utility. Or, for that matter, to any summing up to 100% of fear by adding experiences from the womb to the tomb. Rather, hypothetically at first, and then as certainly as the evidence and logic permit, let us maintain that the human would be fearful and anxious even if he lived a life totally free of frightening experience.

The whole human mental structure appears to be given over to controlling the mind so as to reduce the stress of fear. The poly-self is elected as a governing committee by a central nervous system that was previously under more centralized management. The brain of the hominid loses coordinative ability and in so doing produces the human brain, which imposes a new system of coordination.

The poly-ego, hence self-awareness, would not be present if it were not for the depression and confusion of instincts in humans. What forced the human egos to emerge was the necessity for continuous decision-making and what made this in turn necessary was the delaying of instinctive response.

5. Instinct in Man and Animal

The term "instinct," like "human nature," and "ego," has a suspicious slackness about it. No wonder, given its history. Charles Darwin used it not quite as loosely as he did the idea of "natural selection," S. Freud used it as a workhorse for one speculative probe after another; MacDougall, the social psychologist, pounded it into mincemeat; Tinbergen managed to use it respectably in his study of animal behavior; and Fletcher recently reconciled its ethological and psychiatric meanings usefully.

N. Tinbergen defined instinct as a hierarchically organized nervous mechanism, susceptible to primary releasing and directing impulses of internal and external origin, which responds to these impulses by coordinated movements. The hierarchy is altered by changes in the intensity or by suppression of other instincts. An influential hierarchical order by Rensch gives as instincts sex, deference, feeding, cleaning, and ultimately hunting and collecting. I doubt, however, that there is any hierarchy of instincts in humans except in a group statistical sense, owing to the human ineptitude for specific instinctive response.

Obviously, all writers have had in mind the large fact that animals and men respond automatically when stimulated in certain ways: they blink quickly when about to be struck in the eye, for instance. (Even so, madmen and small boys can teach themselves to control the blink.) This unrestrained reflex is instinctive, as are a great many chemical and motile reactions of the organs and limbs.

As instincts come to require training (the baby can be toilet-trained) or as the stimulus of the instinct provokes a broader response (when struck, the creature dodges, snarls, and strikes back), they enter an area of science that can ultimately merge with speculative philosophy, as when one speaks of an aesthetic instinct in man. Freud's last thrust in the arena of instinct emerged with a death (thanatos) and a life (eros) instinct. This dualism reminds us of entropy and negative entropy, the universal breaking down of motion and material and the countervailing creativeness of life, which, if given optional conditions of sustained full reproduction would soon cover all the stars and the spaces between with organic matter, and then presumably expand the universe beyond even the dreams of the explosive universe theorists. This thought might be taken as an irrelevant comment on the irrelevancy of Freud's two-fold classification. But neither is the case.

In this very book on the pleasure principle, Freud came as close as he ever did to the theory of homo schizo. In the course of denying the domination of pleasure over human mentation, which relates to the anhedonia symptoms adverted to later on, he moves to the question of unpleasure. "Unpleasure corresponds to an increase in the quantity of excitation" that is present but unbound in the mind; pleasure is a diminution of excitation. He thus agrees with G. T. Fletcher (1873) who linked pleasure and unpleasure with stability and instability, in between which lay indifference. And he foreshadowed the behavioral conditioning school of today.

Now Freud, typically pushing ideas to their limits of tolerance (and toleration), makes death a pleasure and then an instinct. "Instinct is an urge inherent in organic life to restore an earlier state of things which the living entity has been obliged to abandon under the pressure of external disturbing forces." Instinct: reversion: death.

Freud's "death instinct," so readily misunderstood, can be shown to make sense in the light of the theory of homo schizo. For we say that man seeks to revert to the animal in order to recapture the instinctive bliss of the single self. That is, man unconsciously seeks his death as a human, and of the human species. This must be very close to what was gestating in the mind of Freud.

I see confirmation of this thought in a cloudy but weighty remark that relates to the dependent clause of the quoted sentence. For he writes: "In the last resort, what has left its mark on the development of organisms must be the history of the earth we live in and of its relation to the sun. Elementary things do not wish to change but are forced to evolve organically by external disturbing and diverting influences." My work on Quantavolution deals heavily with such "influences." What is pertinent here is that, by the theory of homo schizo, the evolved thing, man, wants to rid itself of the burden of the very trait that speciates it, that makes it a unique species, and such is the instinct-delay that creates and maintains its perpetual angst.

Freud's early preoccupation with the sexual instinct is less pertinent, disclosing, as critics have pointed out, an ideological attachment to the worries of well-to-do patients in a bourgeois society before World War l. The varieties of sexuality, it rather seems to us, given the cultural accent upon the subject, indicate a dispersed instinct, a conflict of selves, and an employment of sexual displacements to dispose of existential fear. Love consists of identifying an ego element with people and objects (even a 'security blanket') which reassure one against fear. Love is usually deeply involved with control, and control of course is a heavy motive in sexual attraction.

Affection plays so large a part in nurturing and training an infant that it becomes naturally a well-developed area of fixation for many problems of other instinctive zones besides the sexual. One can understand how affection is attached to all manner of "irrelevant" encounters and objects. It can be plucked out and credited with being the basic drive. But we always

should refer to the human basic drive as self-control, then to other essential interests such as sex and food, and finally to myriad mixed displays of all of these.

Without enthusiasm and with qualms, a definition of instinct may be put forward: instinctive behavior in a species is present when, in the absence of training, a uniform behavior reliably results following upon a definite stimulus. The number of instincts in mammal species subsumable under this definition must be in the hundreds. An important fact is that for every primate instinctive action, there is a human equivalent, ontologically recognizable. This fact is relatively easy to argue. However, the near reverse may be also true, as ethologists and socio-biologists increasingly contend: for every type of human action, defined with increasing specificity, there may be a genetically related primate instinct, with allowances made for training in both cases.

The discussion of human instinct centers about the comparative laxness of instinct in the total behavior of man when compared with the behavior of animals most closely resembling him. Compare the separation of the mother bear and her cub, so simple, with the separation of the human female from her child, so complex, so full of woes, the inspiration of thousands of customs and volumes of literature. And include especially the "exceptional" societies such as those in which the mother is trained like the bear mother, who lumbers away leaving her cub whimpering on the limb of a tree, or the societies employing all- male initiation ceremonies to break the maternal grip, or fascist and soviet societies whose nursery schools are intended to abort family influences deemed incompatible with the ideas of the regime, or societies where the tie is broken by taking up one's first job in a distant city. Humans can come close to, or seemingly go very far from, animal practices.

When giving birth, women in comparison with primate females are more agitated and uncertain, and follow practices not observable among the primates, such as engaging attendants. Again, primate females have a defined rut period when they will accept sexual advances, whereas human females frequently are receptive of sexual overtures most of the time. Kinsey found that a mild rut period is present in slightly over half of a human female population.

The complications in the life of humans introduced by just these two departures from the instinctive norms of the primates are numerous. On the one hand there are the "unhappy" components in the difference of instincts: confusion, doubt, malaise, anxiety, ignorance, ineptness. On other hand there occur some "happy" elements: flexibility in relating to other environmental demands, such as planning hunting absences; reasonable timing; more frequent opportunities to breed; and the possibility of introducing healthy practices; not to mention luckier males. Here are two behaviors, in primates and humans; they "could be" alike. But some mechanism generalizes and renders indistinct the human behavior.

6. POLY-EGO VERSUS INSTINCT

The basic product of the instinct delay is the poly-self. Assuming that several centers of the brain can become seats of an "ego," the delay of instinctive response will cause these centers to develop and exercise influence. The instinct delay produces milliseconds of "hesitation" and "doubt." This is enough for the several centers to sense a problem, that is, nonfulfillment of the instinctive loop of stimulus-response-extinction of impulse, and to react. The general consciousness is supplemented by a superior and dominating special brain center and several inferior but rival ones.

The dominant consciousness now perceives its rivals and the "problem." It casts a pall of fear over the central nervous system, including itself. The problem of non-immediate fulfillment of the instinct impulse is complicated by the sense of competitive decision-making or instruction-giving centers associated with it. Hence the external fear of ourselves is established. Physiologically a low-level of Cannon's fear-flight effect, involving the adrenals, is produced.

Now we have in operation: instinct delay, poly-ego, and existential fear. The person behaves accordingly. He seeks control of the laggard instincts and their wayward derivatives. He seeks to organize his poly-ego into an effective and more comfortable relationship. He tries to abolish his fear, which, after all, is nothing but a continuous play of the fear sensations of animal life and which, mistakenly, he treats as nothing more than an interminable chain of immediate fears.

Man can and would like to fill infinity with his control activities. "... The overriding purpose of the behavior is an attempt to achieve some security and certainty for the person who feels threatened and insecure in an uncertain world. The possibility of controlling oneself and the forces outside oneself by assuming omniscience and omnipotence can give one a false illusion of certainty. Therefore the main ingredient is one of control." So writes L. Salzman on *The Obsessive Personality*.

Control by evolutionary reversion is impossible. Man is unable to reestablish the instinctual basis of existence. He cannot speed up his responses and eradicate their derivatives, except that his attempts at doing so produce the astonishing phenomenon of culture. Nor can he put aside his multiple centers in favor of "one king, one throne, one people." Nor finally can he do away with his fear. The failure to complete automatically his instinctual urges, the dissipation of these urges into bizarre forms, and the conflicts of his "split brain" guarantee a level of fear that would approach panic if it were not channeled into new worlds of activity and location.

7. "You Can't Go Home Again"

Seen from one perspective, the human behavior is homologous to the animals. Seen from another, the human behavior is only analogous. Whether one opts for the former or the latter view is simply a question of whether to regard as important the phases that intervene between stimulus and response: instinct delay— loss of control— existential fear— diffused or diverted reaction. Since these phases create the human, we prefer the view that human behavior as a whole is only analogous to animal behavior, but that man and primate share homologous infrastructure and functions. The human adds a special neural loop to the stimulus- response cycle.

Most ethologists seem presently to agree that the differences between man and other species, while apparently wide, are still differences in degree rather than in kind. This view may be correct on a phenotypical level; as stated, analogous behaviors can be extracted from man and beast. But this is not the crux of the matter. If the human utilizes a mechanism that duplicates animal behaviors but this mechanism is implicated at the same time in other functions which do not accompany the animal behavior, even though in some cases the animal uses other mechanisms for the implicated functions, we have an important structural difference.

The individual is constitutionally unable to reinstinctivize himself. It is commonly imagined that humans can revert to the beast. Not if it is homo sapiens schizotypus whom we are discussing. Mental disease (i. e., schizophrenia) cannot cause such a reversion and does not in fact do so. A person is not genetically capable of becoming (by mental illness or otherwise) a healthy (or even unhealthy) mammal. Such a person can commit every imaginable peculiar or abnormal act, but it will be a human act. One should not be misled by the multitude of instances in which sometimes, in cults, ceremonies and mental illness, a person will play the role of a pig, bear, horse, etc., or for that matter, identify with clouds, angels, chairs, and rocks.

This inherent incapacity to re-mammalize is one of the most persuasive proofs that a genetic mutation occurred in the final transition from hominid to human. For, even if human behavior had changed from the hominid to a new fixed behavior owing to a permanent change in environment, it would be possible to retreat from mental illness in the direction of mammalianism. That is, a way would be found to psychosomatize and build up a new chemico-electrical combination to supply a new type of person.

De-instinctivization is accompanied by another important development in the human, namely, individuation. A social group is forced to tolerate deviations, if only because its totalitarian intentions must founder upon the rocks of its inabilities. The individual must fail, as well, in his desperate attempts to control himself, that is, his alter ego. Much less can he attune himself perfectly to the modal group behaviors. Under such circumstances, the ideal of individualism evolves and prospers in the very presence of the ideal of group conformity. The person becomes ultimately aware that he is different from others, the gaps between the various demi-instincts and the required definite response in actions and habits become filled with his unique character. Each human can be different— and comes to think of himself as different— because he has a unique set of habits or activities to fill the gap between semi-instinctual response and definite practices as the norm.

We conclude that homo schizo produces more behavioral effects than any species, as many as a great many species put together. When operating with a perceived challenge and under scientific rules, he can imitate or reproduce almost every animal behavior. He can reproduce his logical apparatus by computers. He can out-whistle birds, and decoy ducks. He can cut his birth fertility to nil, or stimulate multiple births by chemicals. He is the most flexible animal, the most individually varied, a virtuoso, a polymath, and so on. Prod his brain electrically and an endless flow of free associations and a stream of consciousness is verbalized, though it is neither free nor conscious. And what is in his mind is potentially in his behavior. Considering that his physiology is almost identical with certain primates and that the apparatus used for being human has been hitherto practically indistinguishable from them, we much seek the origins of his uniquely broad and sophisticated outputs in a freedom from instinctive binding. The search is just begun. We go on with it now, in the operations of the brain.

CHAPTER THREE

BRAINWORK

The human skull is an impressive work of natural architecture, especially when bald. It seem to be an apt crown for a creature of will and decision. But look inside this formidable bone casing and see what a disappointing glob of ooze is the master of thought. Undifferentiated, it has a pasty white stuff over a grey matter, spangled with fibrils and fibers, the whole suffused with pink from countless blood vessels, and, if it could be dug into, it would reveal firmer membranes, and a chunky stem that connects to the long spinal chord.

The differentiation of large areas is so vague that some surgical operations are couched in how many grams (of the 2000 or so) are removed or how many millimeters of depth one may safely penetrate.

Unlike other organs in the body, the brain does not usually reject tissue transplants. This feature not only suggests that the chemistry of the brain is generalized but also why specific functions can locate here and there and relocate, too.

Brain operations are delicate partly because we do not know what to be delicate about. Still, successful brain operations must be as old as the oldest settlements of mankind, because a number of ancient skulls exhibit trephinations, penetrations by bores and saws, and a subsequent healing and continued life. What came out is unknown or perhaps the aim was to relieve pressure; there is, after all, little reason to believe that the atmospheric pressure on earth, or the electrical pressure for that matter, have remained conveniently the same.

Anyone who has confused brains with "sweetbreads" at a butcher shop will agree that "the brain has many characteristics of a gland." And R. Bergland and R. Page go on to say: "it contains hormones, it is bathed in hormones, it has hormone receptors, hormones may serve as its synaptic neurotransmitters, and hormones modify the brain's main function, behavior. .. Endorphin and other hormones may be produced in small quantities locally within the brain but transported in larger quantities from the pituitary to the brain on demand." We see one reason for the brain's innocuous appearance now: its innumerable cells are like a massive inelegant hotel, in and out of which hordes of tourists flit. It houses a great many transactions.

This grand hotel is well lit. It is completely electrified. Every neuron is an electrical capacitator. The hormones could not move so readily otherwise. For the brain lacks muscle to lever the hormones about, and it is insulting slang to call a man a "musclebrain." The blood capillaries of the brain are very numerous and carry around the food of the cells and remove their excrement. They also transport hormones to their work sites. If the mechanical pumping system is out of order, even for a few minutes, and the hormones do not get to work, many brain cells asphyxiate and the lights go out forever. The capillaries may burst, too, from time to time; the larger the rupture, the greater the damage, but the location of the stroke is more important; vital faculties may be impaired. Although the brain can switch many functions around its inchoate mass, there is a limit to its versatility.

Its liability to asphyxiations and strokes does not mean that the brain is overworked. A persuasive case can be made for the belief that humans have far more brain matter, especially of the highly touted "grey matter," than is needed for normal functions. J. Lorber found a "socially completely normal" young man with a large cranium, a 126 I. Q., and a first class honors degree in mathematics, but with cerebrospinal fluid taking the place of at least nine-tenths of the normal complement of cerebral tissue.

Highly excited and continually enraged characters are sometimes subjected to leucotomies in which, perhaps emulating unknowingly the ancestral practice, some material of the frontal lobe is cut or burned away like a malignant tumor, leaving the patient afterwards somewhat dulled but relaxed. One wonders whether there is a disproportion between the storage capacity and the practical facilities, among other problems, so that the active behavioral outlets are technologically backward. Perhaps this might account for the displeasurable unsatisfied agitation of people, who sense too much, undergo spasmodic muscular urges, and want to express an impossible number and variety of thoughts.

1. THE ANIMAL BASEMENT

Animals sense their surroundings — smells, chemicals, sights, temperatures and so on. Their sensitivity can be more varied, greater, subjectively "more ingenious" than the human's, or less. Like man, "an animal does not react to all the changes in the environment which its sense organs can receive, but only to a small part of them." (A carnivorous water beetle does not attack and devour a tadpole simply if it sees one, but it will attack any solid object, tadpole or not, if a kind of "smell of meat" arouses it.) Internal operations of the nervous system must contribute "motive" in animals.

Thus Tinbergen goes on to say: There is a mutual relationship between internal and external factors in the sense of an additive influence on the motor response. A high intensity of one factor lowers the threshold for the other factors. A high hormone level increases the responsiveness to external stimulation; if the hormone level is low, very intensive external stimulation is required to bring the total of causal factors above threshold value.

Automatic centers of the Central Nervous System maintain a continuous flow of impulses to central nervous motor mechanisms; but some kind of block prevents an uncontrolled and chaotic discharge of muscles. The discharge requires adequate stimulation by signs typical to the species, whereupon an innate releasing mechanism removes the blockage.

In the thirty years since Tinbergen wrote his book on animal instincts, ethology has run wild and is pressing upon the sacred functions of the human mind. We need to be most cautious in our defense, consequently.

The cerebrum of higher animals directs the hypothalamus to cause the pituitary gland to stimulate by an "adrenocorticotrophic" hormone the adrenal medula to exude catacholamines and the adrenal cortex to emit corticosteroids that plague the motor and nervous system from the brain to the toes for action. The hypothalamus governs the pituitary gland, controls the clocks of the brain, alerts the body to changes soon to occur, and synchronizes the performance of the endocrines throughout the body.

The hypothalamus, buried deep in the brain, is so widespread among animal species that it is unlikely to be the source of distinctive human behavior or to have changed to become so. The fact that it is responsive to cerebral signals suggests that the "human- disease", when it affects the whole system just described, originates in the cerebrum.

The human anatomy offers essentially three brains, it is true — four, if the cerebral hemispheres are accorded autonomous status. One is reptilian (the archicortex), one from the lower animals (the mesocortex), and one from the higher animals (the neocortex).

The cerebral cortex or neocortex activates intensely and overall when stimulated by pain or strong anxiety. Blood flow, metabolism, and electrical fields and discharges increase. The person becomes alert to the self and the world around him. There is no question that the pragmatic philosophers were correct in assigning to anxiety, to the sense of problems, the major role in problem solving efforts, hence intelligence. Man would like to solve his problems by automatic reflexes but he must feel pain and anxiety, both specifically and generally before he can perceive the problem, and while he works upon it and resolves it.

The cerebral cortex, for all the noble brow it presents and the most important tasks assigned to it, results from a slap-dash design. Surprisingly, when the subject is at rest and awake, in a comfortable supine position with eyes closed in a silent laboratory and neither spoken to or touched, the pattern of flow throughout the cortex is not uniform. On the contrary, the flow is always substantially higher in the front part of the cortex than in the central or rear parts. The density of blood vessels is the same hence "the remarkable difference in flow rate suggests that the overall activity level of the front part of the resting brain is about 50 percent higher than that of the rear parts."

Apparently, as expected, the aware human is spending his quiet time "getting his head together." The phrases are: "busy planning and selecting" behaviors; "focused on inner thoughts, particularly on reflections on one's own situation;" "simulation of behavior." Notable is the fact that the perennially excited frontal region of the brain seems to suffer from poor evolutionary logistics, a merely ordinary blood circulatory system. It is as if Washington, D. C., had the same postal system as Detroit, but the employees worked longer hours. The situation suggests that the evolutionary saltation or quantavolution which precipitated mankind, in order to evade helpless confusion, may have selected large frontal brain mass but that this expansion in volume did not result in all-around equalized work assignments in coping with the problems presented.

Widespread among animals and working to all intents and purposes as it does in man, is the neuro-transmission system. In humans, neural impulses are passed from one neuron or nerve cell to another; 10 billion neurons, these with their projecting fibriles— an axon to emit a message, dendrites to receive them — are half in the brain and half elsewhere in the body. No two neurons are chemically the same, because, said Polyak once, "All neurons have different shapes." And they carry differently shaped muscles along their dendrites, too, according to Crick. Between any two neurons exists a gap, a synapse, where chemical neurotransmitters wait like boats to ferry messages among the neurons. There may be a dozen types of boats; a few pick up a message merely to dump it, while the others transmit their messages dutifully.

An increased electrical potential of a neuron, relative to an adjacent neuron, makes it emit a charge which rides a chemical molecule, the neurotransmitter, across the synapse gap to hand it over to the adjacent neuron. The neurotransmitter then breaks down like the exhausted messenger in King Lear. It seems a great waste to lose these myriad molecules when they might be left to ferry many another charge between neurons; perhaps it is to keep the hormone factories humming.

This island- hopping path may be considered "slow" or "fast" depending upon what kind of speculation one is indulging in. Speeds of a mile a minute are common. Delays here may be significant in letting messages go elsewhere. Also the messages may not get through because of the insufficiency of neurotransmitters to carry them and because of sabotage by other boatsmen.

The neurons are exercised to do their part. They beat electrically in the nervous system's rhythm of perhaps ten times a second. "Neurons...

have electrical beats, and large numbers of them beat in place, as armies marching in step. When impulses come in, from the eyes particularly, the neurons begin to scintillate, to get out of step with each other, and the brain wave rhythm breaks up." So says Ralph Gerard. Too much synchronization gives tubular vision, too strict an attention to one thing. Too weak a synchronization would promote inattention and flightiness. Perhaps, thinks Gerard, the strictness would depress attention and be a cause of mental depression; perhaps slackness would elicit anxiety, mania, even epilepsy.

We say, yes, this is an old and common system, but we can see in it some possibilities of human peculiarities. Gerard gives us more food for thought: the thresholds for messages crossing the gaps fluctuate; electron movement, Brownian movement, and other factors vary at any given synapse. If this were not the case, every input would excite exactly the same output. Innovation would be as minimal as with spinal and emotional reflexes. With threshold fluctuations the same, stimulus may vary the paths of its impulses and thus favor innovation. Of course, he suggests, in this case there would be less coherence and more flights of ideas.

2. THE LOCATION OF INSTINCT-DELAY

We wonder whether the synapse may be the location of the instinct-delay that we regard as the basic glory and problem of humans. Gerard points out that the synapse can only be present because it is important not to have a message go through on an express track from one receptor to one effecter. Otherwise, why break the nerve path and cause slowing, the chance of confusion, and all the rest? The synapse permits changeability, allows the units to connect now this way, now that way... Every synapse is thus, in effect, a decision point. A message comes to it from the pre-synaptic fiber; does it go out over the post-synaptic one or doesn't it go out? That is the decision the nervous system makes — at a near infinity of places and times. Many animals have the same system, indistinguishable in detail from the human. Let us grant that here may be the source of animal decision-making, for instance the determination to hunt rather than rest, or simply to rest rather than move. With this, I see two possibilities for the "Human Difference."

We may have a pollution problem: the human system may be dumping so many neurotransmitters and neuro-inhibitors into the synaptic canals that messages cannot pass or cannot pass clean. This would occur, say, if the human endocrine glands were over-busy at a constant rate, whether from mutation or some physiological constant, environmentally induced. Suppose that dopamines, which are neurotransmitters, generally clutter the passages: the results would be a universal set of schizophrenic behaviors. Remove a large proportion of them and we revert to the hominid.

The present interest in dopamine receptors in the brain and elsewhere highlights the electro-chemical complexity of the human being. Dopamine neuroreceptors are more numerous in the brains of diagnosed schizophrenics, especially in the limbic area and the *candat nucleus.* Depressing the receptors suppresses schizophrenic symptoms. The energizers of the dopamine receptors are numerous drugs, some of which exist naturally in the body. Does a particular diet or food do so? Does an atmospheric gas do so? A particle? An ion as attached, for example, to oxygen? Does ambient stress level? Does climate? (hot, cold, damp. . . ?) Winds? Could a combination of these stress the hominid to the point of humanness? And persist permanently?

The body absorbs a continuous supply of small negative ions, negatively charged molecules. There are some 1000 to 2000 ions per cubic centimeter of air over open land, in a ratio of five positive to four negative, according to Soyka and Edmonds. Many reports declare overdoses of positive ions are unhealthy, inducing overproduction of serotonin and emotional imbalance and listlessness. A. P. Dubrov has assembled a volume of studies, many of them from the Soviet Union, on the effects that the geomagnetic field has upon the biosphere, including humans. The two effects— the ion and geomagnetic— are related, and both are implicated in brain activity. At today's levels, they condition the anxiety level of humans, and the behavior of plants and animals.

Like people, who pollute their own environments, the brain is frequently its own poisoner. For, in a coup d'état which may have occurred at the time of humanization, the "higher center" of the brain seized most of the power to requisition the drug supplies of the body and to order the manufacture of more. Even though a hemo-encephalic barrier exists to protect cerebral tissues from most of the drugs going though the body tissues, the barrier can be breached by concussions, intoxication, and "affective storms" which result from the sudden flushing of the brain with certain hormones. The storm could originate from traumatic fear— accident, rape, battle, etc. Acute states of anxiety ensue. If "continuous and successive, they might weaken the hemo-encephalic barrier, causing an increase in permeability, and, consequently, the affective ability of the individual."

The instability of the cortex is rendered more possible by its separation by other barriers from the midbrain extra-pyramidal apparatus. Cortical agitation may itself promote increased stress on itself, also, and on the central nervous system if it breaks up, as for example happens temporarily in hypnosis; its transactions with the midbrain and limbic system through the extrapyramidal apparatus are destabilized.

A baffling problem is presented in that no single substance seems to control any given behavior of the human or his brainwork. The brain has a great many endorphins and peptides, which are identical with hormones found throughout the body. What instructions do they convey? Are they coded or merely combined in their association with other substances and electrical charges? The flow of adrenaline from the adrenal medulla, a neural tissue atop the kidneys, is excited by nerve fibers which can ultimately be excited by the cerebral cortex. Electroconvulsive therapy, for example, applied to the brain, activates the adrenal medulla. Drugs can motivate cerebral activity but also distort it. Since enhanced motivation always presents the problem of its control, the distorted attitude is most likely out of control.

The pituitary gland is associated with the brain, part of it being composed of brain tissues. It emits perhaps a dozen hormones. These affect growth; they stimulate thyroid gland activity, the sexual organs, pigmentation; they influence blood pressure; and so on. These processes and many others in endocrinology are not well understood yet. Substances come from several sources; they may have specific or general inhibitors. They may affect several organs, their quantities do not have well- measured effects. When later we speak of displacements, the multifunctional overlap in behavior affecting endocrinology becomes a factor of importance; it invites confusion (including perversions) in the absence of intense directiveness toward a goal.

If one also asks only which gland or organ is the most important determinant of human nature, one would have to give the traditional answer: the brain, and particularly the cerebral cortex. All else is almost indistinguishably animal and no peculiar human operations have been noted for any function or secretions. We cannot discount the possibility of a constant change of a quantitative nature in the total endocrinal system or even in the adrenals that would place the human in a distinctive drug environment, compelling him to behave differently— to think, to talk, to make war, to have gods. However, if this has happened, it is because of a "decision," a forced and involuntary command, of the cerebral cortex. It orders its own drugs, its own blood supply, its own electrical currents and charges. It can both reduce and increase its orders: that is the vital point. It may appear, all too early, that I am coming to the idea of the World as Will, to use Hegel's expression, so I must say that I am exceedingly aware of the complex interaction occurring inside the human and between the human and his environment.

Homo sapiens schizotypus is not at all the traditional idea of cerebral homo sapiens sapiens. The human cerebrum, we hypnotically repeat to ourselves, is much larger than the primates', even if exceeded by the elephants'. If the human central nervous system, including the endocrine glands, is not proportionately increased in size, we have a situation where electrical and chemical supplies have to be generated or, if not generated, then rationed among a vastly greater number of neurons and synapses.

This, although working against the first mechanism of Human Difference— pollution and excess—would yet have the same effect, of confusion, dispersion and delay by frequent nonachievement of synaptic threshold requirements, overworking feedback signals for more supplies. Far more displacements would occur. The most "ridiculous" and "irrelevant" behaviors and thoughts would be normal. A chronic general anxiety would be present: justified fears of failure coupled with continuous interference in the completion of tasks. This begins to look like the Human Difference.

That the speed of neural activity accounts for differences among species is unquestionable. Alexander von Muralt has called "saltatory conduction" a great advance in evolution. The speed with which nerve impulses are transmitted is lowest in primitive forms and highest in mammals. It is 25 meters per second at 20 deg C in myelinated (sheathed) frog nerves to 100 meters in mammals. The velocity depends upon the conducting mechanism, the diameter of the fibre, the myelinisation of the fibre, and the ambient temperature. The cephalopod nerve must carry a far heavier bulk of fibre and consume much more oxygen to carry the same message as a frog nerve. The frog nerve relies upon tubular sheaths of high- resistant protein, myelin, to concentrate the passing electrical impulse, and upon feeding the impulse at nodal intervals between sheaths with ions and dyes to accelerate it by leaps from one sheathed interval to the next. Mammals have evidently a more efficient system than the amphibious frog.

Experimentation is in too early a stage to distinguish between man and primates with respect to their relative efficiencies in saltatory conduction. That man's conduction velocity may be less, or may put strains on the supply of charges and accelerators is conceivable; humiliating though it might be to possess a "regressive" evolution, this could promote a generally higher level of nervous tension, hence "intelligence."

Holding neural speed constant in all "higher" animals, there would still exist a speed problem with humans. We should inquire whether the human brain expanded coincidentally with humanization or "long before." If the two happened together, humanization might be the effect of slowed responses owing to greater synaptic distances. In the simplest model, two types of distances are involved in a stimulus response. Thus: the left hand transmits a feeling through the central nervous system to the right brain hemisphere, which feels "hot" but must transmit the information through the intervening fibers of the corpus callosum to the language center of the left brain, which then forms the words "it's hot !" for the voicing apparatus to exclaim. First, does the large size of the right cerebral hemisphere make any difference to the speed of the impulse of the heat signal? Second, does the distance traversed within the brain to inform the left brain mean another delay? Third, does the distance from the language center to the exclamation center and then the voice muscles add more delay? The answer in all three cases is probably "yes."

When a hemisphere is performing one of its special functions, high electrical measures register, by contrast with the opposing hemisphere. The activity is evidenced in high average evoked potentials (AEP) and in electroencephalogram beta waves. These results confirm that the experiencing which is taking place in one hemisphere is not occurring strongly in the other. Under such circumstances, there must ensue over time a great many contradictions between the left and right brains, in memory, method, and predispositions of attitudes and behavior. Every new experience therefore requires more preparatory transfers for coordination and planning.

The "ever restless human mind" thus must be more than a metaphor and more than an abnormality of some people. To behave as a whole unity, decisively, with both hemispheres, requires continuous trade-offs of impressions. If enough cannot be done while awake, dreamwork must go on apace. Here, too, is a source of obsession. Transfer dyssymmetry of several types occurs analogous to coordinating two allied armies on a battle front, one can never be sure that all are agreeable, informed, supplied, and prepared for action, and the action carries its own nasty surprises.

But, before going further with the potentialities of the bicameral brain for producing human nature, a brief statement of the situation may be in order. Many studies have appeared in the past few years. An impetus was provided by the availability of persons who had undergone a commissurectomy in which the cord of fibres constituting that giant commissure, the corpus callosum, was severed. With this, the great part of all direct connections between the two cerebral hemispheres is broken. The left side of the brain is not privy to new information or signals presented to the right brain, and vice versa.

The patient is not apparently abnormal; indeed, if the operation were performed to block epileptic seizures, he feels better, because the electric storming of the right hemisphere cannot cross the chasm of severance so as to storm the left hemisphere. Sperry wrote: "Everything we have seen so far indicates that the surgery has left each of these people with two separate minds, that is, with two separate spheres of consciousness," — together, we would add, with the general consciousness discussed above.

The gist of the studies, whether carried out upon normal brain structures or commissurectomized ones, is that the brainwork of the two hemispheres differs. Although either hemisphere can carry on all known mental operations alone, when the two sides are coordinated in the normal manner, each has its special functions and "superiorities." "Asymmetries are in general present at birth or in early childhood or even in utero.. very probably genetically determined.. not absolute, but are distributed along the spectrum.."

The left hemisphere, which, contrastingly, connects with the right side of the body, is called dominant (except that in true left-handers the right hemisphere is dominant), not so much because it specializes in the logical and analytic processes, and verbal and mathematical functions, as because it controls the right hand. The right hemisphere specializes in spatial orientation, arts and crafts, recognitions of whole images, and music and acoustics, including vowels but not consonants. Generally the left brain is more localized, the right more diffuse and prehuman.

Yet something of all of these occurs in both hemispheres. Some of it is culturally induced; fluent Japanese speakers carry their vowels on the left, westerners on the right. The right hemisphere (until the Japanese case came along) was described by some students as feminine, the seat of intuition and artistic taste, whereas the left was labeled rational and correct.

Memory is notably diffused throughout the brain, although a single hemisphere or less could store more memories than one could ever recall. A hemisphere is insensitive to its sources. It does not footnote a datum as coming from outside or from across the corpus callosum.

In fact the brain receives, recognizes and stores information and sensory bits without discrimination. They all become electrochemical transmissions whether they begin as caviar or cacophony. Once they reach the brain the transmissions generate resonances in a number of cells, sometimes widespread, sometimes localized. If they are intensive experiences, they resonate thousands of times on top of the electrical rhythms already present in the cell. They dig in especially where similar circuits already are patterned, and both reinforce, refer to, and learn from (are modified by) the preexisting patterns. Discriminations of sight, hearing, touch, smell and taste are created in the brain. A taste of nectar is a gang of electrically resonating cells with experiences of sweet things from the mouth. A mild electrical stimulation of related cells in the brain might provide an even sweeter taste.

The source of consciousness appears to be still in the brain stem. From there, even in commissurectomized subjects, some alternative— that is both left and right side — operations are controlled. Such operations "are capable of exercising a metacontrol over the higher processes of consciousness." This would be the animal consciousness, not self-awareness.

It must be stressed that specialization in the brain is not complete in any respect, no more than the division of labor in society is ever absolute (there is always a shoemaker or tailor at work despite the great factories). Brain specialization is limited to a dominant ganging or bunching of cells such that they alone respond (or do not respond) unless they are excised, in which event the minor gangs take over their functions, on a reduced level at first, then increasingly so, and sometimes with full capacitation. The union shop, when on strike, so to speak, finds its work taken on by less skilled scabs.

J. Levy argues that differentiation of functions which are lateralized is a result of competition whereby speech and language, e. g., develop in the dominant hemisphere and displace less elaborate psychic processes such as patterning images into the opposite sphere. The source of this pushy competitiveness must be humanly genetic. Anatomical and physiological differences between cerebral hemispheres develop in the human foetus. "Non-human animals have not been demonstrated to possess cerebral specialization in any manner similar to humans, that is, no double dissociations have been reported in nonhuman mammals." The genetic impetus may have originated in a mutation to the large cerebrum, with a neural "weakness for data collection and transfer" in one hemisphere, which required continuous orderly attention.

That these minor locales may be disaffected is not an extreme view; the researchers, perhaps in their exuberance, speak of contradictions. When callossally sectioned, one hemisphere can be led to think and act angrily against the other. Are we to believe that there is no tension between left and right brains in the presence of specialism, when the corpus callosum is exchanging not only sensory information — albeit sometimes traumatic— but novel commands to change itself, give up its habits? Hence sensations of hesitation, doubt, reflection, disobedience may be part of inter-hemispheric relations since a hemisphere does not know the source of a message, it cannot be declared that doubt and disobedience and fear are "external" sensations, incapable of being incited from a source within, namely the opposing hemisphere.

Nor should we ignore another reciprocating effect of specialization. In society as a whole, a tendency to specialize intensifies efforts at coordination. The same logic may apply to inter-hemispheric relations. A great many more messages will flow as the brain specializes. This may occur in a given lifetime and be more cultural than genetic. The civilized capabilities are left-brain and may be at the basis of the larger ever-present anxiety of the civilized person.

3. HANDEDNESS

Handedness, usually to the right, may itself be an important factor in precipitating humanization. It is genetically predisposed. For example, newly born infants turn four times more to the right than to the left. It seems quantitative in its intensity; it can be ranked by how much it dominates a person's relevant activity. It can be altered and reinforced by training. Injury to the dominant hemisphere can of course affect it partly or totally.

Dexterity by its very existence reinforces poly- egoism. Apart from what may be happening in the brain (though never separated from it), the full anatomical laterality, manifest in a thousand ways, makes itself felt as a division between major and minor modes, dominance and subordination, ruler and ruled. The dominant body side is even sensed as heavier, Ornstein has pointed out. When someone slaps his own forehead guiltily (usually with his dominant hand) and says "I could kick myself," it would probably be with his dominant foot.

The brain as such is an insensitive organ, so we feel no contradiction in the left brain dominating the right side of the body. Hence the right (though representing the left brain in action) is obviously authoritative in legend, custom, law, politics, work, and other practice. The right is morally right. Right-sided behavior and authority are connected. Since the right side is authoritative, the opposite of authoritative is antiauthoritarian. Often it is "leftist." We may surmise that also in the individual the left-side is anti-authoritarian. The basic polyego is of a ruler and ruled, but the ruled is frequently antiauthoritarian, a leftist. In authoritarian cultures the left- handed are said to use the "wrong" hand (e. g. Alsace, France).

The specialization of the left brain encompasses speech, grammar, figures, signs, abstract solutions, classical logic, and the dominant right hand movements. These are products, not the *Ding in sich,* the underlying drive of the brain. They must refer to a more basic concept, and I find it in the term "order." "Order" contains nuances of "Truth," authority, goal-setting, completion, instrumental and linear progression. This is all that we would expect from human nature (of course, the left brain simultaneously contains its mammalian routines of half the body). We need only turn over the final card: the opposite of order; what prompts order: confusion, delays, fear, disorder. We need not be amazed and then suspicious at the stupendous analogy with society and social thought, where right and order fight together against anti- authoritarianism and disorder.

In much mental illness and in personal and collective disaster, as Deikman and Parry have indicated, there occurs a takeover of behavior by right hemisphere religious, aesthetic, ecstatic, imagistic thinking and intuitive irrational action. The reader may be reminded of an expression from World War II: "There are no atheists in foxholes." According to A. Shimkunas, in schizophrenia the left hemisphere is overactivated and overloaded, and is accompanied by a highly arousable right hemisphere. Interhemispheric transfers are defective and cannot be processed in the commonly organized manner.

4. MEMORY AND REPETITION

Presumably, unlike animals, the human develops his memory by continual brainwork; that is, memorizing is itself an obsession, transferring and reinforcing memories is part of the overtime behavior of the human mind. The desire to forget is in competition with the fear of forgetting. Who is to judge when memorizing has become obsession, and should cease? Decisions of what to forget and what to remember are "policies" of the "highest" importance to the person and to society.

Memory, obsession, habit, and compulsion all reduce to a single basic operation in the brain: that of repetitiveness. It is for the ameliorators of undesirable symptoms and for ethical philosophers and politicians to make innumerable distinctions of practical conduct. People and cultures can be graded and scored, encouraged and deprecated, in hundreds of ways.

Within the brain, sometimes dealing with itself, at other times transacting with the outer world, a constant busyness occurs which a) experiences by internal and external sensing, b) imprints neurons electrochemically, c) distributes and redistributes charges, and d) emits commands, many to be aborted, to change some internal function or external relation.

Homo schizo's aim in life is to recover his instincts so as to reduce fear. In a roundabout way, the being seeks to control all the ultimately uncontrollable operations to reestablish the tranquility of conscience- less, instinctive behavior. Even when unsuccessful and painful, he persists. Given the options of a blow from outside or an unending succession of self- blows, what creature would choose the way of man and rest content with it? What blow can equal the premeditation of death— a thousand blows to a coward and who is a hero, except the animal, while man suffers the inevitable consequences of identification with the dead, poignant recall, projections into the future and anticipations thereof?

CHAPTER FOUR

DISPLACEMENT & OBSESSION

Personal histology and history cast their images against the screen of instinct delay. A major effect is the human displacement complex. Animals displace, but humans do so with a plenitude and magnificence that lets us be astonished at ourselves.

1. DISPLACEMENT

Every action involves an emotion, which is the sensing of action, therefore an experience. And every other experience involves an emotion. Every experience invites a response, an experience feedback of some affect. Both the experience and the feedback cross the neural synapses and are in the human manner delayed at the crossing. Wherever they may wander, while waiting for their vaguely denominated neurotransmitters, they inspire a new activity, a new experiencing, which is a displacement.

The human has all the instinctive foundations of the animal. But once unleashed, human instinctive behavior can rarely reach its target, if such can be discerned, but expresses itself all over the cerebrum in a splatter of displacements. Tough and hungry and devout, the Aztecs cannibalized their enemies. The Hebrews of Leviticus devised fulsome logic to accompany the slaughter and eating of a beast. I think that only the infinite variety of human displacements lets homo schizo congratulate himself on his large imagination, splendid lucubrations, ingenious associations, and poetic invention. It is important, all-important, "what we live for,"etc., but who says so is ourselves—judges in our own trial. The human is sufficiently depressed instinctively, and thereupon anxious enough, and has enough continuously active positive and negative feedback operating, amidst ample gray matter, to support a world of delusions, no two of them alike.

We can appreciate then how absurd it is to attempt physiological distinctions between good and bad (healthy and unhealthy) displacements and projections, just as it is to divide good from bad (healthy from unhealthy) psychosomatism. Nor can we even speak of true and false displacements and projections. Once the brain casts its affect upon the external world, that world is physiologically real. When a god is perceived as suffusing the starry heavens, or a lover glances at a suddenly all-important stranger, what happens to the brain is as real as what happens when one drinks a wine or receives a blow in the stomach. All are real experiences. Displacement might be conceived very broadly as one's sensing of anything as having effects upon one. Surely it is animal, yet the concept is the same in ethology and psychoanalysis. Thanks to displacements, a great world exists that has no "excuse" to exist.

2. *Projection and Pedagogy*

So it is with projection, which is a common feature of displacement; anything can be a subject of projection. Projection is the animation of the universe. Everything potentially is sensed to have a will with regard to oneself. The breeze sings to one, the birds call one, the volcanoes command one, one's automobile refuses to run, the enemy possesses one's thought. Displacement and projections operate in the thousands in the human mind. While the mammal tends to a few things, the human extends almost unlimited attention to the world, an attention containing affect, that is, psychic involvement. Displacement is accompanied by affect or emotion. The human, and for that matter the animal, does not pay attention to anything unless it invests the thing with emotion and anxiety. This process seems predictable, inasmuch as the reason for the displacement in the first place is to test the capacity of the displacement object to receive a neural load that is not being fully unburdened by an instinctive reaction to a stimulus.

The central nervous system is laced with interconnections of affection. Once the larger world opens to the baby, he must begin to accept those displacements that his attendants point out as the true sources of indulgences and deprivations. His teachers, while pointing to certain nearby objects with a cause-and-consequence nexus fairly obvious even to the inexperienced human are especially interested in indicating to him some very great abstractions as ultimate causes of his well-being or ill- being; they are not at all scrupulous, even if they could be, in pursuing cause-and-consequence in such cases.

For most teachers, logic has an authoritative meaning. The myriad names of gods and spirits are short-hand vulgar logic. So it happens that their obsessions with gods and laws and great natural forces are imprinted early upon the young. A consensus of obsessions is achieved, along with some disrespect for necessary causal connections between the objects of identification and the actual production of benefits and evils.

To inculcate in a child the determination to use only a special pot for his toilet needs can be, depending upon the age of the child and criteria of correct performance, a massive exercise in the transfer of obsessive behavior from adult to child. The displacement of toilet-training obsessions upon many other objects occurs readily, whether the object is an administrative routine (a "clean desk") or the traits of god, so the JudaeoChristian god is not imagined to have an alimentary canal, but many other cultures dwell upon the excretions of their gods, the very word "urine," for instance, being originally from the god "Uranus," who rained many things upon the Earth.

3. TIME AND REMEMBERING

Man practices displacement and projection in creating space and time.

The space dimension is little more than the scope of displacements, to begin with. Measure the area of displacement and one has the boundaries of space. The poly-ego can fill space; put another way, all space can be internalized so that a most remote object - a star, say - or a thought or an hallucination can vie with an insect bite for his attention.

There is a need to sense time, one more empire for the mind to conquer. Memory is of the animal, too, and so is the ability to reach back for the pattern of experiences to relate to an immediate or approaching experience. The distinction of human memory arises from its flexible control of recall. Since man's experience is rich, his memory impressions are richly patterned. It is one more great area to which he can resort for the resources of control. He can play one film against another, like a curator of a hologram museum, until he selects one or imagines a new one that will cope with an ongoing experience. Once more, fear drives him over the stubble field of instincts.

The delayed impulses that are aroused by experiencing flow out electrochemically in all directions through the mind and back and forth between the hemispheres of the brain. They excite the glandular and muscular system. The feedback of "flowback" is voluminous, too. The irrelevance of much of the activity does not embarrass the brain. It highly stimulates it. Some twenty percent of the typical person's oxygen intake is consumed by the brain.

What effect does this have upon the level of fear? It makes the universe fearful; for owing to projection, now the fear which is displaced upon others returns, relevant, and reinforced. But, at the same time, the fear is probably thus rendered more bearable, at the price of weakened identity and a great many unnecessary involvements.

Cultures take over the obsession with time that the individual cannot avoid and *pro bono publico* define the intervals of time that must be mastered. This requires certain schizoid distortions of time. "Pure time," or "absolute time," does not exist save as another delusion yet one of the greatest of all cultural drives since the beginning has been to find absolute time. Lunar time is a mass of obsessive behaviors — rites, lore, and scientific study— surrounding a fairly expectable cycle. Past bad experiences and their anticipated reoccurrences are probably the chief factors in the choice of time clocks and the ways of using them. Disputes over time-reckoning and calendars have precipitated many bitter struggles in human history.

The "rational" student protests: "See the big pay-off from marking time: planting, hunting, saving resources, warfare, rendezvous." No doubt, these are the pragmatic effects (gains) from partially restoring animal instinctive capacities. And they can only come when a culture's people succeed in frightening themselves into observances of certain obsessions. As an *aide de peur,* gods and suns and terrible memories are called upon to assist as aides-mémoires.

Time is also a way of watching oneself, hence watching the world of one's displacements. The sky- watcher is fascinated in part because he can see how "absolute time" is up there and controlling his destiny. As Immanuel Kant once said: "Two things fill the mind with ever-increasing wonder and awe, the more and the more intensely the mind of thought is drawn to them: the starry heavens above me and the moral law within me." There ensue the obsessions of mental discipline, the routines by which life is conducted so as effectively to generate, in individuals and groups, those technologies, artifacts, and means of subsistence that so exceed the capabilities of other species.

The rules of memory and the rules of forgetting are the two sides of the same coin. Who says remember, says select. Who names memory, names forgetting. In the earliest human, memory was a desperate structuring of current events to retain on the surface of the mind what was necessary to be human — that is, a sense of time, a manipulation of symbols and projections, and the practical means of ordering the environment — while relinquishing to the subconscious the impressions so intense that they would catatonize or panic the organism.

The sharpness, detail, and durability (in conscious and subconscious form) of remembering is proportional to the gravity of a trauma, that is, to the deepness and adverseness of its effect upon one or all areas of life. Forgetting speeds up with the intensity of the trauma. Thus a severe memory and its forgetting in all or part go hand-in-hand, and they follow upon the heels of the traumatic event.

The most intense memories occur without being willed. They emerge encased in dreams and myths. The less intense memories ride upon and cover over the more intense ones. Disorders of recall, of forgetting, otherwise unexplainable, can be interpreted as the effects of what is memorable having become willy-nilly attached to the un-rememberable, and often relate to the symbols and affects of the repression of the great disaster, that is, the primal conditions that established the rules of the conscious human.

Memory and forgetting operate like a bookkeeping system to keep the mind in balance. Littleis forgotten, and therefore the balance will continue to show a profit or increase throughout life. Like many bookkeeping systems in commerce, memorial bookkeeping has numerous ways of casting a balance. With the forgotten material, the mind works to create myth and art, even scientific hypothesis. By storing and re-categorizing forgetfulness, the mind achieves its ability to maintain consciousness and behave with instrumental rationality (that is, with a cause and effect logic in relation to practical goals).

Memory, in the specifically human sense of ability to recall at will, is inseparable from the sense of time. One can bring back the past in part voluntarily. This past can be shared with others through signs, symbols, and language. It can also be cast into the future, just as other current and past experiences are cast forwards in time.

The future-thought is born and partakes of the delusional quality of human nature in general. For time is a concept whose only existence is that given it by the time-keeper. Yet its implantation in humans gives them a tool for mental expansion and environmental control (as well as for suffering), not otherwise recognizable in the plant and animal kingdoms.

Culture institutes furious rites to make people remember something that they are forbidden to remember in all of its detail. Saturnalias, the prototype of all anniversaries, famous scenes of joy and orgy, are masquerades literally of the end of the world. They must be compulsively celebrated in order, by aggressive joy and wantonness, to cover up, to ensure the amnesia, of events that cannot be forgotten. People create an elaborate mnemotechnology, to use Friedrich Nietzsche's term, to assure that they do not forget whatever it is that they have forgotten, that is, suppressed. In this same sense, a great paradox emerges: we remember most emotionally what we forget most determinedly.

4. Obsessions, Compulsions, Habits

Stern memories, pathologically called obsessions, are an important part of human memory. Obsession is "excessive," "unstoppable" attention to something (by which, as usual, we mean "anything"). An animal can be trained to "extra" obsessions. Humans are naturally obsessive. They can be obsessed with a pair of shoes, a piece of cow dung, an inner voice or pain, the Second Coming of Christ, or a line of poetry. This "obsession with obsessions" determines what is remembered, what is recallable from memory, what the person will spend his time on, where the important things of life are in his estimation to be found. Moreover, it will give him an enormous capability. A person will be able to abandon all other thoughts and temptations and stick to a task through thick and thin. With this one (or two) abilities, he reconstructs animal instincts with some embellishments. That is, provided that he is compulsive as well as obsessive. Freud speaks of "the compulsion to repeat— something that seems more primitive, more elementary, more instinctual that the pleasure principle which it overrides. . . the pleasure principle — to which, after all, we have hitherto ascribed dominance in mental life."

A simple example of compulsion is the patient who insists upon playing the same chord a thousand times in succession. Or the repetition of nonsense syllables unendingly, which reminds one of religious liturgies that depend for effect upon an obsessive idea and the compulsion to repeat, which in turn must be related to the catatonic wish to stay as one is, forever in place. This again connects with the obsessive compulsive return to origins, to a recollected and imagined primordial set of events, illud tempus (that primordial time when...) which M. Eliade has so well abstracted from primitive ceremonies.

The terms "obsession" and "compulsion" are separate but confused in the psychiatric lexicon. No special neurology is assigned to one or the other. Each can be tied to both behavior and thoughts. Obsession is of the family of memory, planning, and habit. It is a compulsion to repeat. Compulsion is the kind of driven act which is likely to become an obsession. Insofar as all major problems associated with the terms are internalized, they are indistinguishable. Furthermore, they may originate together and act together.

Existential fear, and the need to control it, pursue the logical line of reestablishing the human as an effective mammal. It is not our choice whether to vary from it. We have no recourse; if we seek to follow the smithereens of our explosion of attachments, we shall go to pieces ourselves, psychologically and shortly afterwards as living organisms. We must stay at home while our displacements travel adventurously, and here recompose the hominidal character as best one may: hence, obsession and compulsion, or obsessed compulsiveness.

Obsessions are linked to habit. They are deemed "bad habits". Compulsion may be an urge to shout obscenities upon entering a church. Obsession may be an agonizing repetitive recall of an embarrassing scene, like the time one uttered a string of obscenities in a church. The relation of obsession to habit is clear in this case. The relation of compulsion is not, as in many compulsive behaviors the thought has preceded the deed and has occurred obsessively prior to the occasion when the act is finally committed. Then it is a habit of thought converted into a deed.

Obsession can be viewed as a form of deeply imprinted memory, which repeatedly calls the attention of the self to its selves. If a person suffers a fearful accident, the memory of it may occur with or without volition and despite a will to the contrary. At the same time, an obsession (and even faint memory is in a sense an obsession) is a repetitive trained behavior. Therefore it is a habit. It is also a compulsion, for one is compelled to recall.

Compulsions as acts are distinguishable from habit only by intensity. A drug habit becomes a compulsion, or addiction, when the behavior that is represented in the mind forces itself upon the organism, consciously or unconsciously. It is impossible physiologically to distinguish between a compulsive tic of the eyelid and a compulsion to step on the brake when a deer surprisingly leaps out ahead of one's speeding automobile.

Furthermore, a great many compulsions are consummated repeatedly, especially when uninterrupted by the forces of law, the community, the family, another person, or by destructive reaction or nature of the objects, or by self-destruction as in the compulsion to commit suicide.

In the illumination provided by psychopathology, habit is more readily understandable as an obsession under some degree of control, but at all events as a repeated practice, the regularity of and insistence upon which makes it obsessive, and the social judgement of which makes it reasonable as opposed to pathological. So it goes with compulsiveness very often. One might say that all obsessions are compulsive, but not all compulsions are obsessive; these latter are better called impulsive acts. Still, it would be rare that an impulsive act does not proceed from unconscious obsession, or from an impulsive character, typically given to such actions.

The human is obsessive-habitual because he cannot otherwise cope with existence. His obsession-habits are infinitely variable. They succeed, not precede — although the organic structure is partly in place — the basal human disorder and are the human method of correcting the disorder, with all due limitations and difficulties.

CHAPTER FIVE

COPING WITH FEAR

First of all, the gods created fear in the world." So goes an old Latin saying. The expression can be reversed, for it was also said, as by Statius, "In the beginning, fear created the gods." Let us suggest the primordial condition: human fear and holy dread. The fear is the existential fear of which we speak in this book. It remains, when immediate causes of fear are absent.

1. OMNIPRESENT FEAR

People will agree with the observation that life is spent largely on the problem of feeding themselves. The comparable idea that life is spent largely in coping with fear is met, by contrast, with disbelief. That fear should be central, dominant, and universal in humans is hard to believe, and harder to accept. Indeed one may propose, to begin with, that a cloak of denial is spread over the idea. We are afraid to admit what is forever present and all-determining because the admission, we fear, will simply worsen the condition.

2. PHYSIOLOGY OF FEAR

In terms of the psychology of conditioning, existential fear is to be regarded as continuous self-punition, whether it is called fear, anxiety, frustration, or depression. (Depression, writes J. Gray, is "a state induced by sudden loss of important sources of reward.") The punishment takes the form of inhibiting rewards, the most basic of which is probably the surcease from existential fear itself. According to Neal Miller and his associates, in a conflict between approach and avoidance the animal will come to rest at that point where the forces favoring the simultaneously feared and desired goal equal each other. The "decision" is to come to rest. Then the human, by our theory, cannot come to rest, nor does the wild animal completely, in the presence of the effective extinction of the stimulus. The human continues to live in a heavily displaced world where the avoidance- fear sensation will always find some home and sustenance.

The physiology of existential fear, apart from the brainwork of cerebral conflict, is not structurally or electro-chemically much different from that of mechanical fear (in the presence of accident, blows, threats of punishment, aggression of divine wrath). Nor is it distinguishable from the long-term fear of death. It is clear that it does not constitute a major structural leap in evolution, though its actual effects are quantavolutionary. Hence it operates much like the animal mechanisms.

Especially apropos is the "fight or flight" system, about which we can say, with Tepperman:

> Although people who are disturbed by teleologically 'impure' thinking in biology are sometimes made uncomfortable by Cannon's 'fight-flight characterization of the sympathoadrenomedullary discharge, the fact is that the over- all effect of such a discharge is to mobilize the individual to meet an emergency.

Hans Selye elaborated a model of the fight/flight mechanism, as well as Cannon, generalizing it into a "General-Adaptation Syndrome" or G-A-S. So much of total human activity implicates the G-A-S, that one is compelled to view it as the dominating action determinant, which by the theory of homo schizo, means that once the G-A-S comes into play, once the Central Nervous System is operative, the essence of the response to stress is also functioning. One does not flick off the G-A-S when one is thinking about the rings of Saturn or washing dishes. And Jeffrey Gray employs in his own theory of fear essentially the Cannon-Selye model, "a single fight-flight mechanism which receives information about all punishment, and then issues commands either to fight or for flight depending on the total stimulus context in which punishment is received."

The enhanced release of the adrenomedullary hormones, epinephrine and norepinephrine, denotes that a fight- flight signal has been passed to the central nervous system. In humans, the signal (I would argue) is incessant, or at least intermittent and rapid, and comes from a high level of excitation of the cerebral cortex, which is continually dealing with conflicts - past, present, and future.

Anxiety is moderate and continual fear and, were we dissatisfied with the mechanisms of fear in producing human nature, we should seek the mechanisms of anxiety. But they are the same as those described already. The dozens of physical and mental symptoms of anxiety, if they are not subsumable under the fight/ flight system in operation, are by-products or effects of the system.

The explanations of anxiety are significantly related to the theory of homo schizo. M. Gray mentions five types of hypothesis. That anxiety is a catastrophic response of the organism to stress is advanced by Goldstein. That anxiety is a threat to one's concept of the self is advocated by Rogers. That anxiety arises out of unassimilated percepts is put forward by McReynolds. That it is related to commitment and awareness is Kirkegaard's notion. And that it signifies perceived threat to existence as a personality is conceived by May.

All can be related to the poly-self problem. Going beyond Gray to the sources, one can synthesize from them a concept of anxiety as an unending (because not quite exhausting) flight from oneself occasioned by defeat in containing emotional stress and by inability to face up to the everyday world. These opponents would be never a sufficient cause, however, were there not the ever-present existential fear that the self is not itself, but rather a dissociated confederacy.

And the outer world *per se* is a source of fear because with this discordant confederacy, who can be sure that the world is real and fixed? One need not be a philosopher to sense this fear, but it helps in expressing it. As Schopenhauer wrote, "The uneasiness which keeps the never-resting clock of metaphysics in motion, is the consciousness that the non-existence of this world is just as possible as its existence."

Is anxiety normal? Yes, endemic. Is anxiety part of the fear-flight syndrome? Yes. But for the "flight" part of the syndrome, we must investigate other behavior, beginning with punishment.

3. GUILT AND PUNISHMENT

A taboo forbids some relationship of a group's members to an object or being. It functions as an obsessive phobia. It can also be regarded as an unconscious plan by a social group to proscribe an activity in general but to grant that the prohibition will be violated by "sinners." Despite the score of theories as to the nature of taboos, there seems to be no significant difference between a taboo and the process of law in rationalized societies except in the degree of analytic awareness accorded to the two types of phenomena.

Both condemn a transaction, elicit guilt for its performance; excite reliably the "satanic" impulse of individuals to violate the proscription; thereby exciting impulses in the satanist and, by identification, the society, to feel guilty; punish the guilty and give thereby a sado-masochistic bonus to the group members; and then conclude with the expectation that the cycle will repeat itself. So the taboo, and the law, are techniques of varying awareness for producing routine behaviors, with exceptional cases providing a leavening of guilt and punishment.

A taboo and a law that are never violated are not only unnecessary, but undesirable, and a contradiction in terms. It is for the group consensus that the law or taboo is needed, for "reasons" and "consequences" that are more or less clear but in any event affirmed, if only in order to exercise the ritual guilt and punishment that the human uses to assure that his psyche is under governance and can control its aberrations, both internal and external.

Punishment of the self and of others has the same etiology. Humans who could punish others but not themselves, or vice versa, are structurally impossible. That observers have been misled to believe that people exist who are capable of one but not the other attests to the deluding permutations that are possible in the essential need to deal with the punitive aspects of the gods.

Guilt is an obsession with the conceived need to punish some part of oneself. Guilt and punishment are both "moral" activities, meaning by moral that their processing in the central nervous system is associated with compulsive identifications with symbols or beings of authority. The wish for, or condoning of, such authority is a wish to control oneself and others, thence ultimately existential fear. The origin of morality is frequently the perceived behavior of the gods, whether based on some reality or hallucinatory. The behavior of the gods is an effective instrument for inculcating fear because of their actual behavior as perceived by the delusory and projective apparatus of the primeval human mind.

The individual split-self, comparing itself with others, works to discover a pattern of actions that distinguishes his behavior from that of others, and one category of others from another category. Driven by extreme anxiety, his hypotheses become obsessions. He discovers what he thinks to be such patterns and the relation of such patterns to the response of the gods.

He proposes and follows with obsessive subjectivity and paranoid zeal the line of conduct that appears to promise the greatest benefits and the lightest treatment from the gods (and their representatives— men, animals, plants). He projects his own correlations into the motives of the gods. Depending upon his charisma and the degree to which the god's behavior actually has a consistent appearance to others as well as himself, he makes a consensus of believers, a religious sect.

To strengthen his own self-restrictive behavior and to bargain for control over others, he transfers his authority to the sect. He finds thereby an accommodation that far exceeds, in the security and satisfactions that it brings, even the great benefits being experienced by group cooperation in hunting, farming, and manufacturing. Indeed, this basic security is so beneficial, that it causes the great development of these latter activities, and of art, sex and procreativity. All of these activities, affecting all life, became then suffused with and dependent upon the origination of guilt and punishment.

Punishment takes many forms — of the self and of others, of the world and of the thought, of the body and of the mind, of deliberate action and of subconsciously driven behavior, of mildness and extremeness. All emanate from self-awareness and the reservoir of primeval fear filled by it.

The need to punish is always independent of the demonstrable effects of the punishment. Fear finds its nexus whenever it encounters the obstacle of its own illogic. To sacrifice one's own child to a demanding god is by its own extremity of pain and sorrow the proof that the punishment must be effective. Moreover, the persistent and universal presence of human sacrifice is the mere outcropping of self- destructive and destructive activity, both deliberately and subconsciously conducted, which masses itself beneath the innumerable different cultures that have evolved since mankind originated.

The kind of punition that asks first for the logical connection between offense and punishment, and seeks to follow the crime with correction in the future, has only occasionally and especially in recent times had some impact upon deliberate punishment. Even then, though faced with psychiatric theories attacking primeval guilt and punishment, the urge to punish has tended to go underground into the subconscious, there to expand upwards once again in destructive behaviors that are handled in a context of fear in which punishment occurs as an ever- present instrument of relief and discharge.

Basic patterns of behavior are infused with appropriate modes of punition-catatonism with paralysis for having moved, and by moving, moved the world order to destruction; obsession with numerous rules to proceed in fixed ways on pain of a variety of punishments ranging from mild social disapproval to the most horrifying extirpation that can be devised; sublimation with the torturing of thought, art, and institutional behavior, generally playing out the eternal drama of the sin of being human; orgasm with displays of violence, sacrifice and ecstasy.

The common need for authority, especially one who can punish and forgive, is plain, for punition tends to unite the self, whether it be by the self or by the authority, and a god is therefore much needed. So is the god needed in external relations, interpersonal affairs where, as Martin Buber, the theologian, says, God, the Great Thou, enables human I-Thou relations between one person and another to subsist.

It is plain what an important role the gods play in holding the self together, and why gods are assigned the tasks of punishment. The tenuous self, striving for integrity, for self-rule or selves-government, invites a third party, the god, to intervene, who, by definition rather then logico-empirical proof, restores order in the self by punishing one or more of its components; it matters not which, and matters not why. What matters is that the punished being feels whole, feels "better" afterwards. Such is probably the essential dynamic of masochism.

The transfer of godlike qualities to the rulers of the state - kings, judges, and generals - and of institutions (presidents) and families (parents) lends these real beings authority, which hoi polloi can disregard only at the risk of punishment, whether self-inflicted or imposed by the rulers.

Every life activity displays guilt and punishment - sexuality, food production, tool-making, war and justice, death. The peculiarly human aura of sexuality, like that of other human life areas, is owing to the architectonics of fear, with its arch-instruments of guilt and punitiveness, working its way through the continuous mind-exploding mechanisms of the split self (selective memory, symbolization, control- seeking).

As M.. Gray has mentioned, fright in man is more complex than in animals because "the new dimension that is reached in man can be viewed as symbolic fear." Sex and fear systems are tied together in endocrinal secretions, so that there is little need in psychology to devise a logico-rational explanation of how, beginning with fear, the human mind runs to sex, or vice versa.

Bleuler discovered early that "the idea of intercourse is often expressed by that of murder," that "wars and duels are symbols of cohabitation," and that the idea of being burned is tied up with murder and sex. Often we observe that peacetime anxieties or fear produce sexual impotency, whereas wartime fears, where violence is pervasive, produce lust and rape. Men going about looking for jobs develop impotence; men infiltrating an enemy town or abandoning it seek sexual outlets. Unleashed violence uses sex as a screen for and release of fear. However, regarding murder, say, sex is not the originator. The life areas are intertwined, as are all others.

What the mental strategies do to characterize the sex instincts of humans is also done in the other areas of life. Consequently, if one were to eliminate, in each successive area of life, the generic holistic mechanism working upon continuously renewed and stored fear, one would discover at the end of each process of elimination a root element, the sexual, the grower, the maker, the fighter, the dying person. Although no more complex than the fullness in toto of animal behavior, these areas are elaborated in an exceedingly rich manner within humans by the basic mechanism operating out of fear in unending supply.

4. AVERSION AND PARANOIA

A common element in schizophrenic symptomology is an aversiveness to humans. This includes social distrust, desire for privacy, fear and dislike of others, expectation of being rejected, and the conviction that one is unlovable - all on an intense level. A strong rejection to being helped is common. It is a mutual rejection of the body-selves and a separation from the world and others of the poly-self. It is self-conscious - not animal awareness, but a delusory standing off from oneself, with the whole world (inner and outer) thus revealed, expanded by the sense of time - of recall, memory, and forgetting - and of space.

The "paranoia" comes from the Greek where, anciently, to be "out of one's mind," that is *para* (beside) and *nous* (mind), denoted insanity in general; in American vernacular one would say "I was beside myself with worry." Now, narrowly, paranoia is restricted to projections of threat. The paranoic aspects of primordial and existential fear are well known. Fear fathers destructiveness. Whenever the skies darkened, or certain stars approached, or the Earth trembled, paranoia was stirred up. Legends often openly assert that "because" the gods were destroying the world, men took up arms against each other. Even voices out of highly civilized cultures, such as Alsatian peasant culture, will assure one that wars follow upon great meteoritic showers, as in 1914.

What the gods are intending, as projected, is retrojected back to oneself, to one's group members, and to outside humans. In extenso, it often happens that the gods instruct men to destroy each other. Official psychiatric reports of the great Alaskan earthquake of 1964 describe how blame for the disaster, often couched in terms of an Act of God, was expressed in guilt-feelings and in accusations against others, such as "I changed my church," or "He changed his church."

The inventive aspects of paranoia are considerable. The paranoid is highly energized. He stresses symbolism. He acutely discerns remote analogies as he searches for conspiracies. He prospers upon the unknown and unintelligible. He is obsessive, confident where others falter; he has answers, and insists that others adopt his answers. The dominating role of the paranoid in the origins and history of civilization becomes understandable. His "sickness" was no sickness in primordial times.

Nor, properly rationalized in the language and procedures of existing culture, is the paranoid "sick" in the world of today. For every provable connection of effect with cause, humans resort to hundreds of paranoic assertions. The reason, when experts are asked why, is usually given as "ignorance." But is it not as easy to apply a simple logic on what is known, or to confess ignorance, as it is to learn or invent false accusations continually? The primary ego is compelled to assert its omniscience and, since the function of "knowing" is really to reduce internal fears, accusatory explanations are in heavy demand.

A paranoid is usually beset by ambivalence, the love-hate double face and double mind of the schizophrenic, the abrupt turn from one to the other make relations between schizophrenics and others often more terrifying than consistent hostility.

5. Ambivalence

But ambivalence, we should argue, if we are set to follow homo schizo theory, must be universal in man and culture. And so it is. It is the good and the bad of everything. It is "the two sides to every question" that underlies the judicial systems and many cultures. It is manifest in the fact that people and cultures "choose" to differ in every way that they can, from one another and within themselves. The alter egos must emerge. Cannibals can be divided into those who eat their enemies and those who eat their friends.

Levi-Strauss says that the incest prohibition is the only universal culture trait, a bold and learned claim. But a moment's consideration will put the claim into a revealing context. The scope of incest rules differs: some stop with parents, other with nuclear families, others with uncles and aunts, other with clans and so on in various combinations. By their existence and limitations, the rules implicate contrary wishes, hence ambivalence.

Furthermore, societies have rules about everything that can be the subject of rules; incest is no more specific a concept than murder, and all societies have rules about murder; further, all societies "by right" are totalitarian implicitly if not explicitly endorsing the old army saying that "there is a right way and a wrong way to do everything in this Army." (There is an alternative rendering which is, "There is the right way, and there is the Army way; you do it the Army way!") Finally "rules are made to be broken;" without infractions there would be no rules which means, more subtly, that it is desirable to have infractions, if only to gratify the punishers, but, more than that, in order to let what is right be known.

Today the gods are less frankly present in the operations of the "normal" mind and institutions; but the schizoid meaning of the primordial gods is clear. It is not enough to say that the first human mind could imagine the gods and imitate their imaginings (projection and retrojection). It actually did so on the basis of more reality than is fashionable to admit nowadays. The human mind was born with the gods, that is, with those terrible entities of the sky that are said to have wrecked the world, and that were hovering above and swooping down over a period of thousands of years, ever refreshing the reservoir of fear in the human mind.

The animation was imagined, or so our modern logic insists. No matter, for every symbol stood for a memorable sign, every myth represented an event, then came imitation — prompt, unquestioning, and illogical. Above all other relationships in the world was the identification with bodies that both hated and loved humans on a massive scale; these were the gods who turned upon another, castrating (Saturn-Uranus, Jupiter-Typhon, etc.), maiming, tearing off heads and limbs, hurling mountains and cosmic discharges at each other and at men.

"Black magic" sees a hand in everything that happens. The paranoid often sees the same. Comets and meteors readily simulate the hand. The prehistoric caves contain the hand in great numbers. The "Hand of God" is frequently reported stretching out from the heavens to the agonized populace. At one and the same time, it is the image of terror, of inevitability (the hand finally grasps), of help and of punishment. He also "Holds the Whole World in His Hands."

But the hand is only a part of the rich assemblage of forms that natural bodies can assume, all of which contributes to the original and continued reification and anthropomorphizing of the gods. Schizophrenics, more commonly than controlled normal schizoids, reify the outer world. Insofar as the schizotypical human has been in the forefront of human development, the gods as "humans writ large on the skies" are unending.

What they did to men was beyond modern belief and was deeply suppressed. But it was never truly forgotten. The myths of old, the dreams of the normal, and the autistic reveries of the schizophrene are basically alike in structure and purpose: to manage the unmanageable. Whoever survived had to believe that they were the chosen people of the gods. Further, all the new aspects of their environment (like the manna and the ambrosia from heaven) that helped them to survive, kill enemies, give birth, and carry on creatively (arts and crafts) made them believe in the love of god. It is the famous "double bind," which social environmentalists attribute often to a mother who works out a hate-love relationship with her child, and which makes the child schizophrenic. But going back generation--to-generation to the earliest times the mother, and her mother, and so on, can only deal in hate-love ambivalence, because they were so dealt with by the primeval divinities.

6. ANHEDONICS

After ambivalence comes the pleasure-phobia, the symptom of anhedonia, which Meehl has called "one of the most consistent and dramatic behavioral signs of the disease" of schizophrenia. It is a "marked, widespread and refractory defect in pleasure capacity." Man has been mistakenly called a hedonistic or pleasure-seeking animal. The psychiatrist would say that this is correct only if self-destructive and other anhedonistic behaviors can be termed pleasurable. If whatever the organism seeks becomes, by definition, pleasurable, then it is hedonistic.

The hedonistic theory is inherited from the Benthamite school of early nineteenth century England. The psychology of hedonism had taken over a commanding position in Western societies in the guise of democracy and socialism. Certain philosophers — ancient Epicurus for instance — certain societies, not many — the United States for example — take up the idea enthusiastically. Or, at least, so imply the advertisers. But, say the critics of American civilization, the attempt at hedonism fails miserably.

In reality, the pursuit of hedonism is always a secondary social aim. No society has ever been founded upon the pleasure principle. All societies are ideologically committed to the principles of anhedonia: pleasure is evil; pleasure is impossible; pleasure brings punishment, suffering is good; pleasure is a release from disciplined suffering; pleasure is to be tolerated only upon the celebration of a disastrous anniversary, as an orgy inviting repentance. The orgy is a complex of smile and snarl— as in dogs we have known, and in humans we should wish to investigate.

It is from the human person that society is constructed, and where anhedonia is born. To prove this, one may begin with deduction. Pleasure-phobia is logically implied in the theory of the fearful polyego. The mental construction of the human is fundamentally unsettled. The frustrations of existence — to satisfy needs and evade the blows of nature— write an undulating pattern over the basic unsettlement.

Now if one is permitted the irony, is it not strange that pleasure should be regarded as natural and in this day people who are anhedonic should be regarded as mad, when it is the hedonist who perhaps should more naturally be taken to be mad? It is significant therefore that the definition of pleasure itself is the greatest weakness of hedonism as a philosophy. The hedonist begins by thinking of pleasure as a commonplace idea: "eat, drink, and be merry," "a car for every family," "peace and plenty," "a chicken in every pot," "free love," etc. That is, the hedonist turns out to be a superficial psychologist who has a rationalist uni-dimensional view of people.

However, most people seem intent upon rejecting this kind of pleasure in part or whole. As soon as one asks of the hedonist, "What gives people pleasure?" he must reply: "People get pleasure from whatever they wish to do or have done to them." Pleasure, then, in a word, is "voluntarism," as in the old-fashioned expression, found in several European languages as a mode of address to superiors, e. g. "What is the gentleman's pleasure?" Under the Ancien Régime in France, the King signed all of his promulgations with the phrase, "Car tel est notre bon plaisir," which phrase would be related upon every reading of a law or judgement, as when a man was condemned to death: "For such is our good pleasure."

But of what does Voluntarism consist of in primeval humanity? Assuredly it is in fulfilling those needs that are animal in kind: feeding, fornication, fighting and fleeing danger. More than that? Yes, hanging around where one can feed and fornicate and feel free from danger the next time on each cycle.

What else, that is typically human? To refuse food, to refuse sex, to fight instead of flee. Even more, to do all that can possibly be imagined to use these elemental desires in ways that will establish and secure the most wanted triple-control of the self, others and the gods.

Let us look at the earliest legends of mankind. What do they have him doing in these regards? A great many unpleasurable things. He sets up a host of taboos against the most plausible kinds of enjoyment.

He eats only certain food, and at only certain times. He eats what is bad to eat when good foods are available. He eats in a certain way, giving a portion of the best of his food to the gods, as hungry and insecure as he may be. He fasts. He eats his gods and his enemies. He builds an oral literature on what to avoid eating and subsidizes priests to tell him what not to eat, when, and how to prepare what he does eat, ulcer or not. Would it be permissible to suggest that two out of every ten humans who have ever lived have died prematurely from pursuing irrational eating habits? (Meaning by irrational: ingesting or avoiding because of non-dietary reasons what is severely prescribed or proscribed.)

Sexual behavior, too, is thoroughly permeated by restrictions and impositions. Name an animal that lingers in coition. Other than man, of course. That exults in creating a female orgasm. Whose orgasms are compared with death itself? Not "I am born," but "I am dying" is the sometimes climactic ejaculation by the sexual partner, with subsequent relief and relaxation from having met with death and survived. And again creates a philosophy, as the Hindu, for one instance, that explains how you weaken yourself unto death by sexuality.

So, too, conflict. To defend is "natural," to flee is natural; but not to attack deliberately with one's guts roiling, in row upon row with bayonets fixed, with death ahead, yet death from behind upon whoever falters. Whence the awards go to those who have suffered most joyously, doggedly, "without questioning why, but to do or die." Verily who seeks pleasure seeks its own reward; who seeks pain and suffering is exalted before oneself, before man, and before god.

The earliest glyphs and scripts of mankind are pleasure-phobic to the degree to which they are sacred. Coming from the Egyptians, Hebrews, Sumerians, Mayans, Chinese, Icelanders, Myceneans and Greeks, these coherent patches of history and morality stress without exception that the pains of existence must be and should be, hardly ever do they set up mammalian or sublimated pleasures as a human ideal.

The anhedonism of primordial and schizophrenic humans is understandable: existential fear demands not pleasure, but relief. And this relief results from a broad spectrum of activities that are hardly pleasurable: self-mutilation, sacrifice, cannibalism and exhausting ritual. The ambivalence of the gods and of the self, too, warn against pleasure.

T. Reik has argued that the original sin was not sexual but rather of hubris, the imitation of the god's power, such as the seizing of the God's fire. It was alright for the Homeric heroes to address the gods as "blessed and happy;" but calling themselves happy was an invitation to disaster.

Far more institutions have been created in ancient and modern times for the suppression of pleasure than for its enjoyment. But as Freud intimated in his *Civilization and Its Discontents,* without suppression of the instincts there would be no civilization. Herbert Marcuse has more recently elaborated upon the thesis in *Eros and Civilization.* Even these trenchant criticisms for culture now seem superficial. They have but carried forward another version of "the noble savage" of eighteenth century philosophy. One cannot discover, nor properly induce from pre-history, a human culture that sought to provide pleasure except as it might be incidental to relief and escape. Anhedonism is imprinted upon human nature, the individual schizoid psyche, which is pleasure-phobic more than pleasure-prone, abets and invents the anhedonistic institutions. And institutions "hate pleasure," whereas they cultivate suffering.

Usually, pleasure-phobia is an aversion to interpersonal and "animal" pleasure rather than to cognitive and aesthetic pleasure. The schizo-type can evince aesthetic and intellectual hypercathexis without the fears and guilt of interpersonal pleasures (i. e., human pleasures). Anhedonia is not apathy, but is pseudoapathy, a tense and anxious state; its final form is catatonism, which is really "playing possum" with the gods.

7. CATATONICS

Manu, the Noah of the Hindus, "practiced severe and great self-mortification." Wearing a bark shirt, his hair matted, "while he stood on one foot with his arms raised, with bent head and eyes unblinking, he performed awesome austerities for 10,000 years." He was chosen by order of the gods to recreate all creatures after the Deluge. "By virtue of his very severe self-mortifications the manner shall be manifest to him." So the model of man is taught the greatest knowledge by the greatest suffering.

Freud in one place quotes Kaempfer on the taboos governing the Japanese Emperor of old:

> In ancient times he was obliged to sit on the throne for some hours every morning, with the imperial crown on his head, but to sit altogether like a statue, without stirring either hands or feet, head or eyes, nor indeed any part of his body, because, by this means, it was thought that he could preserve peace and tranquility in his empire; for if, unfortunately, he turned himself on one side or the other, or if he looked a good while towards any part of his dominions, it was apprehended that war, famine, fire, or some other great misfortune was near at hand to desolate the country .

Gurdjieff reports an experience from Central Asia. There in a village a religious sect was playing games of magic circles. A girl was frozen in the circle that other children had drawn around her. She could not move out nor could adults from the sect drag her out. Persons who entered and left the circle remained in a catatonic state for many hours. It seems that one entering the circle grants her vital force over to an outside being as a bribe to prevent harm to oneself. Thus emptied of vitality, one cannot move.

There is more than an analogy with some psychological problems in a rigid bureaucracy. The incumbant at his desk gives over his life forces to an outside being - in this case the inanimate collective representation that is the agency or bureau. Whereupon the employee becomes inert, immobile, and cannot direct the very forces he is employed to manage.

What is the fear of change of habits, customs, society and human relationships, even of aging? Is it based upon some pragmatic calculus of cause-consequence; that is, is it based upon the experience of change, discovering that change is always for the worse? And why does the feeling vary so greatly, then, among individuals, so that, for example, a social psychologist such as Lowell could divide politicians and public opinion into two categories: the optimistic and the pessimistic?

The fear of change derives from the anxiety over the potential loss of an ego stability, which is markedly worse the less the poly-ego is stabilized to begin with. May this not explain the phenomena of personal and social conservatism and stagnation? The process in society, as in the person, is self-aggravating. As a society destabilizes in revolution, whether social, industrial, or political, far from people becoming habituated to the change, they become more desperate. Even though the change may be rationalized as beneficial, any material improvement and a formal lift in dignity become inadequate consolations for the failures of individuals to compose new poly-selves for the new times. This should constitute a lesson for non- Marxian revolutionaries.

A catatonic patient, like Manu, is far from "despairing" of control of the world. He may hold a position for months, but if moved, flex and adjust so as to maintain the assumed position. His ability is unconscious in that it far exceeds any normal ability. Other patients (Bleuler called them "waxy cataleptics") were without spontaneous movement but maintained any position in which they were placed. Another type went on repeating motions that were supposed to have begun and stopped. Often the cataleptic exercises himself in an "affirmation of negativism" that requires great muscular energy and coordination.

Catatonism tells the gods or other authorities: "I do not move, lest when I move I am noticed, and the world, too, will move." Catatonism is a common response to the shocks of primeval and historical disasters. It is seen in every accident ward and especially in military hospitals. In a great disaster, such as Hiroshima, catatonism is a major behavioral response.

Non-rationalized cultures (mistakenly termed "primitive")) simulate catatonism when reenacting the earliest days of creation. Members of a certain Jewish sect must remain throughout the Sabbath in the same posture that they were assuming when the Sabbath began. Children present during traditional religious ceremonies are warned to be particularly silent and immobile while the priest reenacts the primordial end of one world and beginning of the next. Wherever the authority of religion has descended upon secular institutions— be it a library or the mausoleum of Lenin— a "respectful" silence is maintained. Gestures become restrained. Clothing becomes "appropriately" somber, unobtrusive. All that is individually outward is suppressed, and it all "happens naturally."

Short of catatonism and beyond anhedonia exists the realm of apathy. Schizophrenics, the psychiatrists say, are so contradictory: sometimes excessively voluble, hyperactive, frenzied, and then again and for prolonged periods apathetic. Nothing interests or excites them. They may be occasionally aroused, and then retreat dramatically into torpor of manner, posture, and speech. They are impossible to arouse. No one knows what stimulus will once more perhaps excite them. Meanwhile no indignity inflicted upon themselves or others, no injustice, no deprivation seems to matter. Electric shock treatment is resorted to, a horrible experience, a repetition of primeval bolts of Jupiter. Occasionally, a cure of apathy results, but note that it is a worse pain, a greater fear of the lightning discharge, that arouses the patient, not a pleasure; no pleasure will ever do the job. Pain, not pleasure, is the route to resuscitation, a recovery of a status as a punished one, a way to please the self, others and the gods.

Self-deception of anhedonia and catatonics can be carried to the expected extreme of self-destruction by suicide, just as with the other delusions of man. Karl Meninger sees in all mental illness a core of self-destructiveness. Freud's last paradigm had "Thanatos," the death instinct, juxtaposed to "Eros," the life instinct. There is no use in trying to draw a line to include only the mentally ill. Self-destruction in its most obvious forms, leading to death, is an extreme anhedonism, by definition, which is to say that other self-destructive behaviors that are not so obviously leading out of anhedonia are tied into anhedonia for the simple reason that man seeks self-control holistically and hologramatically; the symptoms are interrelated. Oblivion in catatonism and death is the ultimate control of the self by surrender; the self is no longer divided or in disarray or scattered.

8. Orgies and Holocausts

But a place must be made for orgiastic behavior. Whereas one kind of violence emerges from the discipline and sacrifices of "law and order" or obsessive social forms and institutions, a second kind of violence entails the tearing down of structures and institutions. Sometimes this occurs in Saturnalian orgies where deliberately, for the nonce, all social forms are turned upside down; masters serve slaves, equality reigns, contracts are broken, wealth is burned or otherwise wasted. Amidst the Assyrian terror, Isaiah cries out: "Let us eat and drink, for tomorrow we shall die." (Is 22; 13) Ecclesiastes does him one better: "Eat, drink, and be merry..." At other times, orgiastic violence becomes warfare and social purges, such as witch hunts and political reigns of terror.

"The conscience for which [Hitler] stands," wrote Lasswell in 1933, "is full of obsessional doubts, repetitive affirmations, resounding negations, and stern compulsions." Mass death followed. The obsessional and the orgiastic work hand in glove. Human sacrifices, an ancient Pharaoh inscribed, will purge you to the satisfaction of the gods. Warfare, for the Aztecs, was continuously needed to provide prisoners whose sacrifice was demanded to keep the world orderly and the sun regular.

The orgiastic event commemorates the end of the world or of one world. "Better a terrible end than an endless terror," said the Nazi slogan. Again, it is a replay of the primeval times of disaster, a carrying out of the will of the indignant or new god. Political force represents phylogenetically the force that overturns the earth; therefore, necessarily often, political force is adored, is not to be restrained, and when abused, remains still genealogically right. There are many analogies in the human mind between natural and political violence; Shakespeare, as Irving Wolfe demonstrates, interchanges social, personal and nature's language in a shower of metaphors. Holocausts are demanded. "The beast within us" is called forth. But, of course, it is not a beast; no beast acts so; it is the human within us that is called out.

A common phrase in writings about repulsive practices is "Even as late as..," as if mankind had been on an upward track of moral conduct. "Even as late as the Roman Empire, infant sacrifices to Saturn occurred... etc." "Not until the Spaniard arrived, did the Aztecs cease their regular cannibal sacrifices of thousands of persons," or "the Jews of Czarist Russia and the Armenians of the Ottoman Empire were not freed from the dread of pogrom and massacre until 1920," or "As late as the 1880's the Shawnee Indians sacrificed a maiden upon the near approach of planet Venus."

However, relief is not in sight. Progress is a matter of a few years. Somewhere, at any given time, a massacre or deadly persecution is probably taking place. The human "holocausts" of the German Nazis, Russian communists, Chinese communists and nationalists and of the lesser Balinese military, Cambodia communists and other groups of the past half-century took perhaps one hundred million lives. Wars and starvation by neglect killed another hundred millions. And the world arms, while it talks in terms of killing more hundreds of millions of people.

It is not difficult to prove that homo schizo is nearly as far from "killing only to eat" as he ever was. Nor has any social invention appeared that might promise a definite end to such catastrophic behavior. This, more than anything else, ought to motivate the study of human nature. We know, do we not, that the historical modes and inventions of systems of social cooperation have failed all critical tests?

There is an automatic anhedonia about modern holocausts. The pleasure-phobia sacrifices others to reassure itself: "The only good Indian is a dead one." But the phobia then goes farther to hide the emotionality of the deed by bureaucratizing (routinizing) it. The self-suffering (which is not rationalizable to others) is translated into sublimated self-suffering that makes sense to others. In cultures where religion provides infinite legitimated anhedonia, this is an easy matter; every impulse to suffering can be indulged.

9. Sublimation of Fear

Nevertheless, older religions (theocracies, specifically) set up channels of suffering that lead to astonishing aesthetic and intellectual products. Especially in cultures deviating from heavy religious norms, self-suffering by sublimated activity is given an individual or scientific-bureaucratic base, and it is here that the schizophrenic is identified; since he is not incorporated into bureaucracy and rituals, his activity is exposed as psychopathic; then wonder is expressed at the great aesthetic and intellectual product that emerges from his suffering mental state as in the case of the composer Schuman, the sociologist Max Weber, the novelist Kafka, and so on.

Aesthetics and invention are displacements and transsubstantiations of interpersonal suffering, conveyed in ideas and symbols. Thus frequently the schizoid patient surprises his keepers by contrasting behaviors, on the one hand completely uncoordinated and erratic where people are concerned, but on the other hand competent, cool, logical in the pursuit of music, mathematics, and zoology.

In such pursuits, the schizoid leaps over his uncontrollable anxieties of the other self, the other person, the living nature and gods. Then he (or she) doubles back upon the percepts and cognitions left behind, purging them until they form a colorless fabric of abstractions which he drapes comfortably over himself.

In movies, novels, and journalism people (normal) express surprise and anger at how many institutions of love and priests of love behave contrarily; these are unfeeling, inhuman, interested mostly in abstracted aspects of people and things. Such frequent behavior of "welfare officials" is partly their anhedonism and partly their personal aversiveness, whether expressed by them characterologically or as typical representations of institutions.

Man cannot lift himself by his own bootstraps. The human has been and is always in some combined state, varying cyclically in intensity, of self-punition, aversiveness, anhedonia, ambivalence, paranoia, catatonism, and orgiasm, as well as obsession. The social structures are but an extension to help him control these embodiments of anxiety. It may be hopeless to seek exceptions via cultural anthropology or special religious sects.

Fear phenomena can be manipulated, handled, understood, balanced, but not erased. Otherwise we should have a vegetable or simple animal, for the animals closest to man have also a problem of eternal angst. A normal scientist (and to be a scientist implies an abnormality), admittedly a superman of rationalized controls, is *ipso facto* allowed and trained to mistrust his senses, to mistrust the word of others, to regard everything as important (that is, engage in the most remote displacements), to check and recheck anxiously, to be obsessive about his subject, to avoid personal identifications and emotions, and to suffer self-punition over extended periods of times. What is science, indeed, but a capitalization of instinct delay and the heavy anxiety-alert to develop and exploit them as actual products?

Chapter Six

THE GOOD, THE TRUE, & THE BEAUTIFUL

The good is what one wants; the true is how to get it; the beautiful is a mask of the good and true. Such is the frame of the argument here to come. The repugnance that should be aroused by it is one which I can feel as sensibly as the reader, for do we not appreciate that the good is what we Good Ones want? And the truth is more than mere means, but ought to be *good* means? And the beautiful is the subtle expression and adornment of our good means and good ends? But let us hide our light under a bushel and speak of others.

How does human mentation work on these matters? Most mentation, we should have to admit, is a muddle, testifying as a whole to the frustration and futility of humankind, but this is so widely known that we need not take the time here to describe it. Mentation, like human behavior generally, seeks to recapture instinctiveness and, by so doing, to hold its poly-ego in a comfortable balance as near to automatism as possible. The closest it has come to this Nirvana is a state in which the rules for quickly achieving goals are routine and effective, and the sublimation of the instinctive ends and means is at a minimum. This is ordinary scientific and rational behavior. Man chooses art, and whatever else is blessed as voluntarism and beauty, because he cannot attain instinct directly, or finds himself stranded in the muddle.

1. THE MUDDLE OF MENTATION

A most apparent excrescence of the human mind is egotism. The human emits a plentitude of ejaculations, demands, and wishes, which he believes are reasonable simply because they emanate from *himselves.* To all of his positive identifications he ascribes a good, to all negative ones (and many are ambivalent) an evil.

He characteristically emits denials of whatever would appear to oppose his good, quite aside from the rules of logic or reason or justice, although to these he may even subscribe. His capacity for denial of the opposition extends to non-perception and non-recognition. The very sensing of things by eyes, nose, taste, ears, and feeling is broadly prejudiced. Nor is this selective sensing a "logical condition for survival" or "a preference — *de gustibus non disputandum est*"; it is a severe effort to destroy what threatens.

The human possesses a rudimentary notion of cause which labels whatever he dislikes as the cause of the evils he perceives and whatever he likes as the cause of the good. Guilt and blame are displaced liberally upon his negatively construed objects.

Everyday thought exhibits an abundance of what psychologists term "erratic cognition." Some Hindus say that "The Sun and the Moon rise and set only because the brahmin recites the Jayatri." The Aztecs, who were butchering and eating an estimated 200,000 persons per year when the Spaniards arrived upon the Mexican scene in the sixteenth century, claimed that, without a gift of human organs, the sun would not rise.

In a related type of case, a disaster occurs; it is "normal" to believe that it happened because the victims had been bad. "They deserved what they got." The reasoning is to be discovered both in the Bible and in present-day Christian communities, whether in Alaska during the 1964 earthquake or in the Wilkes-Barre, Pa., flood of 1973; it permeates every culture's ideology. Nor is this universal reasoning simply a product of the lazy human mind, which does not seek scientifically for the antecedents of a disaster. It is an effort to control the world. It gives humans a great collective responsibility, an intolerable one. In order to believe in human control, it may be necessary to believe that humans can bring the world to an end!

Human mentation is normally preoccupied with the great battle for control of fear. This is and has been of much more interest and concern to the organism and society than the pragmatic concerns of the several areas of life— work, sex, science, health. All of these life areas —the immense structure of civilization— emerge from a "madman" trying to control his head. That pragmatic behavior is neglected in the frenzy for control is normally observable in human thought and behavior.

Governments operate in the same muddle as individuals, whether they be democratic, communist, military, traditional authoritarian, theocratic, or tribal. It is quite clear among them that the good is what they want, the truth is how to get it, and the beautiful is what adorns their good and true. The documentation of this statement is so profuse that I can only allude to it here. A century ago, Ratzenhofer expressed the consensus of the most discerning political scientists when he suggested classifying politics as a branch of psychopathology.

Raison d'état carries properly its cryptic original sense: whatever the state wants is reason enough. "Do not ask questions; it will do you no good." The governments consist of the men who run states; these men are basically similar to those whom they rule; they are constrained by attitudes, ideology and capabilities. They have the same or slightly more of the schizoid traits of split personalities, fear, obsessions, paranoia, and immersion in symbolism that are observable in ordinary people.

2. *The Omnipotence of Ought*

Man exercises from his gouty toe to the heavens above what Freud has called "the omnipotence of thought." He builds in his mind an image of a new reality that is under his control and, further, will believe and act as if the corresponding events are occurring as planned. It is a kind of megalomania.

Still the indulgence of omnipotent thought is ordinary, even usual.

The notion of "free will" in morals and law is an example. A person is expected to claim voluntarism and, if he does not, the community claims it for him. It is appropriate behavior. He must practice affecting himself, others, and the world with his mind, and he is abetted and encouraged often in imagining his own spheres of power; so heaven is attainable, as is hell, good as well as evil, world destruction and world renewal, the triumph of the Ideal, of Truth, of Order.

What grammarians say "ought to be" is obsessively regarded as "is," that is, achieved: "God is on our side." "Work is fun." "Parents are good." The distinction between a preference and a fact is overridden, not because some humans cannot comprehend it, but because they cannot tolerate the world that exists. In the beginning, and in the case of every infant, the "is" is painfully segregated from the "ought" so we should not be surprised at the universal recidivism from "is" to "ought."

The overriding drive to control the self and everything else is the "should be" of all "should be's." The gods are made "to be" as they "should be," friends of ours, of our tribe, or perhaps even of humanity. No matter that they are also believed to have repeatedly destroyed the nations and that they will do so again; they remain, in Homer's cliché, "the source of all blessings."

Wishful thinking, in matters both small and large, is universal and practically ineradicable. Megalomania is but an obvious pathology, slightly out of step with the fundamental human delusion of making a wished world out of a real world. Masses of people who are helpless and frustrated by lack of control achieve miracles by prayer— for the cure of illness, salvation of the soul, and the recovery of popes and presidents who have been shot.

That will and morale are powerful agents cannot be denied. Not so long ago, in the dust bowls of empiricism, these words brought shudders. The development of psychosomatic medicine, the infiltration of the West by Hindu Yoga, the respectabilizing of consciousness- raising and, paradoxically, of suggestive techniques, brought them into new prominence.

Hilgard discusses the experiments of Spanos and others on suggestion. Hypnotized subjects believe all the more readily that their arm is becoming stiff if beforehand they have been supplied with fantasies, as that the arm is in a splint or made of wood or iron, and therefore cannot bend. Spanos calls these "goal-directed fantasies." We regard them also as a type of psychosomatic conversion. We see in them evidence that willpower can become an operational concept, even in practical affairs, after a century of ridicule and obloquy. Will- power in politics, religion, sports, or business must be an exertion upon external objects of the same physiological system that accounts for psychosomatism up to the point of the system impacting on the body tissue, which completes the operation if internal, but carries the operation only into an external activity if involving a displacement removed from the organism.

There is little question but that homo schizo can mobilize his mind for remarkable feats of organs, mentation and behavior. We note this, too, in the connection between so-called visceral learning and yoga, and the illusion of omnipotence of thought. All serve to validate the very old supposition of homo schizo that he could do anything if he only wanted to do so badly enough. The fact that our contemporary world is so extreme a chaos of wills and wants obscures the enormous potential that this age-old idea possesses when harnessed to modern psychology. "We demand a character for which our emotions and active propensities shall be a match. Small as we are, minute as is the point by which the cosmos impinges upon each one of us, each one desires to feel that his reaction at that point is congruous with the demands of the vast whole— then he balances the latter, so to speak, and is able to do what it expects of him."

Once again we must allude to the enormous impact on the world of the drive for control genetically engendered in homo schizo by the failure of animal instinct and the fearful balkanisation of the human self.

3. Rationalization

In this age that is dominated by a belief in "rationalism," much that is believed not to be rational is gathered together in the concept of "rationalization." Rationalization is supposed to be finding persuasive arguments for doing what one thinks one wants to do. It therefore depends upon the sophistication of the persuaded and upon the demands that the rationalizer makes upon himself. Rationalization is assigned to linguistic emissions whose purpose is to conceal real mentation by describing it in acceptable linguistic, moral and logical forms. Thus, as in the case of the anarchist, G. Zangara, a man hates his father, displaces his hatred upon a remote authority, the president, and tries to kill the president, believing and asserting that the sole source of his action is in the policies of the president. The idea of a separate process of rationalization characterizes all human communities, but especially modern communities, where all behaviors are supposed to become "rational," tied in as cause-and-effect with appropriate and approvable community conduct, or at least with an ideal ethic recognized as such, even if opposed, by the community.

In the theory of homo schizo, rationalization is nothing but a pandemic mode of discourse; it is the "rational," but defined and shaped by whatever level of rationality that the community manifests. The most exalted philosophy, as well as the lame excuse of a malingering schoolboy, are equally rationalizations. The human does little but rationalize its wants; it does not do something extra and special called "reason."

Where is the line to be drawn between rationalization and rationality? Out of the clouded mental sky do not some few stars of intelligence shine? If intelligence exists, we would say, it is a rare ability to continuously compress mental operations according to symbolized rules along a track of highly correlated "cause-effect-cause-effect... n" ties, and to make many track-switching associations, as we shall soon discover. But it is worth investigating.

4. THE DISSOLUTION OF LOGIC

The digital (or linear) and analog logics can perform all mentation thus far ascribable to "reason," it would appear. As I would portray these logics, in homo schizo theory, the digital or linear is a coding to take care of "elapsed time" on delayed instinctual reactions, while the analog is a coding to utilize the displacements engendered by the same glitch. In both cases a language appears, which when working "as it should" accomplishes "rational thought."

"Rational thought" is defined as appropriate public symbolic behavior aimed at a solution. The public may be anyone or everyone. A trite lesson in logic goes: "all men are mortal; Socrates is a man; therefore Socrates is mortal." (All X is Y; S is X; ergo S is Y.) Presumably, the lesson is based upon reality. We characterize many men, and then observe that

Socrates shares their modal characteristics. Further, one observes that, among other happenings, they all die; whereupon Socrates must die, too. It helps to observe that Socrates did die, unfortunately. Obviously, all depends upon whether Socrates is accurately placed as a member of the human species and whether any exceptions to death occur. It is clear that Socrates can be deviant from all human norms except this absolutely inclusive norm of death. The problem is first one of analogy, then of algebra.

The procedure is called an Aristotelian syllogism and has long been regarded as the classic deductive proof, but, even as William of Occam surmised in the late Middle Ages, it is an emanation from the structure of the mind, not a quality of reality, that is being processed. The lioness who has seen and hunted many antelope knows that all antelope are mortal, and that the antelope she sees now is mortal, and she can expect it to die by her claw and fang like the rest. The philosopher has to prove it symbolically, cutting through a mass of confused human neurology before putting the major and minor premises together in a conclusion.

Modern psychology and pragmatism have pushed much of Aristotelianism into a corner and occupied its premises otherwise as well. Its three basic laws have become tautologies: that a thing is itself, 'A' is 'A'; that a thing cannot be both itself and something other than itself, 'A' cannot be both 'A' and 'not A', and that a thing must be either itself or not itself, 'A' being either 'A' or 'not-A'. These statements are suppositions of narrow utility, overwhelmed by the multitudinous demonstrations of modern psychology and anthropology that 'A' may or may not be 'A'; 'A' can be both 'A' and what 'A' is not; and 'A' can be either 'A' or not 'A' or both. That is, no thing, no occurrence, no process, no 'A' exists but exists holistically, in the company of its opposites. Causes and anti-causes cohabit. This is the actual operation of the human mind, acting out of its structure. It is also homo schizo theory, which is non-Aristotelian and non-Cartesian.

The question of whether this is the actual condition of the real world rather than of mind alone might not appear germane to the present discussion. However, inasmuch as homo schizo seeks to control the universe because he is displaced throughout its time and space, he will presumably seek to know it for control purposes. Therefore, he will wish to elaborate and perfect whatever human apparatus is best adapted to that end. All of his efforts at controlling the divine and the mundane, objects and existence, will be pragmatically judged. All of the non-logical and logical procedures generated in all of human history are so tested.

The modern age has proliferated not only forms of non-Aristotelian logic to this end, but it also witnessed occult ideas, cults, therapies, and countless other modes of confronting reality. Every nook and cranny of psychiatry, philosophy, mysticism, magic, behavior, life as art, group configuration, and of Siberia, the Caucasus, Egypt, South America, the Caribbean, China, Indonesia, Rumania, Iceland, Tahiti, Africa and India— the whole geographical and ideational world and outer space, too — have been poked, prodded, pierced into and opened up for a better way. All are driven by the hope of discovering and seizing upon a procedure that will give the longed-for control and set the human mind once and for all at ease. After reason has failed to prove reasonable, it is every man for himself, *sauve qui peut.* Internally, the "appearance" of the linear and analog logical forms must be "messy." That is, nothing is as clear in neurological language as it is in public language; this is a truism, since public language has to pursue a clearly communicative format. When a few scientists first began to speculate about the brain as a computer, John von Neuman remarked that probably "it is futile to look for a precise logical concept, that is, for a precise verbal description, of 'visual analogy. " It is possible that the connection pattern of the visual brain itself is the simplest logical expression or definition of this principle." Pribram, with the hologram image in the vanguard of his work, can today supply much of what was missing then.

In ordinary cases, training maintains its grip on the external communication. The mind selects the arithmetic and analog rules which, it has been thought, are acceptable manipulations of terms. The mind "makes sense," publicly. The degree to which its public demonstrations of logical mastery grip the mind, and influence it, can vary greatly. The display logic may convey little of the "true" thought processes; it may conceal them and in any event express only some part of them. But, we stress, this display is what excites much of the response in the transactions between external minds. So two people deal in a currency that scarcely measures the internal values of the exchange.

In general, what makes for intense memories in people also makes for obsession with "correct" logical expression and for following compulsively the dictates, or solutions, provided by the logic. People in logical or rational communication must convey what they intend to convey in all critical circumstances, whether football players or bankers or scientists, or else the language breaks down.

Plato's *Timaeus* as interpreted by Taylor argues also that those who cannot do a sum take fear when the planets show oppositions, occultations and reappearances. This is not the only indication from ancient legend and science, nor from modern psychology and behavior, that numbers are a security device as is measurement, hence astrology, astronomy and astrophysics. The universally observed magic of numbers and the superstitions of numbers support this hypothesis.

Numbering may have originated in the fight against fear; numbers and measures may ultimately have become logical-rational procedures; but they may originally have been methods of fighting fear. Counting, ordering, measuring, have in them a fear therapy. So children are told to count sheep in order to fall asleep. "Hail, Mary's" are recited by soldiers until panic passes. Parachuters count before leaping. The count-down before space-vehicle launching is a public ritual of prayer, wish, and suppression of last frightened thoughts.

5. Causation

For thousands of years, the leading forms of philosophy of truth have been directed generally at the destruction of commonsense truth. The analysis of discursive symbolism among the ancient Greeks affected human communications with these questions: What do you know? (perception and cognition) ; how do you know it? (logic and proof); then, do you know yourself? (Socrates). The three questions cast much of what passed (and still passes) for knowledge into the realm of the non-rational.

The classical Greeks were neither first nor last to go through the act of first constructing natural laws and then of finding out how to evade them. It would appear that this constructive-destructive process is characteristic of high periods of mental development, whether in the Arabic enlightenment, the high Middle Ages, the Renaissance, the Eighteenth Century Enlightenment, or the past century. The more man subjectivizes, the more he can control the outer world, but also the more he can see of the limits of his truths.

So it happened that in more recent times, we have seen the destruction of Euclidean geometric space as an absolutely existent phenomenon; "time" has been reduced to a relative, generically impelled habit and coincidence, and the "caused" has become a "function."

Needless to say, the outlook of science has been forcibly affected by this relativism, but ordinary life has proceeded on its commonsense paths, carrying on faithfully the belief in absolute truths of knowledge while of course carrying along its full complement of illusions and delusions. Yet ordinary life has been always affected by the relativism of time and space, and the incidentalism of causation.

The concept of causality has caused philosophers infinite headaches, leading up to its final denial; but the path remains to support the commonsense belief that stress causes pain. Analysis of alleged "causation," freed of commonplace prejudices, quickly arrives at the conclusion that in any given case of "causation," everything in the vignette can be termed a casual factor, and that what is called "the cause" is whatever the judge deems it to be.

Quantum mechanics also would destroy ordinary causal theory, lending as it does an indeterministic element to the "decision" of a causally potent condition as to whether or not to actuate, that is, happen. A given electron may or may not "choose" to leave its radioactive atom, for example. Or, in order to discover the momentum of a given particle (which is at the same time a wave), one foregoes by the conditions under which this can be observed and measured, the chance of discovering its location— and vice versa. With quantum theory and the Heisenberg principle of indeterminism (uncertainty), the following must be foregone: certainty; predictability; causation; space and time; Aristotelian logic (see above); nor can the quantum-uncertainty principle be proven true or false empirically. What remains, however, are statistical probabilities governing aggregate behavior. As we cannot ask an explanation of the basic fact that "inertia is", neither can we ask why there is a state of indeterminism.

As if this were not enough, what we see in causality in the human mind is a spasm of incompleteness between two events that it is felt ought instinctively to happen in sequence. Here again is the instinct glitch. Anything once delayed builds a secondary displacement circuit or hologram. The circuit continues to be excited by analogous events and the analogous sets become grouped into perceived causal classes, as, for example, "stress causes pain.”

6. TIME AND SPACE

The world, it may be agreed, is essentially vacant of time, space, or causality. So is the human mind. No time-clock as such registers impressions and expressions of the central nervous system. However, what comes to be sensed as time is the neurological superposition of halos imprinted upon neurons as they occur. Then, typical left-hemisphere operations ensue, ordering impressions by digital logic, cleaning up inner time for incorporation into external and especially cultural time schedules.

The past tense of time is perceived as one's recall reaches for lower figurations in the "stack" of impressions. The "lower," the older. When a woman tells a man, "you remind me of my father," perhaps tens of thousands of circuits are retrogressively lighting up. In her mind, a coded representation of his behavior is vigorously seeking analogues (holograms). If, is as likely, she is engaging in wishful thinking, non-analogues and distasteful analogues are being censored.

As with practically every other human trait, a rudimentary time sense is invaluable to the communication of animal instincts, and the storage of time in memory as well as the projection of time are readily observable. We resort, as usual, to the human glitch and the splatter of displacements to account for the rich human display of temporal effects.

No matter what philosophers may say in derogation of time, every cell and every species, even every grain of sand and atom, enjoys its big-sequences and big-rhythms. Not only are all things in change; they are also changing in patterns, and uniquely, hence a kind of triple paradox of change, pattern, and uniqueness occurs.

"Our fearful mind anticipates the future but we can only understand what was in the past," declared Kierkegaard. Something has been said earlier of the sense of dread regarding death, the divine, and the future. To the schizophrenic, writes Meerloo, "the past is something demonic. The feeling of unbroken homogeneity with the present has been lost... 'The world clock stands still, ' says one patient." Mendel perceives the dissolution of historicity and with it the future as a major characteristic of disease.

Exaltation, which can be viewed as an agitated nervous crisis of the present moment, of which the use of the historical present in literary style partakes, is frequent in mental states pronounced insane as well as divine. It collapses both history and the future. It also reverses time. The ancient prophets — the Hebrew Isaiah, Saint John the Evangelist, the unknown Egyptian author of the Ipuwer papyrus concerning the destruction of Egypt, for example, used the future tense to say what had happened in times past. Leonardo da Vinci, whose genius included a set of *Profetie,* also refers in these prophecies to historical materials using the future tense. Time becomes like the chain that propels a bicycle, going backwards as it moves forward. The faster a person or culture moves, the more its future and its history are changed. But a culture denies that it can change, it denies the charge of hubris, and celebrates continually its very beginnings as if they were today and in the future. Time, man's great tool, is repeatedly and deliberately destroyed. Time is projected memory. To control himself, man must control his projections both past and future. Time, to him, is an event, a fact, that must be controlled along with every other happening.

Left hemisphere brain damage interferes with the perception of sequence but right hemisphere damage does not. If our theory that two types of logic form in the brain, one analog, the other additive or digital, and that the first is right-brain or bilateral while the second is left-brain, the inability to perceive sequences may be attributable to a disturbance of time-counting by digital sequence coding. The analog contribution to sequencing would be by superposition of images. It may operate like holography, as was suggested earlier, allowing replicated images to be stored in large numbers, so that the excision of even a great many holograms in either or both hemispheres would not disturb the detection of sequences.

Future tense arises out of obsessional expectations of the return of an event. For ordinary and minor events, the future is helped to emerge by the transference of analogous major obsessions. The anniversary complex has deep roots in the human mind: an intense private celebration, when congruent with a public consensus, forms the peak type of human memory event. When an anniversary is forgotten, whether private or public, it is because of dread of the reoccurrence. Ordinarily, a deep enough trauma is quite suppressed and is celebrated only unconsciously, with depression or psychosomatism or displacement behavior. Thus any of these neuroses may befall a woman upon the anniversary of the painful death of a dear mother, especially if her age approaches the age of her mother at death. Public trauma can be bifurcated to give a mourning occasion followed by a saturnalian release, or a joyful one, as before Christmas, or at Eastertime, among Christians and Jews, for different reasons, the Judaic mourning and joy in these cases relating to the great Deluge of Noah and to the Passover of Exodus, respectively. Thus we both remember and forget.

Space may be dependent upon time. That is, without time, space might be inconceivable. An animal is master of the space around it, more than the human, who probably suffers more awkward bumps and falls in the house he has built than do the animals that may share it with him. But only the human is driven to conceive of, measure, and manage all space, so that one must guess that he is the jack of all space for being master of none.

Human concepts of space are perhaps built upon an infrastructure of time. Once the sense of time is developed, space can be calculated as elapsed time between the self and the displacements of the self, where they were in the memory, where they occur now. Thus we think of primitive space as distance in time from an object or event to the experiencing self. Direction is also a point of reference from one's body— front, behind, right, left, up, down.

Accurate mapping of space within the confines of experience becomes possible. What is beyond direct experience— over the mountains, the stars above— may be naively construed as far away, but much less far than they really are. The celestial bodies are measures of space—" I live a day's walk from you," or "six hours by airplane." But, thus, too, is time.

7. Science as Instinct

If theologians and philosophers vainly sought certainty in order to displace fear, are then scientists merely at another stage of displacement or sublimation? Would the counting of binary star systems be such, too? Might the rejection of the holistic term "human nature" be a collective schizotypical symptom of depersonalization among psychologists? Would perfecting solar energy systems fall in a similar category? Yes, we would say, but all with a notable difference. Truth or the rational is how to get what one wants. Then "howling for bread" is true if it brings bread. So is prayer. Yes. What works, what is effective, is considered rational and true.

Would not bread be more certainly forthcoming if one farmed wheat and baked bread? Perhaps...in some cases... yes, on the whole... etc. The sparrows don't look for the morrow, said Christ; trust in the Lord. The infant howls and is fed. The mob riots and is fed. The primitive band gathers nuts and fruits and herbs. But we applaud the ancient inventors of the science of agriculture. Now they could feed more mouths while resting in place.

Science — and reason — are suppressors of unruly processes. They discipline the fearful selves to follow rules which, they assert, will reliably bring desired consequences. "Get rid of the excess and costly baggage of superstitious behavior: don't chant and dance around the growing crops; hone your spades, plant more seed, dig deeper; (then, reluctantly) it may help if you play music for the crops to grow by."

The rules of science and reason are simple. All things are sensible. They must be given exclusive denotations. These must be acted upon in exclusively denotable ways. They can then be grouped within a closed system of logical counting which is not so empirical. The process and the consequences are to be watched and confirmed by others. At no stage of the process should "wishing" be admitted and given any weight. Scientific procedures give homo schizo controls to add to his kitbag of controls. To some extent they are more reliable controls, though often for things that he wants to let be un-controlled or cares little about, or are not what he wants most to control. Behaviorally, what homo schizo has done, which has come to be called "rationality" and "reason," is to select out of his experience certain operations whose traits are that, first, they give success (by test) in naming, and in the transfer of naming (also by test), and, second, the names can be counted and the count tested. Success in testing is validated inasmuch as the names and their manipulation produce psychic and material effects deemed favorable. Why does man select these operations, and make so much of them that a wonderful science ensues? He finds their effects reliable; he can make easy and gratifying obsessions of them.

Is this all? This is a great deal. It makes the difference between uncontrolled fear and a bearable equilibrium, between helplessness and ruling the earth and all of its denizens. It is the difference between a blank gaze and a child counting apples, or even more, a computer guiding a spaceship out and back to earth. By temporarily giving up his chaotic mentation, by submitting to the controls determined by others, by obsessively dedicating his mind to the proven possible and proven practical, he can gain a share of control. The rules, the identifications, the promises, the secrets of language and experiments, the mystique and authority of science — all help him feel comforted and less fearful. The myriad displacements augment the normal complement of animal foci of attention and sources of stimuli.

The responses are the fantastically engendered capabilities of the human. Satisfactions emerge from a perceived coping with the stimulus by the selected responses. Science takes homospecific urges, applies species-specific responses and obtains species-indulgent effects. To a foreign intelligence, none of these would make sense, much less truth. What the process resembles, in a sinister shocking way, is where it all began, the home that it never left, amidst the unselfconscious breeds of life.

The closest I find to this idea of the search to recapture instinct as inherent in pragmatic or operational science is in Mead's essay on the biologic individual. After describing the physical world that faces the human as a biological and instinctive organism, Mead says: "Just in so far as we present ourselves as biological mechanisms are we better able to control a correspondingly greater field of conditions which determine conduct. On the other hand, this statement in mechanical terms abstracts from all purposes and all ends of conduct." Then he compares modern scientific method as a way of moving into the "now" to test reliability and truth.

On the whole, homo schizo would prefer more direct and easy methods of reaching the good, the true, and the beautiful, than science has thus far afforded him. If he were always comfortable, he certainly would not be rational; but then, if he were always comfortable (that is, had his quota of the values of sex, respect, health, knowledge, affection, and power) he would not be homo schizo, for there is no quota, only endless discomfort.

Paraphrasing Heraclitus, we could say, "One can never bathe in the same river of truth twice." But homo schizo would hate this truth, even though he has had to live by it. What homo schizo would most desire, because it would bring him immediate surcease from his existential agony, would be to become once more a generalized mammal, whose mind fit its body, able to act decisively, unconsciously, instinctively upon the presentation of a stimulus. The good, and the truth that leads to it, depend upon reestablishing the unitary ego, defining stimuli, and affixing their specific responses, all to occur together, holistically. Linear and analog science help. They build a bridge over the stimulus-response chasm. If only science might find a holistic way of bridging the chasm, of healing the instinct glitch, man would feel even better. The clue to this is in the incessant human attempt to embrace the good, the true, and the beautiful in one holistic motion where what is called ethics, science and art have no place and little interest, where he feels at one with himself and the world, an intuitive well- being. Although stoicism and Buddhism and Taoism and many other formulas of conduct prefigure this kind of confrontation that brings comfort and surcease from fear, they cannot manage reliably to control the "reality principle," that is, the persecution, hunger, massacre, frustration and demands visited upon them by the unregenerate homo schizo outside the cult.

8. THE COST OF LOSING MAGIC

James Fraser, a century ago, explained the practice of magic by two principles. According to a law of similarity, like produces like; a magician can produce an effect by imitating it in advance. By sprinkling water, one can bring rain, providing other matters are attended to also. Second, the law of contagion maintains that things once in contact continue to interact ever thereafter. A spot struck by lightening is forever sacred to Jupiter. Such was the belief and practice of the ancient Etruscans.

These principles are found in current as well as ancient and tribal thought. Science often proceeds by imitating nature and unravelling strings of consequences. In them one can locate the operations of analogy that seem to be naturally produced in the brain. Analogy took the primary and more powerful role in the development of Greek science from magic and myth.

We do not change brains or develop new organs in going from "falsehood" to "truth." We get rid of the "hocus-pocus" that accompanies magic. We make the magic public (open display and repetition of experiments). Then we increase the validity of the analogies and sequences until they become reliable.

The history of science shows us many a relation in tandem between magic, religion and scientific practice. Astrology as astronomy and magic is perhaps the most famous. The ancient Chinese could foretell eclipses, a major achievement of scientific observation and logic. But at the same time they applied rituals and emergency policies to quell official and public fear of eclipses and to repel astral invasions. Mesmerism and hypnotism are another example, from the nineteenth century, of parallel evolution of cult practices and scientific method.

The discipline involved in the change from magic to science is intense, obsessive, and costly. An experiment by Liam Hudson performed upon students of history and engineering involved interrupting their sleep upon observing signs of dreaming and asking them to report their dreams. The engineers reported more frequently that they were not dreaming at all or could not recall the dreams. "The engineers' inhibition in dealing with 'primary process' thought— with ideas and images that have not been ordered in a conventionally rational way — is not a superficial aspect of their thinking; it is an integral part of the way in which their minds work."

What we are observing here and in primitive magic are lesser and greater degrees of the conversion of obsession into bureaucratic and scientific habit and showing that, like a form of psychosomatism, the specialized disciplined worker overdevelops a *point d'appui,* working from the conscious into the unconscious along a narrow band. This is not science as new theory or hypothesis, not science as poetry, which is an altogether different mental operation, distinguishing two types of scientists as night from day.

So must the routine administrator or bureaucrat be distinguished from the organizational innovators of the type of Epaminondas, St. Therese, I. Loyola, Thomas Jefferson, Henri de Saint-Simon, Mussolini, Trotsky, Henry Ford and Gandhi; a great many unnamable persons have produced the largest number of inventions— there were, after all, engineering students who did recall their dreams. The rational is the routine, true, but ought one not permit the term for the creative? But, then, the creative is non-rational. So it is both rational and nonrational, a contradiction if both are the same.

Therefore do we propose discarding the term "rational" or letting "rational" mean the ability to obtain what one wants, namely, "truth." We must disagree with those who, like Arthur Koestler, assert of the human dilemma: "No matter how much the symptoms vary, the pattern of disorder is the same: a mentality split between faith and reason, between emotion and intellect. " Faith and reason subsist cheek by jowl in the mentating process; we must abandon this medieval dichotomy if we would understand human nature.

So, too with emotion and intellect: emotion is intellective and intellect is emotional. At the turn of the past century, George Mead was lecturing at the University of Chicago that: "It would be a mistake to assume that a man is a biologic individual plus a reason, if we mean by this definition that he leads two separate lives, one of impulse or instinct, and another of reason.. On the contrary, the whole drift of modern psychology has been toward an undertaking, to bring will and reason within the impulsive life." There is little neurological or pragmatic basis for the words. Nor do they help in programming policies for humanity.

Close in outlook to Mead, John Dewey, too, was long engaged in combat against traditional logic and psychology. In 1929, in his book *The Quest for Certainty,* he devoted a chapter to "Escape from Peril," where he continued his attack upon the philosophers' search for the immutable, the truth, by way of "pure knowledge." "The quest for certainty," he said, "is a quest for peace which is assured, an object which is unqualified by risk and the shadow of fear which action casts." We would add that homo schizo normally wants to escape his perils and invented first historical religions, then theology, then philosophy, moving outwards into abstractions to develop the sense of certainty that would relieve his anxieties. But each further stage of abstraction displaces him farther, too, from the origins of existential fear in his inability to act like an animal.

The Other Books by Alfred de Grazia

Michels, Roberto, *First lectures in political sociology.* Translated, with an introduction, by Alfred de Grazia. Minneapolis: University of Minnesota Press, [1949]. And Harper & Row, 1965.

Public and republic: political representation in America. New York: Alfred Knopf, 1951.

The elements of political science Series: Borzoi Books in Political Science. New York: Knopf, 1952. And second revised edition: *Politics and government: the elements of political science* Vol 1: *The element of political science* and Vol. 2: *Political organization.* [1962]. New York: Collier, 1962– . And new revised edition, New York: Free Press London: Collier Macmillan, 1965.

The Western Public: 1952 and beyond. A study of political behaviour in the western United States. Stanford: Stanford University Press, 1954.

The American way of government. National edition. New York: Wiley, 1957. There is also a "National, State and Local edition".

Foundation for Voluntary Welfare *Grass roots, private welfare: winning essays of the 1956 national awards competition of the Foundation for Voluntary Welfare.* Alfred de Grazia, editor. New York: New York University Press, 1957.

American welfare. New York: New York University Press, 1961 (with Ted Gurr).

World politics: a study in international relations. Series: College Outline Series. New York: Barnes & Noble, 1962.

Apportionment and representative government. Series: Books that matter. New York: Praeger, c1963

Essay on apportionment and representative government. Washington : American Enterprise Institute, 1963

Revolution in teaching: new theory, technology, and curricula. With an introduction by Jerome Bruner. New York: Bantam Books, [1964] (Editor, with David A. Sohn).

The Universal Reference System. Political science, government, and public policy: an annotated and intensively indexed compilation of significant books, pamphlets, and articles, selected and processed by the Universal Reference System. Prepared under the direction of Alfred De Grazia, general editor, Carl E. Martinson, managing editor, and John B. Simeone, consultant. Princeton, N.J.: Princeton Research Pub. Co., 1965–69. "Plus" nine more volumes on the subjects of: *International Affairs; Economic Regulation; Public Policy and the Management of Science; Administrative Management; Comparative Government and Cultures; Legislative Process; Bibliography of Bibliographies in Political Science, Government and Public Policy; Current Events and Problems of Modern Society; Public Opinion, Mass Behavior and Political Psychology; Law, Jurisprudence and Judicial Process.*

Republic in crisis: Congress against the executive force. New York: Federal Legal Publications, [1965].

Political behavior. Series: *Elements of political science;* 1. New, revised edition. New York: Free press paperback, 1966.

Congress, The First Branch of Government, editor, Doubleday – Anchor Books, 1967.

Congress and the Presidency: Their Roles in Modern Times, with Arthur M. Schlesinger, American Enterprise Institute for Public Policy Research, Washington, 1967.

The Behavioral Sciences: Essays in honor of George A. Lundberg, editor, Behavioral Research Council, Great Barrington, Mass.; 1968.

Old Government, New People: Readings for the New politics, et al., Scott, Foresman, Glenview, Ill., 1971.

Politics for Better or Worse, Scott, Foresman, Glenview, Ill., 1973.

Eight Branches of Government: American Government Today, w. Eric Weise, Collegiate Pub., 1975.

Eight Bads – Eight Goods: The American Contradictions, Doubleday – Anchor Books, 1975.

Supporting Art and Culture: 1001 Questions on Policy, Lieber-Atherton, New York, 1979.

A Cloud Over Bhopal: Causes, Consequences, and Constructive Solutions, Kalos Foundation for the India-America Committee for the Bhopal Victims: Popular Prakashan, Bombay, 1985.

The American State of Canaan – the peaceful, prosperous juncture of Israel and Palestine as the 51st State of the United States of America, Metron Publications, Princeton, NJ, 2009 LCCN 2008945276.

Autobiography, Poetry, etc.

The Babe, Child of Boom and Bust in Old Chicago, umbilicus mundi, Quiddity Press, Metron Publications, Princeton, N.J., 1992.

The Student: at Chicago in Hutchin's Hey-day, Quiddity Press, Metron Publications, Princeton N.J., 1991.

A Taste of War: Soldiering in World War II, Metron Publications, Princeton, N.J., 1992 & 2011.

Passage of the Year, Poetry, Quiddity Press, Metron publications, Princeton, N.J., 1967.

Twentieth Century Fire-Sale, Poetry, Quiddity Press, Metron Publications, Princeton, N.J., 1996.

Films

La Roccia di Sisifo, directed by Pilar Latini, Bergamo, 2004.

Il Gene della Speranza, directed by Kinokitchen, Pisa, 2006.

Made in the USA
Monee, IL
06 November 2026

68083938R00329